GIS: A Visual Approach

ruce E. Davis

ONWORD® PRESS

GIS: A Visual Approach

By Bruce E. Davis

Published by
OnWord Press
2530 Camino Entrada
Santa Fe, NM 87505-4835 USA

SAN 694-0269
First Edition, 1996
10 9 8 7 6 5 4 3 2 1
Printed in the United States of America

Library of Congress Cataloging-in-Publication Data

Davis, Bruce E.
Geographic information systems : a visual approach/Bruce E. Davis — 1st ed.
p. cm.
Includes index.
ISBN 1-56690-098-0
1. Geographic Information Systems. I. Title.
G70.212.d38 1996
910'.285—dc20
96-8029
CIP

Trademarks

Warning and Disclaimer

About the Author

Bruce Davis is a geographer, with a B.A. from the University of California, Santa Barbara, M.S. from the University of Southern Mississippi, and Ph.D. from the University of California, Los Angeles. He has taught in various schools throughout the United States, primarily in the Southeast. With over twenty-five years experience in remote sensing and GIS, he has served as director of several centers, as a GIS consultant, NASA visiting scholar, applied practitioner, a Fulbright Scholar to the University of the South Pacific (USP), and teacher. Dr. Davis is currently a Senior Lecturer in Geography and founding Director of the GIS Unit at USP in Fiji.

Acknowledgments

The production of anything worthwhile demands the labors of many folks, too often unseen and unappreciated. My sincerest gratitude goes first to Mr. Kevin Schultz, a long-time and valued colleague who helped start this text; many thanks, Kevin. I also owe thanks (and apologies) to the many students, particularly at the University of the South Pacific, who suffered through various trials and tribulations of refining the material. Mr. George Saemane, dear friend and close colleague at USP, deserves all the appreciation, and more, that I can offer. To the staff at High Mountain Press I am indebted for the opportunity, patience, cooperation, and hard work that helped get this accomplished. To Erin and Eric, who survived the all-too-frequent periods of absentee fatherhood, even when I was there, and compensated for my irresponsibilities. And last, few men accomplish much without the proverbial good woman behind, pushing and pointing and suffering and tolerating; to Tina, my closest friend and love of my life, who knows me better than I ever could, I give of myself. I take full responsibility for any errors or confusions in this book, but nothing positive could have been achieved without these and other magnificent compatriots.

OnWord Press

OnWord Press is dedicated to the fine art of professional documentation. In addition to the author, other members of the OnWord Press team contributed to the making of this book. Thanks to the following people and the other members of the OnWord Press team who contributed to its production and distribution.

Dan Raker, Publisher and President
Gary Lange, Vice President
Janet Leigh Dick, Associate Publisher/Director of Channel Marketing
David Talbott, Director of Acquisitions
Daniel Clavio, Director of Market Development and Strategic Relations
Rena Rully, Senior Manager, Editorial and Production
Daril Bentley, Project Editor
Carol Leyba, Senior Production Manager
Kristie Reilly, Production Editor
Beverly Nabours, Production Artist
Kate Bemis, Indexer
Lynne Egensteiner, Cover Designer

Contents

Preface

I have watched GIS and its derivative technologies transition from the esoterica of academe into the practical world, where most people live and work. GIS has become one of the few modern tools to live up to, and beyond, its early hype and promise. It has spread from the centers of advanced technology to the emerging worlds and remote corners of our planet, to places where computers are still a wonder. But my frustrations have grown, also. What once could be excused as too complicated for all but the advanced is today a technology being used by barefoot villagers who reside out of the global mainstream of information and cyberspace. Yet practically all of the literature is still aimed at the over-involved practitioner or focussed entrepreneur with intellectually padded backgrounds and access to a wealth of the latest technology. Very little written, perhaps nothing, exists for the rest of us; we must plod our way through stilted advanced English (usually written as advanced Stilted English!), inundations of words upon words, and inappropriate levels of technology and applications in order even to begin learning GIS. But I know that it should not be this way.

The comment "I would like to learn GIS, but it is too advanced" is heard often. Actually, that's wrong; GIS can be learned rather easily if only a bit of attention is given to an appropriate approach for teaching it. Most of the world is not in graduate school, and a truly introductory GIS text should have a place in the field. This book is an attempt to fill that need. Having worked and traveled over much of the "Third World" and now living in the South Pacific, where, at best, English is a second language (ESL), it is conspicuously evident that the standard western approach to teaching many subjects is usually inappropriate. Western academia does a rather poor job of reaching the majorities, preferring to spend its vast intellectual resources merely talking to one another. This text is meant for students and professionals everywhere who wish to learn the rudiments of GIS without being overwhelmed by the difficulties of language and culture. We don't have to spend a thousand big words to explain a fifty-word simplicity, and this introduction to GIS was developed to reach those who prefer to learn without unnecessary struggle. Using a "visual set approach" (my term) that presents an explanatory illustration, accompanied by brief and uncomplicated text, I hope GIS can be learned easily and effectively. If successful, I hope this approach is noticed by other fields so that learning what previously had been relegated to the unreachable can truly become accessible for all those who are interested. Thank you for reading *GIS: A Visual Approach*.

Introduction

GIS a rapidly growing field, but most, perhaps all, of the introductory texts to it are too advanced for the true beginner. These books are typically filled with jargon and assume considerable knowledge of computers and technical advances. Particularly disadvantaged are those who have first languages other than English. This book is directed to the many students and professionals in both the developed and developing worlds who need an efficient introduction to GIS. To reach the uninitiated and the ESL (English as a second language) reader, a new approach is used in this book: a blend of illustrations and text that support each other at the appropriate reading level and that present a true introduction to GIS. Toward this end, the following considerations went into the writing of *GIS: A Visual Approach*.

◆ GIS is a visual technology and is best learned by illustration of its concepts and operations. This book presents simple, clear graphics combined with brief text to achieve this goal.

◆ This book uses simple words and phrases wherever possible. Jargon, computerese, and academic language are avoided.

◆ All aspects of GIS are not covered; a complete review is not presented. Too many concepts and too much technical detail tends to confuse and defeat the primary purpose. The aim is to provide a foundation for basic understanding and for growth within the field.

◆ A principal objective is to focus on the user, not the technician, academic researcher, or computer specialist.

◆ Chapters are not exclusive; they are intended to be cumulative, building a sequence of principles and operations for comprehensive understanding. Many points and examples may be useful throughout.

◆ Typically, there are several ways to accomplish most GIS tasks. The illustrations and examples used here are meant to be simple, "generic" representations of GIS concepts and operations in general.

◆ Unfortunately, today's computer technology makes difficult demands regarding operations and procedures. Although GIS is becoming more "friendly," the user must still understand both computer systems and geographic applications. This book tries to make GIS easily learned by simplifying procedures where possible. This is a first introduction to GIS; it is not intended as a "hands-on" manual.

A beginner's book can not replace other forms of learning, particularly hands-on experience at the computer. The reader is encouraged to search other sources of information, especially real applications and projects.

1

GIS and the formation Age

Introduction

›ter is an introduction to and review of GIS: ıts major elements, its general importance, ›f its uses. It begins with a discussion of the ıture, and value of information in society, lefines GIS and, in general terms, describes orks. Included are some of the questions l by GIS, showing the GIS approach. A few adigms (philosophies, models of thinking) › GIS conclude this chapter.

Note: *The word paradigm combines the concepts of philosophy, how we think, influences on thinking, how things work, and models of ideas. The term is used in this book because it best expresses the notion that GIS affects not only the actual procedures of computer operations, but how we think of them as methodology.*

Point to remember: GIS is a new technology for a very wide range of applications; it is not just a tool for academic research.

The value and acceptance of GIS in the professional, working world is one of its major strengths. We are constantly developing new personal and institutional paradigms as a result of GIS.

Information and Change

Post-industrial Information Age

This edited quote by Saul Cohen, 1990 president of the Association of American Geographers, says that times are changing, and that our methods of meeting the needs of the times are also changing. The world is entering the "post-industrial Information Age"—a time when information is becoming a major product of, and foundation for, progress. Increasing emphasis on data management is apparent and necessary.

INTRODUCTION

- VALUE OF INFORMATION
- GIS DEFINED AND DESCRIBED
- PREVIEW OF GIS
- TYPES OF QUESTIONS GIS ADDRESSES
- BASIC PARADIGMS

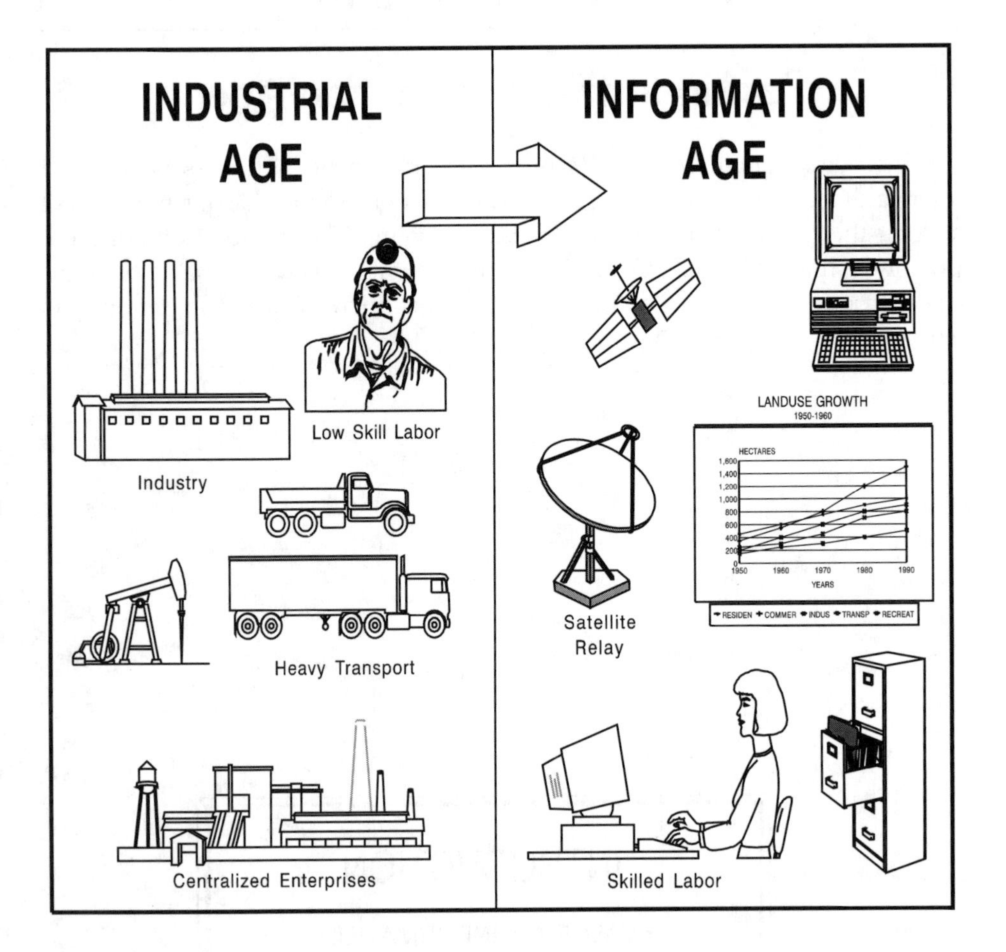
INDUSTRIAL AGE
INFORMATION AGE
Industry
Low Skill Labor
Heavy Transport
Centralized Enterprises
LANDUSE GROWTH
1950-1960
HECTARES
1,600
1,400
1,200
1,000
800
600
400
200
0
1950
1960
1970
1980
1990
YEARS
RESIDEN
COMMER
INDUS
TRANSP
RECREAT
Satellite Relay
Skilled Labor

Industrial Age to Information Age

Illustrated is the transition that Cohen considers. Western society has changed from a time when economies and culture depended on heavy industry, usually petroleum-based and using heavy transportation. Large, centralized factories and semiskilled workers were the norm—the basis for a way of national economic life.

Modern times are undergoing a passing of the old ways into a period in which sophisticated technology is changing the traditional power base. Computers and other electronic equipment (such as satellites) are the machinery whereby economies and cultures are being redirected and sustained. Data and information are the "currencies" of the Information Age, and skilled, technology-trained workers are becoming the labor force. GIS is an important part of the Information Age.

The Developing World ("Third World") is faced with the very heavy burden

> *"Today's focus in geography [or any other discipline] on data handling and data manipulation reflects the national [and international] emphasis on meeting the needs of the post-industrial information age."*
>
> Saul Cohen, 1990

of moving quickly into the Information Age if it is to keep up and avoid being overwhelmed by this global evolution. Even though the Industrial Age has yet to be fully realized in many nations, the Information Age will not wait patiently for slow and careful industrial development. The pressure to catch up is high for much of the world.

NEED FOR BETTER INFORMATION SYSTEMS

"The gap in the availability, quality, coherence, standardization, and accessibility of data between the developed and the developing world has been increasing, seriously impairing the capacities of countries to make informed decisions concerning environment and development."

Need for Better Information Systems

This quote—from a report of the 1992 Rio de Janeiro, Brazil, Earth conference—points to the need for better information and information systems. GIS is a central component in the world's environmental information structure, and it will continue to play a primary role.

> The gap between the Have and the Have-Nots of the information world is increasing; there is a huge difference between the information capabilities and wealth of the developed Western world and that of the Developing World. The gap should be closing, not becoming wider. Developing nations need good, reliable information for survival and progress.

Consider each point in the quote in terms of needed data and information.

- *Availability:* Does it exist, and where is it?
- *Quality:* Is it any good? Can we depend on its quality?
- *Coherence:* Does it agree with or correspond to other data?
- *Standardization:* Are we talking the same "data language"?
- *Accessibility:* Can we get to it, and can we afford it?

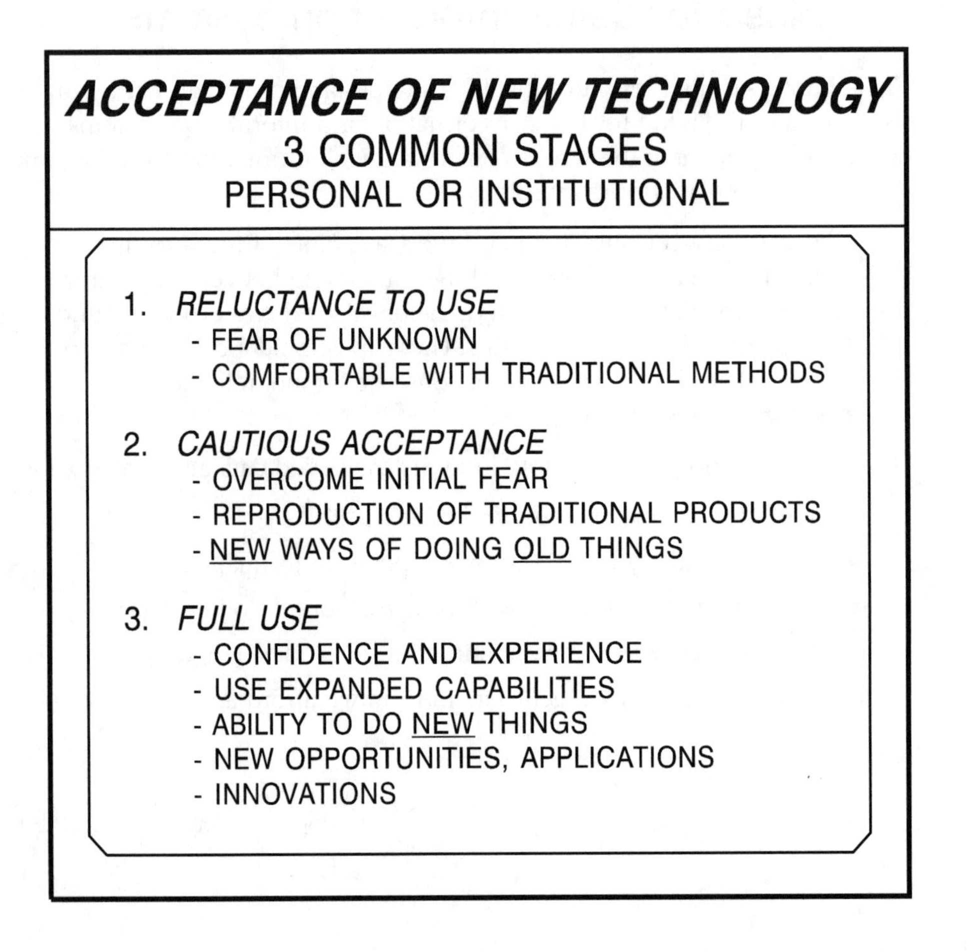
ACCEPTANCE OF NEW TECHNOLOGY
3 COMMON STAGES
PERSONAL OR INSTITUTIONAL
1. RELUCTANCE TO USE
- FEAR OF UNKNOWN
- COMFORTABLE WITH TRADITIONAL METHODS
2. CAUTIOUS ACCEPTANCE
- OVERCOME INITIAL FEAR
- REPRODUCTION OF TRADITIONAL PRODUCTS
- NEW WAYS OF DOING OLD THINGS
3. FULL USE
- CONFIDENCE AND EXPERIENCE
- USE EXPANDED CAPABILITIES
- ABILITY TO DO NEW THINGS
- NEW OPPORTUNITIES, APPLICATIONS
- INNOVATIONS

Stages of Acceptance

Almost any new technology, whether a simple tool or an advanced electronic machine, usually goes through three stages of acceptance. These stages may be on an individual, personal level or institutional (e.g., school or government) level. These stages are

1. *Reluctance to Use It:* We reject the new technology, often out of fear, and tend to stay with traditional methods. That fear may be from simple ignorance of what is involved in the use of the technology, fear of learning something so new, or fear of the changes it may bring. We are comfortable with our current state of technology and ways of doing things. Persons and institutions react to and accept change slowly, having to be convinced that the change is productive.

2. *Cautious Acceptance:* After overcoming initial fears (maybe because we see other people using it), we cautiously begin to accept the new technology and start to use it. The first uses are not particularly advanced, often being simple reproduction of traditional products in a new way—new ways of doing old things. It is very exciting when we begin seeing the potential wonders and value of the new technology.

3. *Full Use:* The most exciting aspects of acceptance are in the full use of a technology and the recognition that our capabilities can be expanded. We begin seeing new things that can be done, and new ways of accomplishing things we only dreamed about previously. New modes of analysis, for example, begin opening up new opportunities. The truly innovative aspects of the technology are now appreciated.

GIS is one of the most important new technologies to arrive in recent years. Its users may experience these three stages.

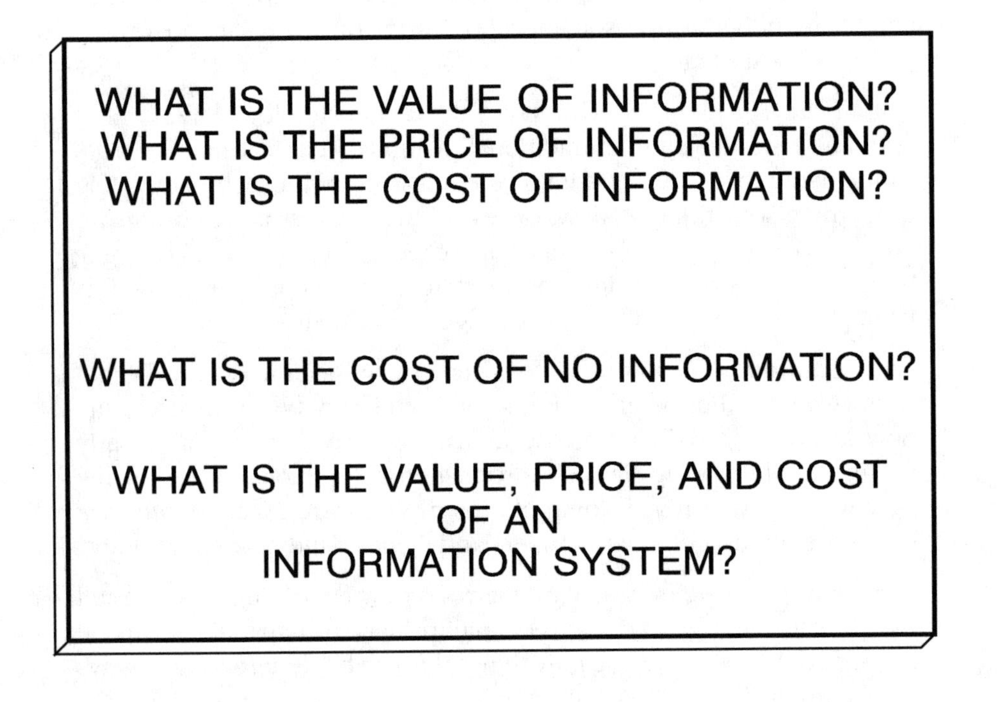
WHAT IS THE VALUE OF INFORMATION?
WHAT IS THE PRICE OF INFORMATION?
WHAT IS THE COST OF INFORMATION?
WHAT IS THE COST OF NO INFORMATION?
WHAT IS THE VALUE, PRICE, AND COST
OF AN
INFORMATION SYSTEM?

Why GIS?

Value of Information

As the world moves into the Information Age, the major "currency" is becoming meaningful data. A critical question of our time concerns the value and the use (and misuse) of data and its ultimate form as "information." We ask about the value of information: How important is it?, What can it be used for?, and What are its benefits?

The price of information deals with how much money (or equivalent) we pay for it. Often the price seems very high, but it might be low in terms of its cost-effective benefits. (A satellite image may have a high price, but it is data for a large area.)

The cost of information involves much more than price, and might include the machinery needed to process information, the people using it and their support and organization, and many other elements. Cost is the sum total of investments: price, people, facilities, effort, time, and others. These "ingredients" constitute the "infrastructure" of a GIS implementation and maintenance.

Perhaps the more important question concerns the cost of *no* information. We may pay dearly for not having critical information. For example, critical "informed decisions" in regard to the infrastructure previously mentioned may be impossible without useful information.

What is the value, price, and cost of an information system—the major tool of the Information Age? Perhaps the price and cost are very high, but an information system is an excellent investment, considering its value, and GIS is a major information system.

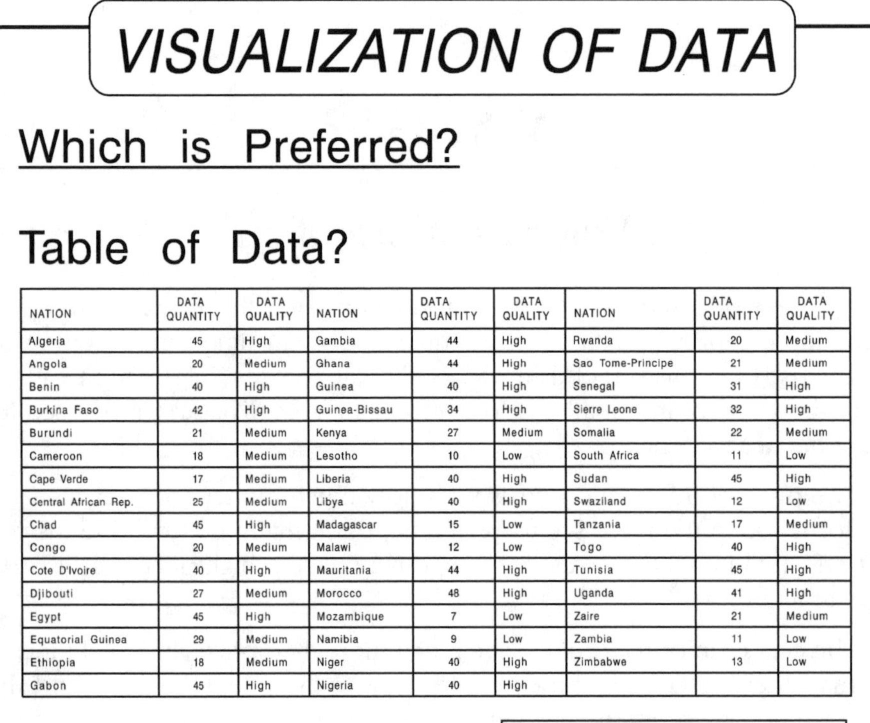

VISUALIZATION OF DATA

Which is Preferred?

Table of Data?

NATION	DATA QUANTITY	DATA QUALITY	NATION	DATA QUANTITY	DATA QUALITY	NATION	DATA QUANTITY	DATA QUALITY
Algeria	45	High	Gambia	44	High	Rwanda	20	Medium
Angola	20	Medium	Ghana	44	High	Sao Tome-Principe	21	Medium
Benin	40	High	Guinea	40	High	Senegal	31	High
Burkina Faso	42	High	Guinea-Bissau	34	High	Sierre Leone	32	High
Burundi	21	Medium	Kenya	27	Medium	Somalia	22	Medium
Cameroon	18	Medium	Lesotho	10	Low	South Africa	11	Low
Cape Verde	17	Medium	Liberia	40	High	Sudan	45	High
Central African Rep.	25	Medium	Libya	40	High	Swaziland	12	Low
Chad	45	High	Madagascar	15	Low	Tanzania	17	Medium
Congo	20	Medium	Malawi	12	Low	Togo	40	High
Cote D'Ivoire	40	High	Mauritania	44	High	Tunisia	45	High
Djibouti	27	Medium	Morocco	48	High	Uganda	41	High
Egypt	45	High	Mozambique	7	Low	Zaire	21	Medium
Equatorial Guinea	29	Medium	Namibia	9	Low	Zambia	11	Low
Ethiopia	18	Medium	Niger	40	High	Zimbabwe	13	Low
Gabon	45	High	Nigeria	40	High			

OR:

A Map?

HIGH
MEDIUM
LOW

Visualization of Data

We start with a simple question. The illustration presents two sets of data. Both say the same thing, yet one is presented in detailed, numeric, tabular form, whereas the other is a map. The question is: Which one would you prefer to help make the data more understandable? The map gives the best initial impression of what the data mean. The table is best for detailed analysis.

Visualization is the presentation of data in graphic form. Whereas tables and lists of numbers are usually difficult to understand without careful study, visualization is used as a convenient and effective way to communicate complex information. Computer technology supports visualization by offering sophisticated techniques of changing data into pleasing and understandable display. Greater importance is being placed on data visualization today, and GIS is a leading technology in this movement.

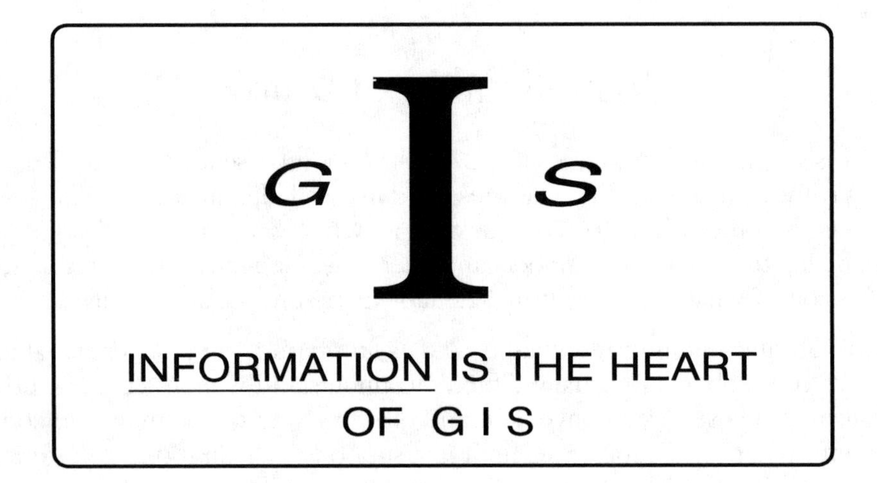
G I S
INFORMATION IS THE HEART
OF G I S

Information: The Heart of GIS

GIS may have three parts, but information is its heart. Without the I, the G and S are unrelated and disconnected. It is easy to concentrate on dazzling systems, for example, but in information resides the real work and value of GIS applications.

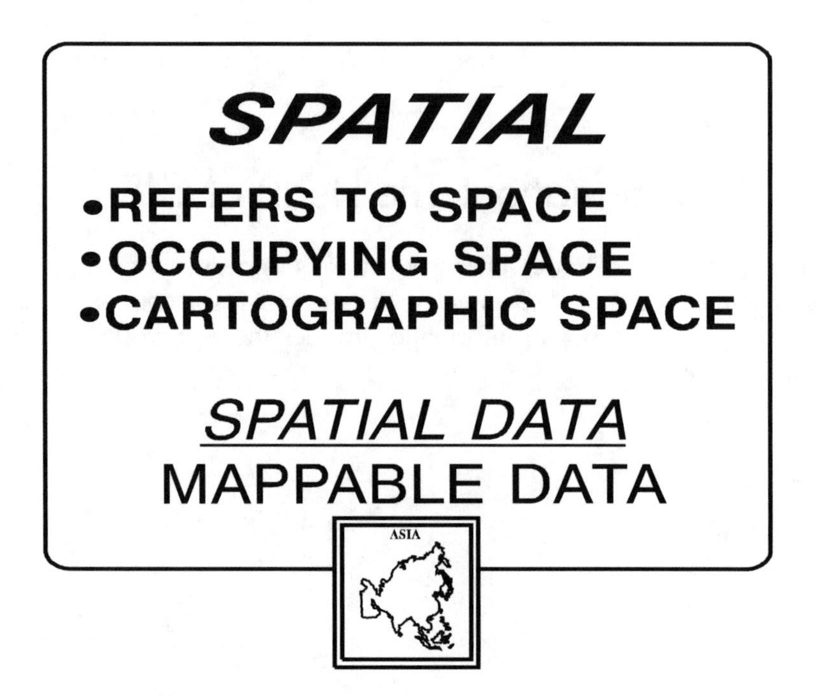
SPATIAL
•REFERS TO SPACE
•OCCUPYING SPACE
•CARTOGRAPHIC SPACE
SPATIAL DATA
MAPPABLE DATA
ASIA

Spatial Data: What GIS Uses

A common term throughout GIS discussion is *spatial.* It refers to geographic space, or the space that something on the earth occupies. Cartographic space is location or position on a map. Therefore, spatial data refers to mappable data—data that can be located in space.

The term will become more clear in Chapter 3, but the initial point to remember is that GIS uses spatial data, which normally can be placed on a map. Nonspatial data is not location specific, but can be used with GIS. An example of the difference between spatial and nonspatial data is that the address of a house is spatial information, whereas its color or the name of its owner are nonspatial. (See also the glossary for definitions of these two terms.)

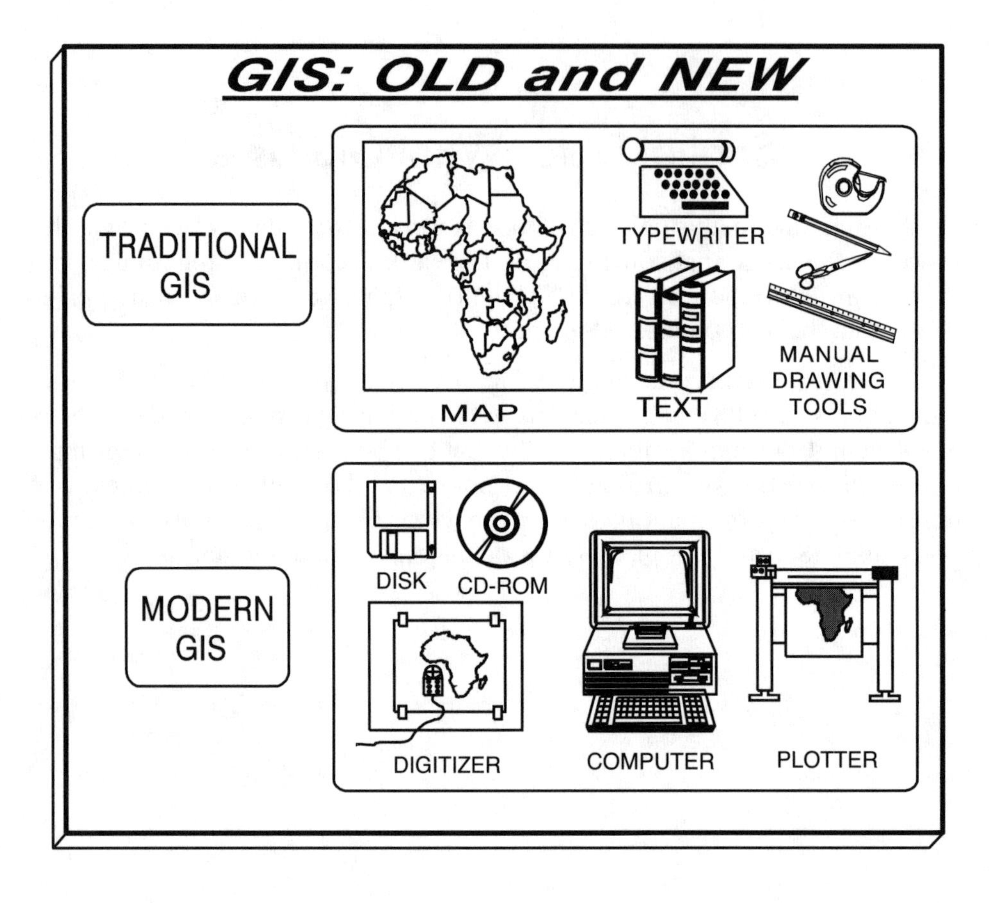
GIS: OLD and NEW
TRADITIONAL GIS
MAP
TYPEWRITER
TEXT
MANUAL DRAWING TOOLS
MODERN GIS
DISK
CD-ROM
DIGITIZER
COMPUTER
PLOTTER

Old and New GIS

Spatial analysis has been around as long as maps have been used for purposes other than simple location and navigation. Traditional GIS has included paper maps and manual tools such as typewriters, drawing implements, and supporting text. This was the "BC" era—Before Computers. The G and the I of GIS were the same as today, but the S was rather elementary.

Modern GIS arrived when computers became powerful, more easy to use, and more affordable. GIS as we know it today involves the use of computer technology to accomplish the traditional tasks of yesterday. The basic goals of computer use and most types of data haven't changed very much, but the methods have altered considerably.

Compare the illustrated equipment. Computer disks hold tremendous amounts of data, the digitizer table converts paper maps into digital data, and the electronic plotter produces maps with very fine detail and in a much shorter time than what once was hand-rendered art. The computer is central to modern GIS technology and is truly changing our professional lives.

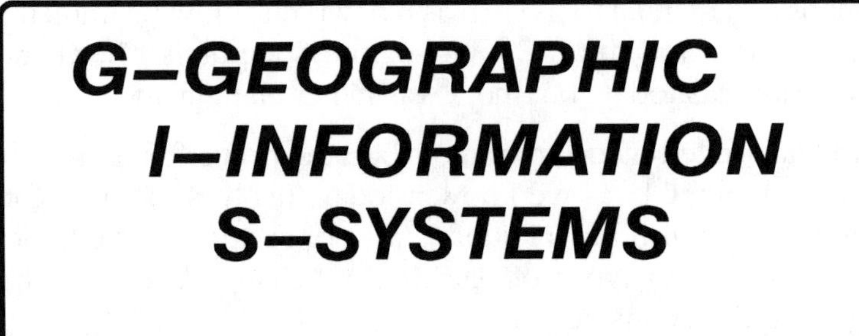

GEOGRAPHIC: REAL WORLD; SPATIAL REALITIES
INFORMATION: DATA AND THEIR MEANING
SYSTEMS: COMPUTER TECHNOLOGY

What Is GIS?

GIS Overview

GIS refers to three integrated parts.

- *Geographic:* Of the real world; the spatial realities; the geography.
- *Information:* Data and information; their meaning and use.
- *Systems:* The computer technology and support infrastructure.

GIS therefore refers to a set of three aspects of our modern world, and offers new methods to deal with them. You will learn later that GIS is much more than a computer system; it is also a methodology (an approach) in science and applications, a new profession, and a new business.

Notes:

- The term *GIS systems* is valid when referring to the computer systems that support GIS.
- The correct plural of GIS is GISs.
- There is no difference in meaning between geographical and geographic; it is correct to use either one. Geographic is used more commonly in the United States, and geographical elsewhere in the world. Whichever you use, be consistent.
- GIS is not pronounced as the single-syllable "jis." The three letters are usually spoken, even though they are considered a single word. (Who says English is easy?)

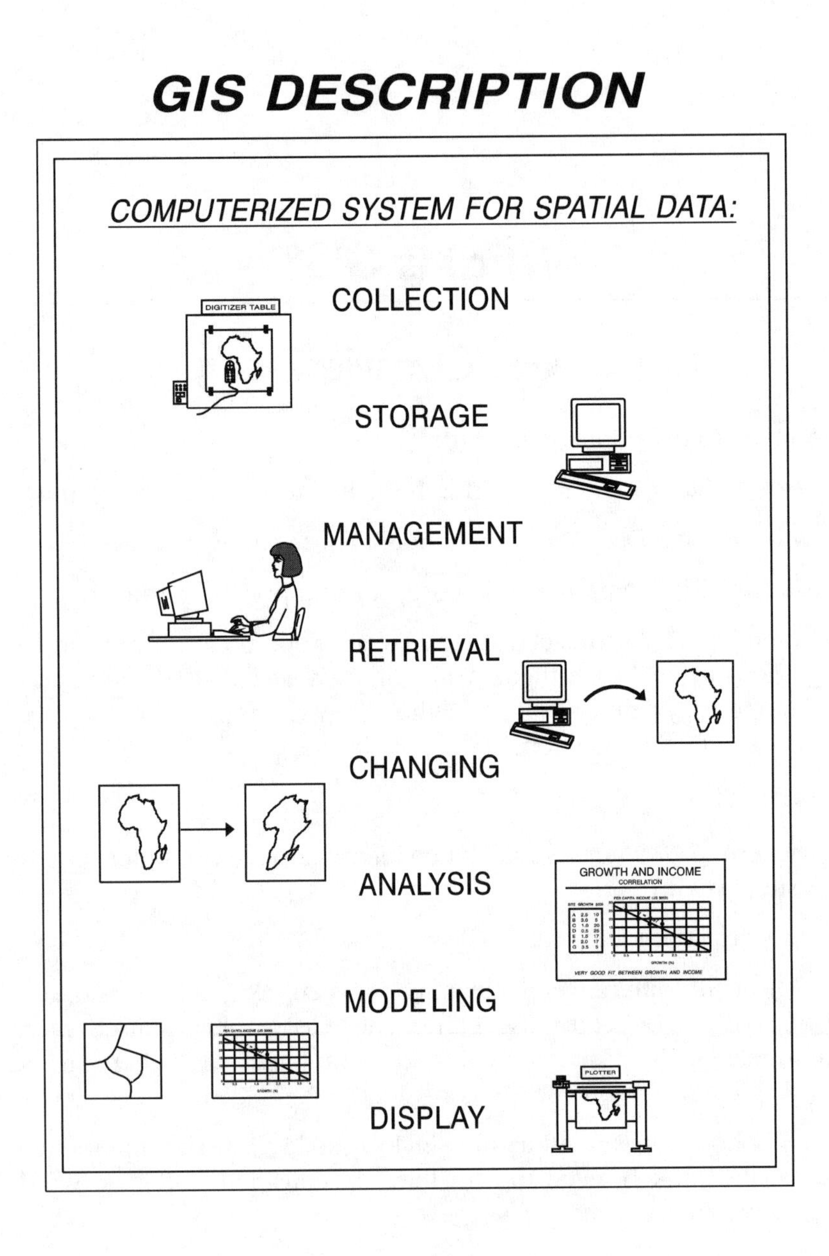
GIS DESCRIPTION
COMPUTERIZED SYSTEM FOR SPATIAL DATA:
DIGITIZER TABLE
COLLECTION
STORAGE
MANAGEMENT
RETRIEVAL
CHANGING
ANALYSIS
GROWTH AND INCOME
CORRELATION
MODELING
DISPLAY
PLOTTER

GIS Description

A very brief definition and description of GIS is that it is a computerized system that deals with spatial data in terms of the following:

- *Collection:* Gathering data from many sources. The illustration shows a digitizing board for converting (by tracing) paper maps into computer maps.
- *Storage:* Efficient digital storage.
- *Management of Data:* Administering and keeping track of data, including integration of various data sets into a common database.
- *Retrieval:* Easy and efficient selection and viewing of data in a variety of ways.
- *Conversion:* Converting data from one form to another; that is, conversion from one geometric projection to another, rescaling, and other computer "tricks" to make the data more useful. Changing one map file to match another.
- *Analysis:* Manipulating data to produce insight and new information.
- *Modeling:* Simplifying the data or the world and its processes to understand how things work.
- *Display:* Presenting data in various ways for easy understanding; for example, maps or reports.

These and associated elements are discussed in detail in material within this chapter.

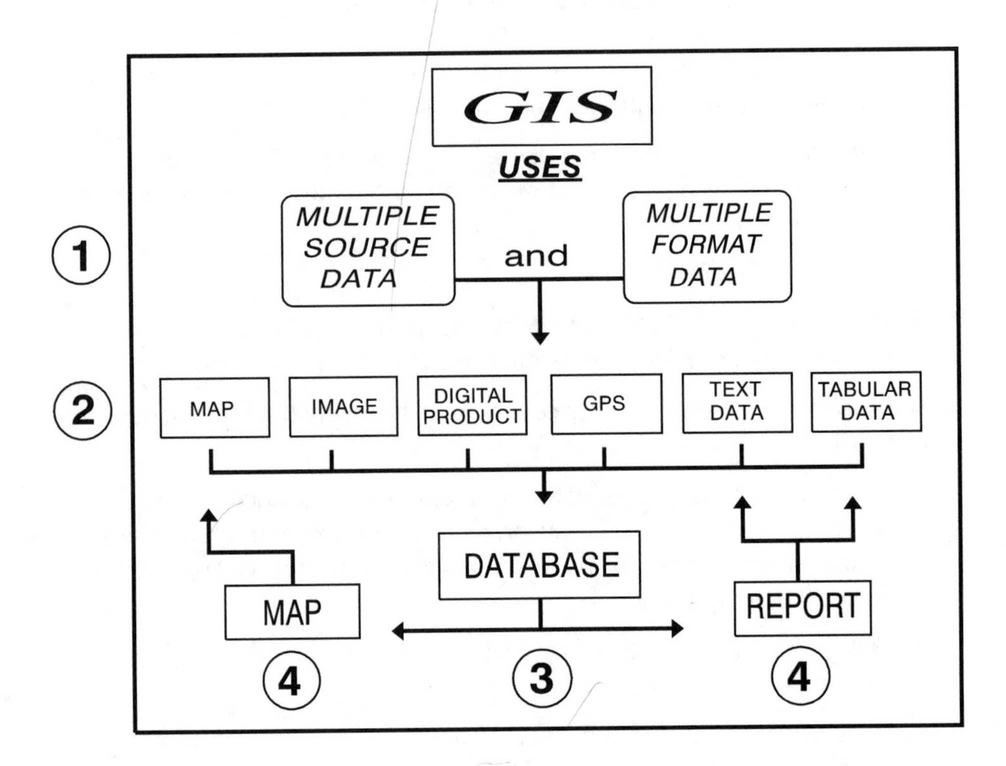
GIS
USES
MULTIPLE SOURCE DATA
and
MULTIPLE FORMAT DATA
1
2
MAP
IMAGE
DIGITAL PRODUCT
GPS
TEXT DATA
TABULAR DATA
DATABASE
MAP
REPORT
4
3
4

Organization

GIS is a "complete system"; it performs many integrated functions. The illustration demonstrates one way of viewing GIS's basic organization and operations. Follow the numbers on the diagram.

1. GIS accepts data from multiple sources, which can be in a variety of formats. In other words, GIS is very flexible in the types and structures of data it can recive and integrate with other data. This is a special strength of GIS that few other technologies share.

2. Data types include the following:

- *Maps:* The most common type of spatial data in existence. Perhaps in the near future digital or computer format will be the most common form of spatial-data storage.
- *Images:* Pictures and digital data from aircraft and satellites; normally termed remote sensing data.
- *Digital Products:* Data sets already stored in digital form that come on computer disks, tapes, CD-ROM disks, or even through telecommunication networks.
- *Global Positioning System (GPS):* A special satellite system that provides highly accurate locational data anywhere in the world, along with other types of information, such as speed of movement and elevation. This can be very useful GIS data, particularly for field work.
- *Text Data:* Reports and text dealing with spatial subjects.
- *Tabular Data:* Lists of numeric data, such as census data.

3. These and other types of data are combined and integrated into GIS in the form of a database, a program or system to hold data conveniently for use in a variety of ways.

4. From the database, products are made, such as maps and reports. These in turn can be recycled back into the database as additional data.

This is a simple organization of GIS. More functions and capabilities will become evident.

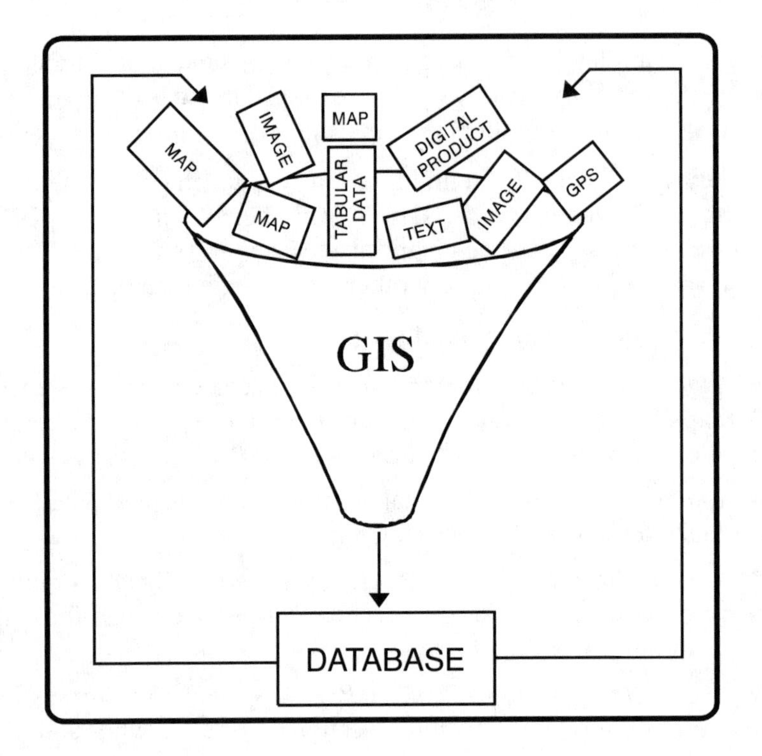
MAP
IMAGE
MAP
DIGITAL PRODUCT
TABULAR DATA
MAP
TEXT
IMAGE
GPS
GIS
DATABASE

GIS Funnel

Another way of viewing GIS is to think of it as a funnel. In effect, GIS acts as a large digital funnel of many types of data in constructing a database, which can then return data and analysis back into the system for subsequent use. GIS is not just a computer mapping system.

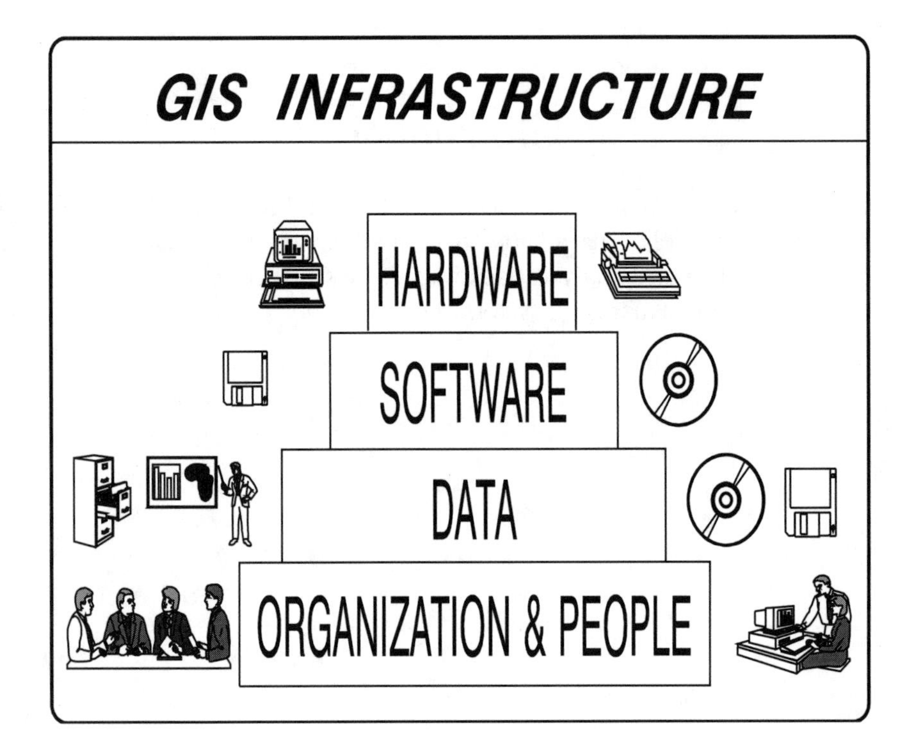
GIS INFRASTRUCTURE
HARDWARE
SOFTWARE
DATA
ORGANIZATION & PEOPLE

Infrastructure

The illustration indicates that the four primary components of GIS can be arranged in a pyramid, in order of importance from the bottom up. This is a basic GIS infrastructure.

- *Organization and People:* The most important part of a GIS structure. Without organization and people, GIS will not work. GIS is too powerful a tool to be considered just another piece of equipment to be used once in a while. To be useful and successful, it must be part of the organization and must have people and facilities dedicated to it.

- *Data:* Without data, there is no meaning or purpose. The focus of attention should be on data. In fact, most work will be devoted to data input and management. Data (information) is the foundation of GIS applications.

- *Software:* The computer programs needed to run GIS. There are many GIS programs available, from low-cost and low-performance packages to expensive and very powerful ones. This also includes support programs, such as statistical, word processing, graphing, and others.

- *Hardware:* The machinery on which GIS operates—computers, printers, plotters, digitizers, and other types of equipment.

Unfortunately, many organizations develop GIS in the wrong order, from the top down, paying more attention to the dazzling hardware and software rather than trying to fit their data to the computer technology. Also, there may be little regard for the people who are to run the system, and the organizational structure required to support it. There are many implementation (start-up) issues that must be considered when developing GIS in an organization. Some consulting businesses exist just for this purpose.

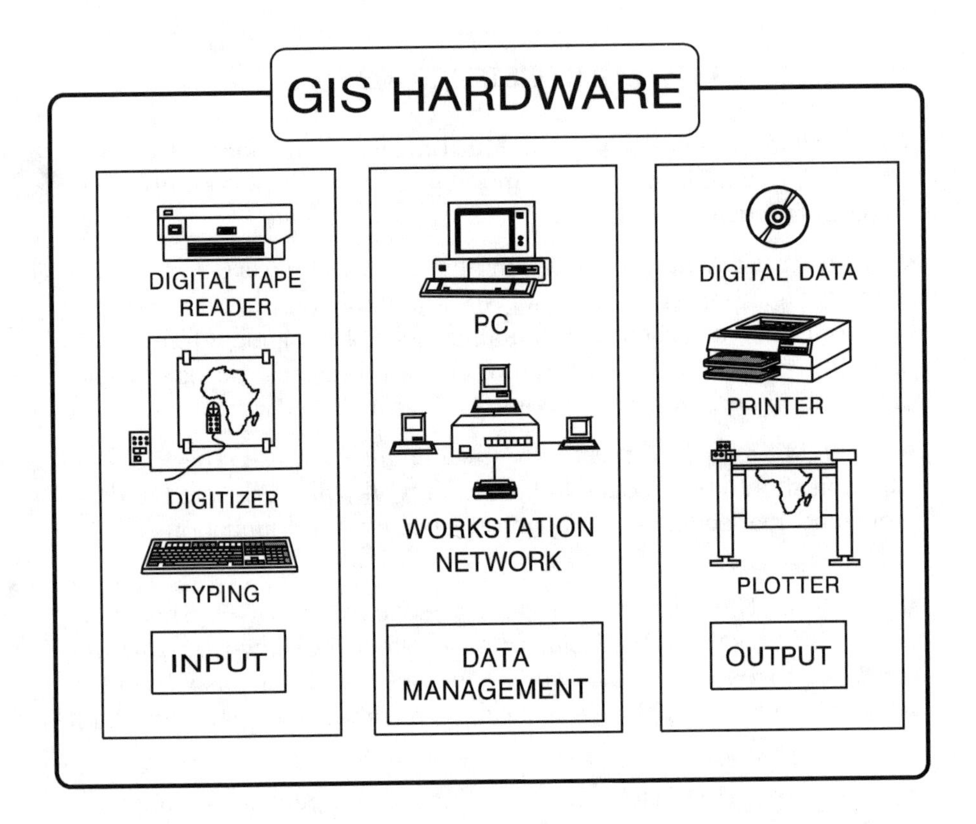
GIS HARDWARE
DIGITAL TAPE READER
DIGITIZER
TYPING
INPUT
PC
WORKSTATION NETWORK
DATA MANAGEMENT
DIGITAL DATA
PRINTER
PLOTTER
OUTPUT

Hardware

A typical GIS hardware infrastructure can be divided, according to purpose, into three basic groups, as shown in the illustration.

- *Data Input:* Data are entered by reading previously prepared digital data (from prior work or from outside sources) on CD disk, tape, or floppy disk; manual digitizing (tracing maps on a digitizer table); or typing from a keyboard.

- *Data Management and Analysis:* The core of a GIS is computer work to manage and analyze data. Illustrated in the middle box are the two common computer "platforms": the personal computer (PC) and the workstation network (a central server and several "satellite" monitors or stations). Other types of central computing equipment are used, and other associated structures may be attached, such as the Internet.

- *Output:* GIS can provide a variety of products. Monitor display is the most common because looking at what is on screen typically precedes actually producing an output product. Output data are usually presented as maps, such as from a plotter or printer. A plotter gives very high-quality output. Accompanying statistics (perhaps tabular data listing records) and reports (perhaps detailing the project) are also frequently printed. Digital data can be output directly to tapes, disk, or a network and then input into another GIS. For these reasons, digital products are becoming more commonly used with GISs.

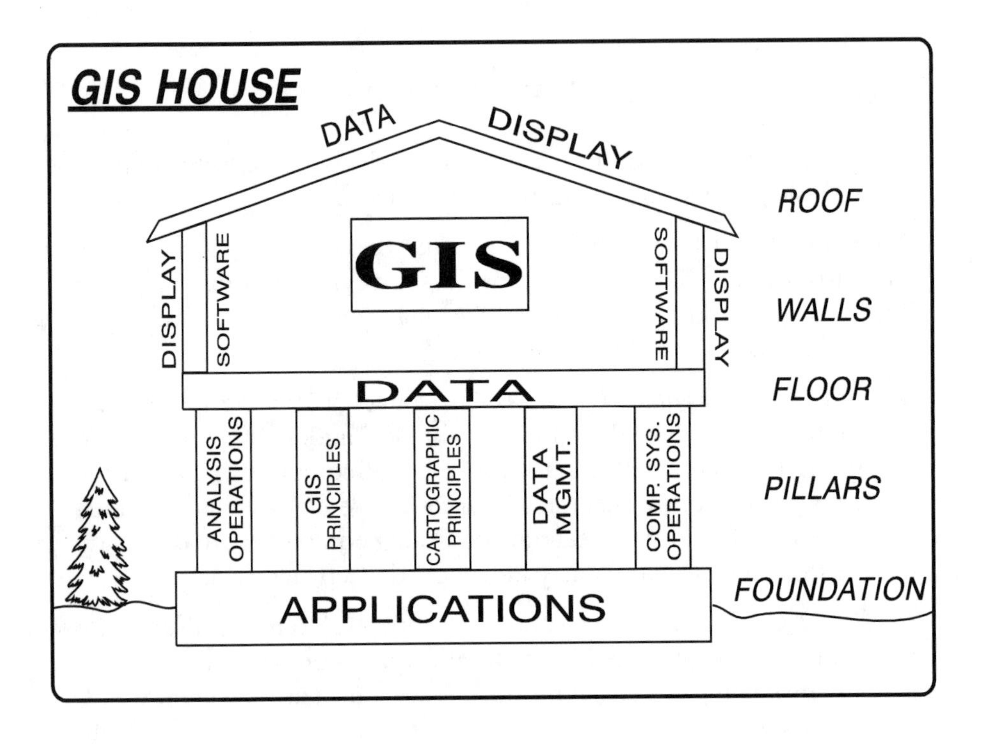

GIS House

Let's think of the GIS organizational structure as a house and its design. Successful operation of GIS normally requires a breadth of knowledge, usually involving several staff members who operate various parts of the structure. The house is built as

- ◆ *Foundation:* The base level. In GIS, the fundamental base, or major driving purpose, is applications (its uses). The most important part of a project is its purpose. Without a good knowledge of the application or project goals, GIS can be without direction, and failure can result.

- ◆ *Pillars:* The "support structures" (specializations) linking the Applications foundation to the rest of GIS. Five examples follow, though others exist.

- *Analysis Operations:* Knowing the application and how to analyze its data are essential. Someone has to know about analytical techniques if more than simple mapping is desired.
- *GIS Principles:* It is necessary to understand what GIS is, how it operates, its strengths and weaknesses, and the many concepts that make it useful to various applications.
- *Cartographic Principles:* Mapping must conform to cartographic standards, such as using or excluding specific map items, or making sure the theme of the end product is easily interpreted. Someone has to know the principles of constructing and presenting maps, reports, and other output products.
- *Data Management:* Managing data can be a full-time task, often requiring a dedicated professional. If data are lost or in poor condition, the project faces significant difficulties. This person is part computer technician and part applications specialist.
- *Computer Systems Operations:* Because GIS is a computer technology, a technician is necessary to perform equipment repairs and system maintenance.

◆ *Floor:* Supporting the "living space" of GIS is the data. It is the base (or "floor") for analysis and information.

◆ *Walls:* Defining the living space are the walls—the software programs that manage, analyze, and present project data. Software surrounds the data space. Most GIS work is done with the software, and most training is aimed toward its efficient use.

◆ *Roof:* The outside walls and top of the GIS house are Data Presentation, where the most visible part of the project emerges. The roof, supported by software walls, is the "cap" on the project.

A GIS consists of many parts, each of which is important to the form, strength, and "livability" of the structure.

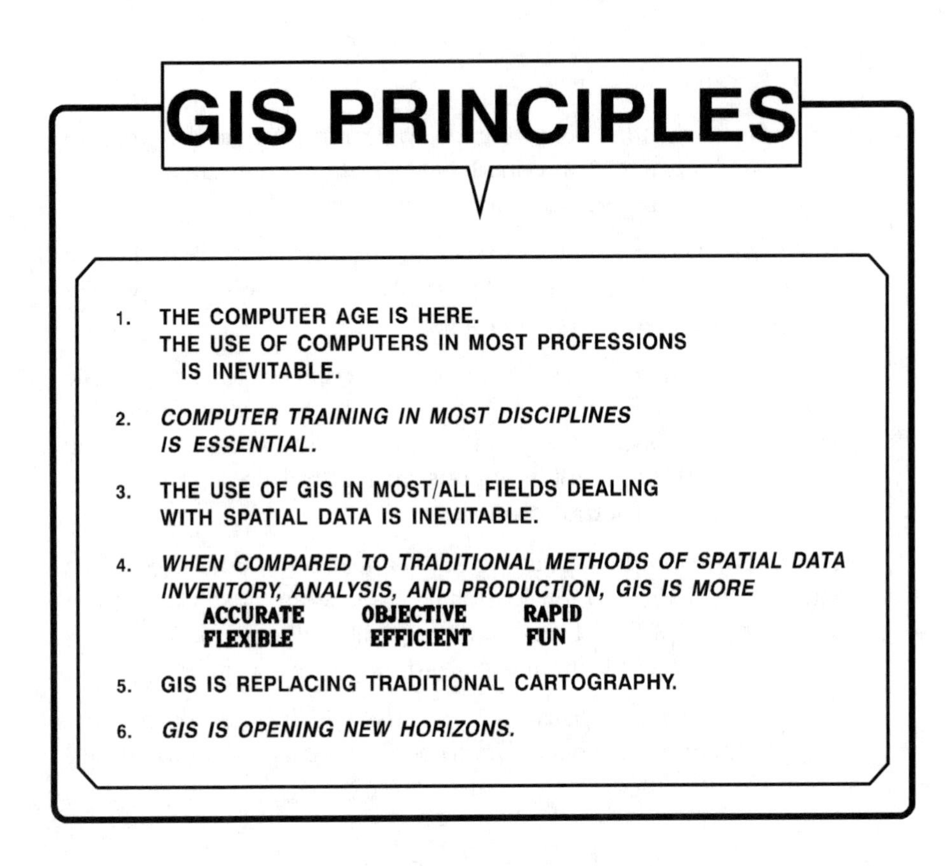

GIS Principles

Computers and GIS are creating significant change. We can recognize six key principles important to an understanding of this new technology.

1. The computer age is here and it cannot be ignored. Many, perhaps most, professions are using computer technology and computerized methods. The word *inevitable* is valid here. There is no doubt that computer technology will continue to be a necessary element in almost all professions.

2. Computer training in most academic and professional fields is essential. Without it, students will be handicapped in their chosen professions and undoubtedly will learn computers on the job, without proper preparation.

3. GIS has become the accepted and standard means of using spatial data. The use of spatial data is growing very rapidly in diverse fields. Phone companies, banks, advertising firms, emergency services, and many other enterprises, public and private, have adopted GIS as a major support technology. The adoption of GIS by many other fields and professions is inevitable.

4. GIS has in most cases proven to have many advantages over older methods of making use of spatial data. When compared to traditional methods, GIS has numerous benefits, including the fact that for many it is simply more fun to use than outdated techniques.

5. Much of traditional "pen and ink" cartography done by skilled draftspersons and artists is being replaced by GIS. Hand-rendered cartography, other than that desired as art, is becoming outdated because it is expensive, is not easily altered, is difficult to store, and is expensive to reproduce.

6. GIS is opening new horizons. It is not simply a computerized version of traditional means of accomplishing tasks, but is an innovative area in which new modes of analysis and new applications are constantly discovered. GIS is a catalyst (or stimulus) for many advancements.

This chapter has introduced GIS and explained how important data and information are to it. We've seen what GIS is, its components and parts, and a bit of how it functions. Remember that people and organization are the most important parts of GIS, and that data and applications are the critical guidelines for its use. The very high value of GIS should be apparent. Let's now turn to a more in-depth view of what GIS does.

2

What Does GIS Do?

Introduction

We have reviewed the basic components and structure of GIS and now need to see what it can do. This chapter presents a basic preview of GIS by discussing some of the generic questions it can consider and answer. These types of operations and applications are presented in more detail in later chapters.

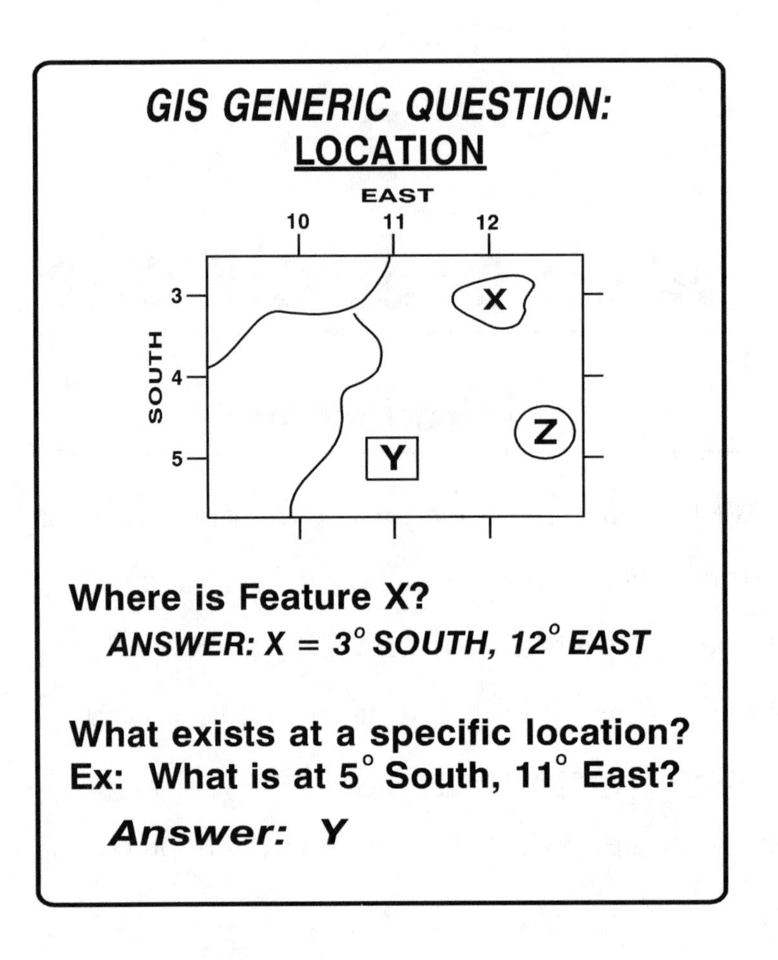
GIS GENERIC QUESTION:
LOCATION
EAST
10
11
12
3
4
5
SOUTH
X
Y
Z
Where is Feature X?
ANSWER: X = 3° SOUTH, 12° EAST
What exists at a specific location?
Ex: What is at 5° South, 11° East?
Answer: Y

Preview of GIS Functions

Location

Spatial data deals with location, and GIS addresses questions concerning the "what" and "where" of data. The illustration shows a simple map, marked on its edges with latitude and longitude "ticks" (small marks) representing coordinate values in degrees. On the map are several features: two types of lines (perhaps paved and unpaved roads), and three shapes (X, Y, and Z, which might be agricultural areas).

The first question asks where a particular feature is, usually in terms of a coordinate system. "Where is feature X?" is answered by making horizontal and vertical coordinate points and answering according to the given system: 3 degrees South and 12 degrees East. On the flip side of the question, the location might be given and the feature is to be identified. The location asked is 5 degrees South and 11 degrees East, which, when followed to where they cross on the map, arrive at feature Y.

Location is the fundamental characteristic of spatial data, and GIS considers position directly (site) and indirectly (situation). It gives the site (a specific place) in terms of a coordinate system, and the situation (relative position) using the surroundings. Location is a principle aspect of GIS data.

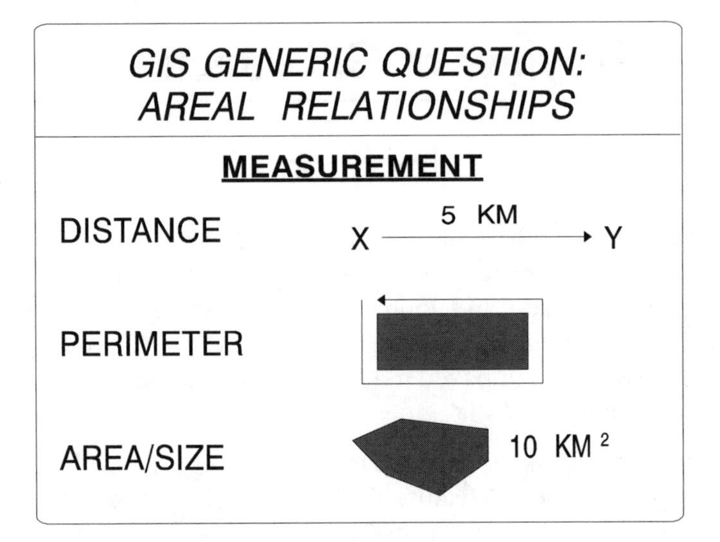

GIS GENERIC QUESTION:
AREAL RELATIONSHIPS
MEASUREMENT
DISTANCE
5 KM
X
Y
PERIMETER
AREA/SIZE
10 KM 2

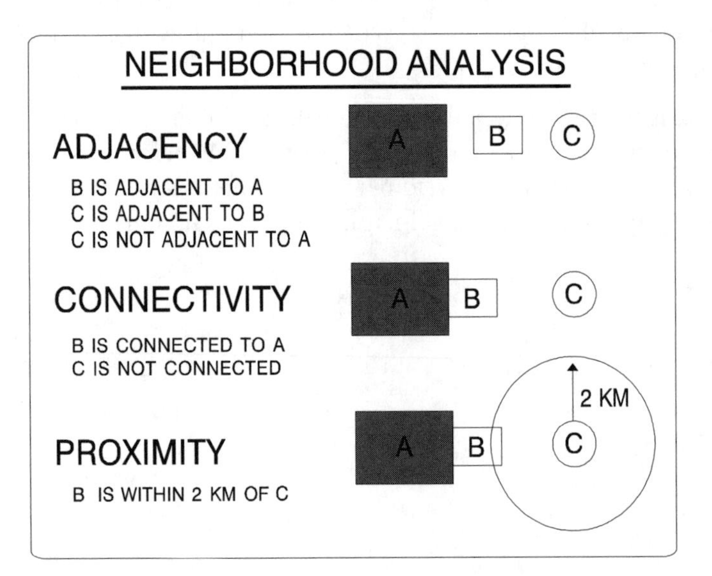

NEIGHBORHOOD ANALYSIS
ADJACENCY
B IS ADJACENT TO A
C IS ADJACENT TO B
C IS NOT ADJACENT TO A
A
B
C
CONNECTIVITY
B IS CONNECTED TO A
C IS NOT CONNECTED
A
B
C
PROXIMITY
B IS WITHIN 2 KM OF C
A
B
C
2 KM

Measurements

Measurement is relatively easy in GIS, and several types of spatial "measurement" analysis can be achieved, typically using automatic (or built-in) functions of the software. These are shown in the top portion of the illustration.

◆ *Distance:* Measuring the length from one point to another usually requires only definition of the points; that is, pointing to the beginning and end. Most GISs give real-world distance (such as kilometers), but some low-end systems need the operator to translate internal units to real-world equivalents.

◆ *Perimeter:* Getting the distance around features may be very easy, often requiring only selection of the item and reading the value from the database. Some GISs automatically calculate perimeter and have it stored in the database from the outset.

◆ *Area and Size:* Like perimeter, some GISs automatically calculate and store the spatial extent of each feature. Low-end GISs may offer indirect areal (area) measurement, requiring the operator to translate internal units to real-world area. In either case, these types of measurements should be easy to obtain.

Other types of spatial measurement or relationships may be available. These deal with "neighborhood" characteristics and distance. Three of the most common are presented in the bottom portion of the illustration.

◆ *Adjacency:* What is next to or near a given feature or category of features. In the illustration, feature B is adjacent to feature A, and C is adjacent to B, but C is not adjacent to A. These features could be almost any type of data found on the map.

◆ *Connectivity:* Physical connection is an important spatial characteristic for many applications. What is actually connected to features can be calculated. In the illustration, B is connected to A, but C is not connected to anything. Can you think of a good application for this information?

◆ *Proximity:* As in adjacency, distance is an important spatial factor of the positions of features. The minimum distance that constitutes proximity (spatial association, or "nearness") can be defined. Therefore, features within certain distances and positions are calculated. In the illustration, the distance from C has been set at 2 km, and only feature B has the required proximity.

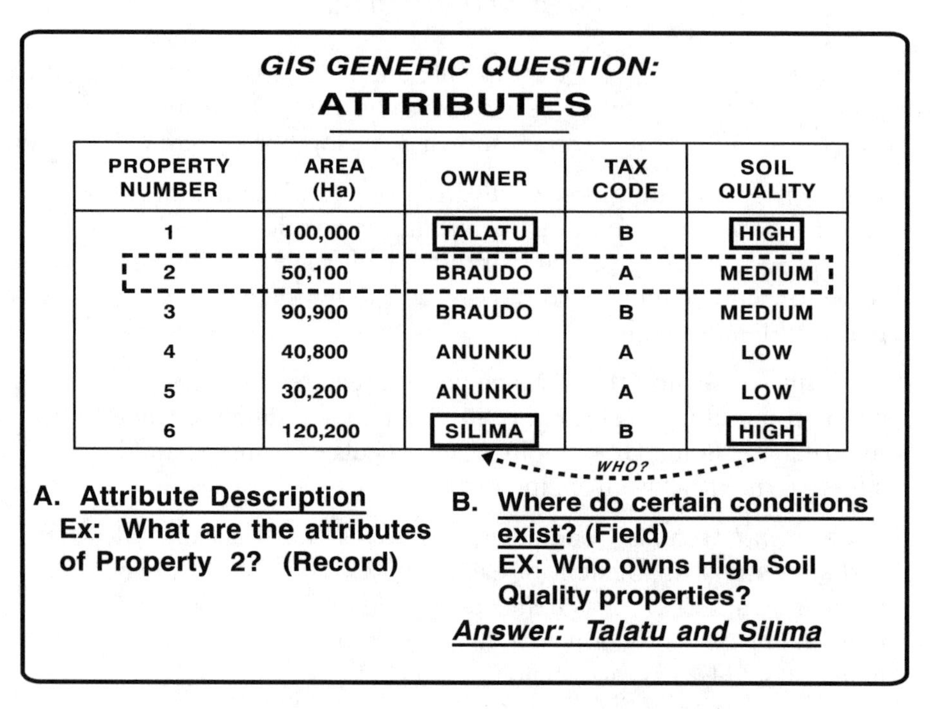

PROPERTY NUMBER	AREA (Ha)	OWNER	TAX CODE	SOIL QUALITY
1	100,000	TALATU	B	HIGH
2	50,100	BRAUDO	A	MEDIUM
3	90,900	BRAUDO	B	MEDIUM
4	40,800	ANUNKU	A	LOW
5	30,200	ANUNKU	A	LOW
6	120,200	SILIMA	B	HIGH

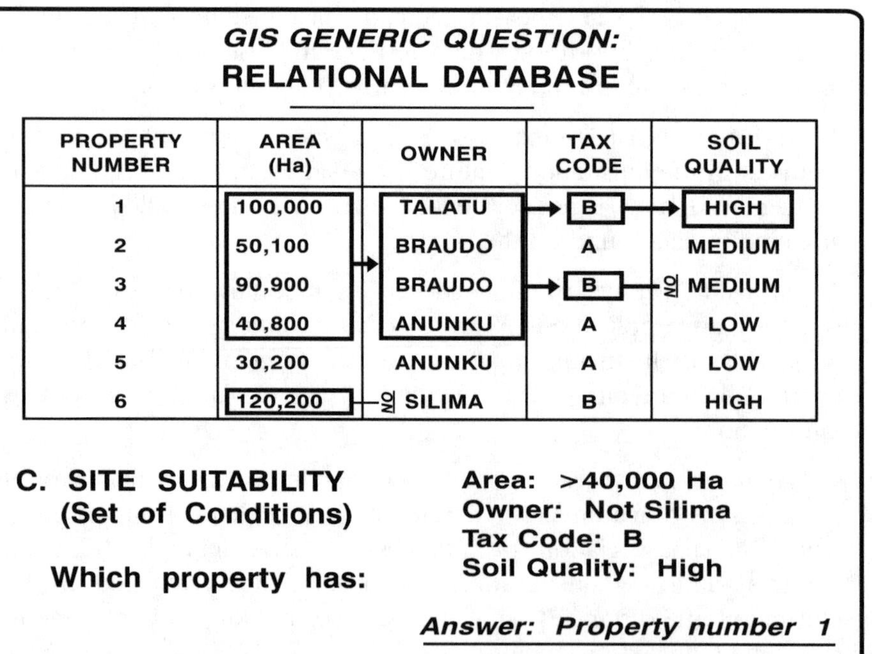

PROPERTY NUMBER	AREA (Ha)	OWNER	TAX CODE	SOIL QUALITY
1	100,000	TALATU	B	HIGH
2	50,100	BRAUDO	A	MEDIUM
3	90,900	BRAUDO	B	MEDIUM
4	40,800	ANUNKU	A	LOW
5	30,200	ANUNKU	A	LOW
6	120,200	SILIMA	B	HIGH

NO
NO
C. SITE SUITABILITY
(Set of Conditions)
Which property has:
Area: >40,000 Ha
Owner: Not Silima
Tax Code: B
Soil Quality: High
Answer: Property number 1

Attributes

An attribute is a description of a feature—a characteristic of it. An attribute can be a nonspatial aspect of spatial data, such as name, classification, or color. It may be quantitative (numeric) or qualitative. The upper illustration shows a small database (in tabular form), with six agricultural properties (numbers 1 through 6) and their attributes of area, owner, tax code, and soil quality. A database is a very convenient and efficient way to manage attribute data. In a database, rows are called "records" (in this case, for each property) and the columns, or attributes, are "fields."

A common question (called "generic" in the illustration) asks the attributes of a given feature, such as "What are the attributes of property 2?" The answer is outlined in the upper illustration by the dashed line, with its area of 50,100 hectares (ha), owner Braudo, tax code A, and medium soil quality.

Another easy question concerns where certain attributes or conditions exist: "Who owns High Soil Quality properties?" The procedure is to "browse" the Soil Quality field (column) and then match all "Highs" with the Owner field. Talatu and Silima are the only High Soil Quality property owners.

The lower illustration is the same database. It shows a special type of question, called "site suitability," which addresses sets of conditions to find the most suitable site according to a list of criteria. Here, the task is to find the most suitable property or properties that have the following attributes:

◆ Area over 40,000 ha

◆ Owner who is anyone but Silima (we wonder what is wrong with Silima)

◆ Tax code of B

◆ High soil quality

This is a relational database, with the powerful capability to relate one field or set of records with another. In this site suitability example, the database selects those records meeting conditions in one field and then moves to the next field, relating the previous selections to the records in the new field.

The program begins with Area, and from that list selects all records meeting the over-40,000-ha condition. It then searches owners to locate and exclude Silima. From that list, B tax codes are selected, followed by selection of the High Soil Quality attribute. Only property 1 meets all of the criteria, making it the only suitable site.

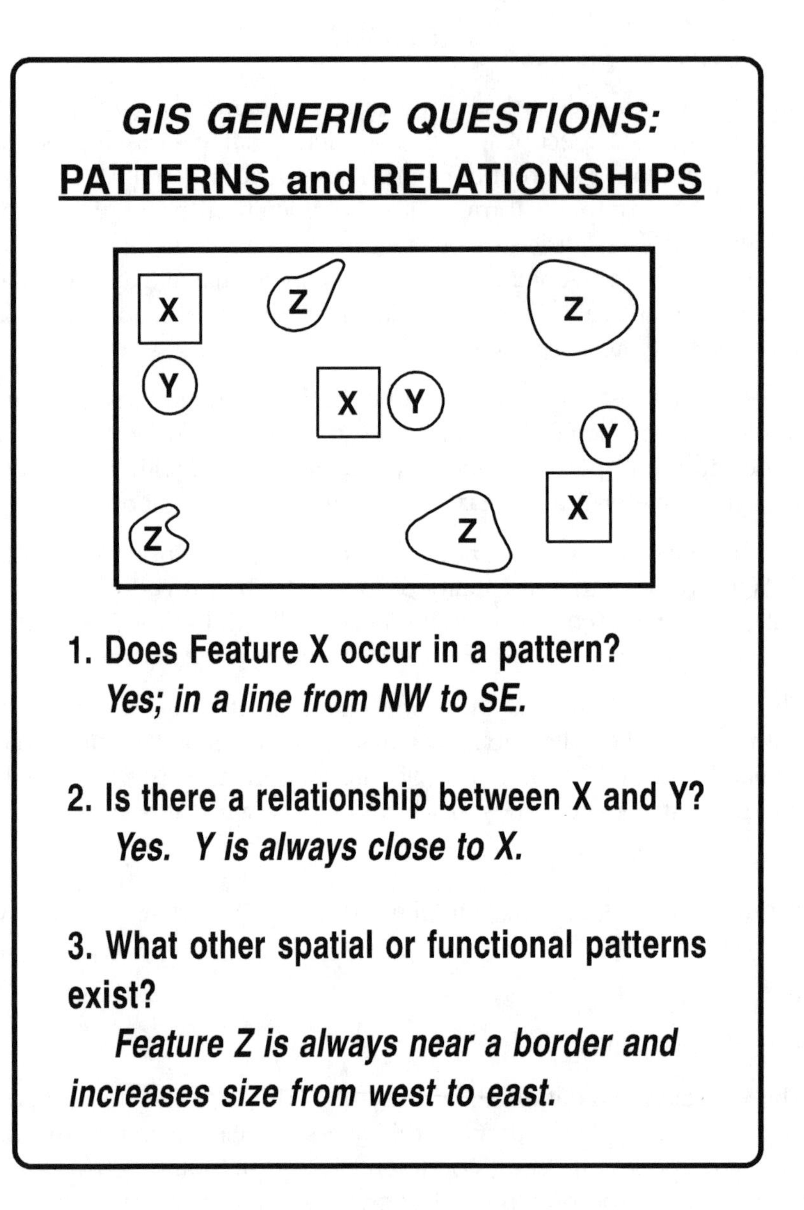
GIS GENERIC QUESTIONS:
PATTERNS and RELATIONSHIPS
X
Z
Z
Y
X
Y
Y
X
Z
Z
1. Does Feature X occur in a pattern?
Yes; in a line from NW to SE.
2. Is there a relationship between X and Y?
Yes. Y is always close to X.
3. What other spatial or functional patterns exist?
Feature Z is always near a border and increases size from west to east.

Patterns and Relationships

All map features have spatial relationships of various types. Some are easy to see with the eye, whereas others require GIS work. Three relationships are presented in the illustration. Some GISs may answer the following types of questions directly, but others require work from the operator and several steps to extract results. (Numbers correspond to the illustration.)

1. *Does feature X occur in a pattern?* Careful visual observation shows that feature X (could be a vegetation type) occurs in a line (pattern) running from upper left to lower right. It may be easy to see the pattern in this simple map, but consider a complex, multiple-feature landscape, where sophisticated assistance might be needed.

2. *Is there a relationship between X and Y?* Again, the eye reveals that Y is always very close to X. Also, it may be that the Y position changes in a pattern from one X to the other. Can you see this spatial relationship?

From these spatial relationships, the operator may learn functional associations. For example, if X and Y are always together, there may be an important reason for that linkage. Such information often leads to other questions, which eventually lead to discoveries or questions that had not previously been considered.

3. *What other spatial or functional patterns exist?* We can see that feature Z is always near a border, and in fact increases in size from west to east (or decreases size east to west). Also, it seems to have a zigzag (up and down) location pattern. Again, there may some functional reason for this pattern other than mere coincidence, and GIS might help to discover that reason and its process.

Spatial patterns and relationships are important characteristics suitable for GIS exploration. Where traditional cartography and spatial analysis were limited in their abilities to extract data and to see complex relationships, GIS uses computer technology to investigate the landscape and to discover things the eye cannot discern.

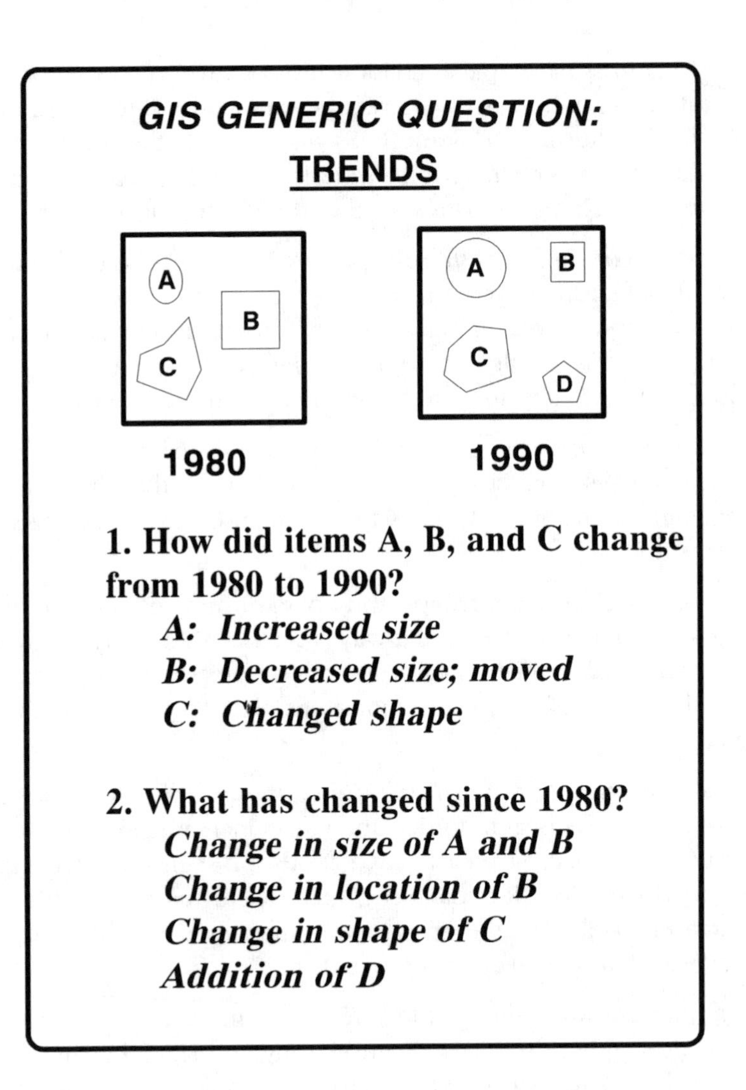
GIS GENERIC QUESTION:
TRENDS
A
B
C
A
B
C
D
1980
1990
1. How did items A, B, and C change from 1980 to 1990?
A: Increased size
B: Decreased size; moved
C: Changed shape
2. What has changed since 1980?
Change in size of A and B
Change in location of B
Change in shape of C
Addition of D

Trends

In the previous illustration some spatial trends were discovered. When time is added, temporal (relating to time) trends can be considered. If something changes in a predictable manner in a given amount of time, it might change in a similar way in another, identical time increment. In the illustration, two landscapes are viewed over a ten-year period (1980 to 1990). Spatial changes are to be expected, and GIS can reveal them. How did items (features) A, B, and C change from 1980 to 1990?

- A increased size.
- B decreased size and changed position.
- C changed shape (but size remained about the same, though that can be determined exactly by GIS if needed).

Another view of the question: What has changed since 1980?

- A and B changed size.
- B changed location.
- C changed shape.
- D was added.

In summary, we can use GIS to see how spatial features change through time, or to see how time affects spatial features.

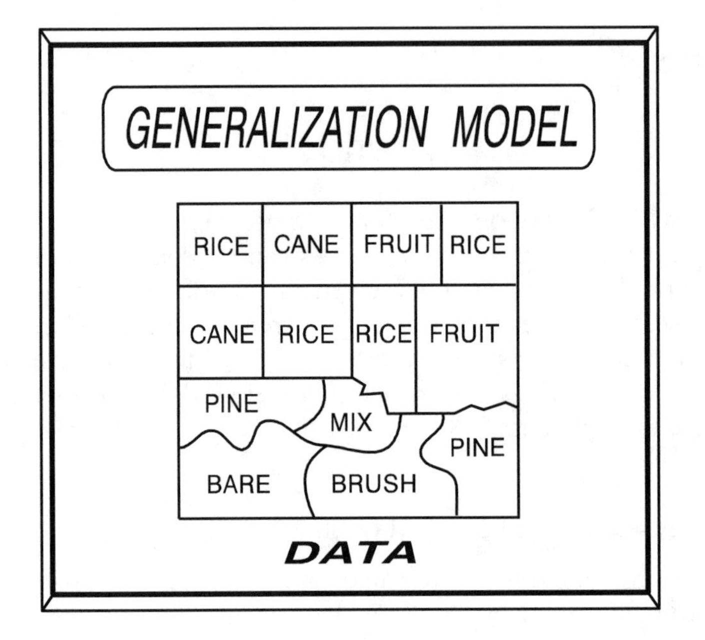
GENERALIZATION MODEL
RICE
CANE
FRUIT
RICE
CANE
RICE
RICE
FRUIT
PINE
MIX
PINE
BARE
BRUSH
DATA

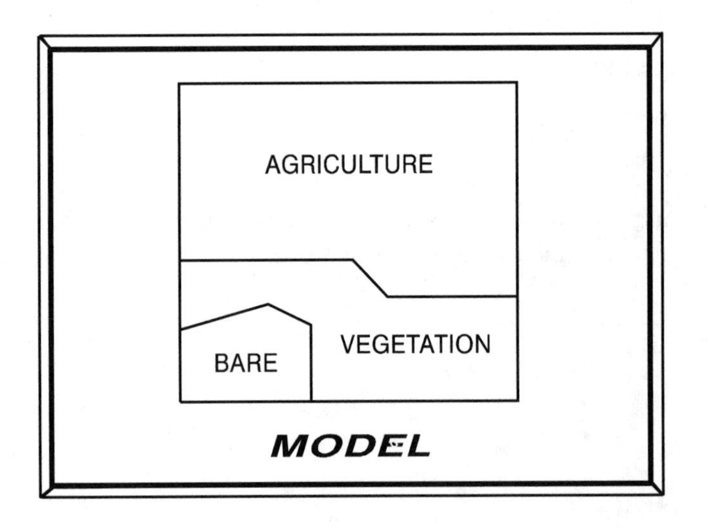
AGRICULTURE
VEGETATION
BARE
MODEL

Modeling

What Is Modeling?

The high power of computers now permits the advanced analysis called "modeling." Models are representative versions of the world and its processes. Models summarize data and analysis to make general statements on the way things exist and operate. They help to reduce data clutter and to reveal generalizations.

Illustrated is a simple, generalized model of an agricultural area. At the top of the illustration, titled Data, is a map of what the initial data reveal: specific crop and vegetation types, and detailed spatial configurations of the fields. At the bottom of the illustration is a simplified model: generalized land cover and spatial patterns. All crops have been recoded and combined into the category Agriculture, and all nonagricultural plant growth has been collected under Vegetation. The small twists and turns of boundaries have been smoothed to general shapes. This generalization serves to simplify the landscape for additional comparison and analysis.

Both versions of the landscape may be useful. The original data are beneficial to detailed and highly accurate study, whereas the generalized model is useful for other purposes, such as categorizing broad areas of land.

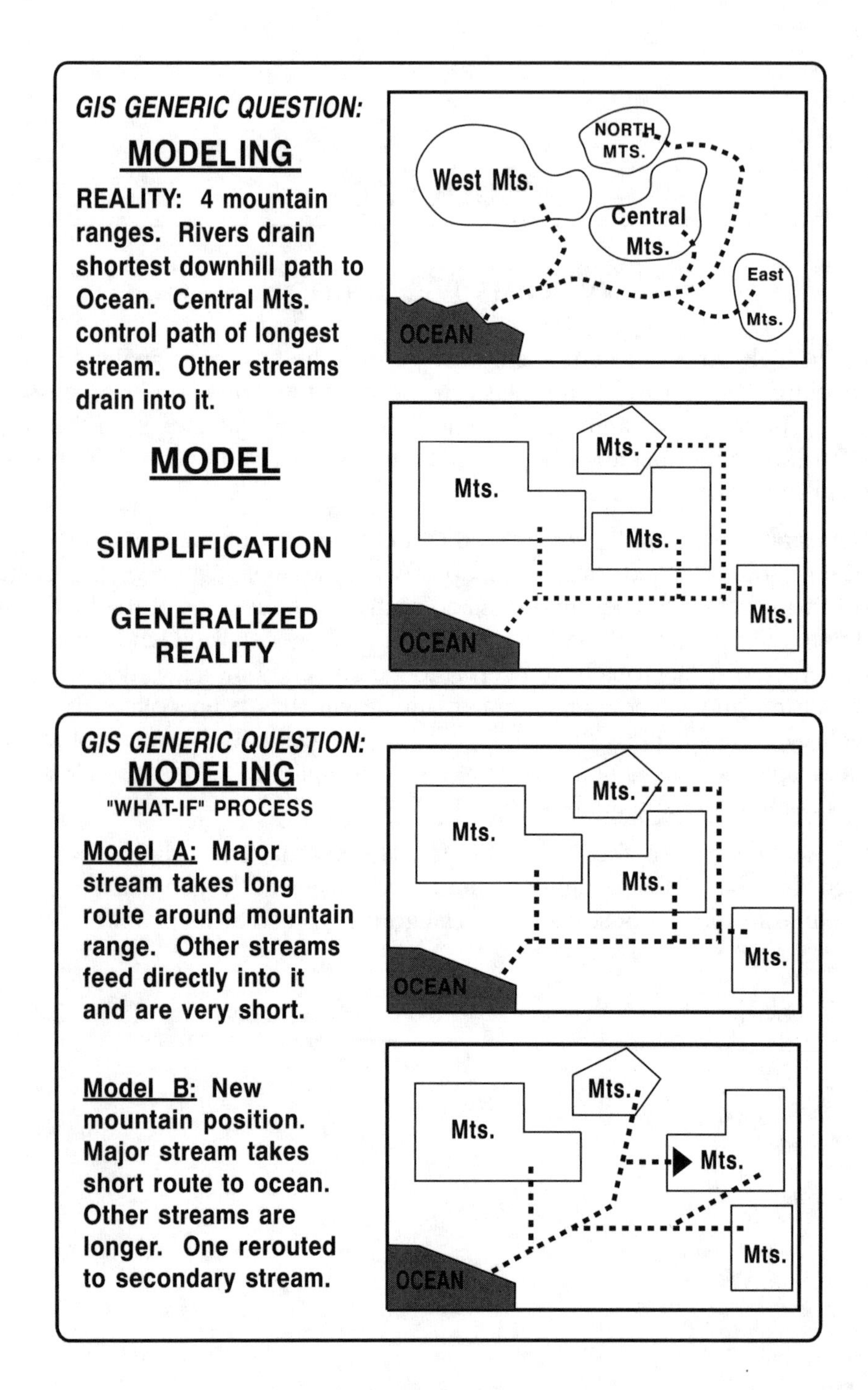
GIS GENERIC QUESTION:
MODELING
REALITY: 4 mountain ranges. Rivers drain shortest downhill path to Ocean. Central Mts. control path of longest stream. Other streams drain into it.
NORTH MTS.
West Mts.
Central Mts.
East Mts.
OCEAN
MODEL
SIMPLIFICATION
GENERALIZED REALITY
Mts.
Mts.
Mts.
Mts.
OCEAN
GIS GENERIC QUESTION:
MODELING
"WHAT-IF" PROCESS
Model A: Major stream takes long route around mountain range. Other streams feed directly into it and are very short.
Mts.
Mts.
Mts.
Mts.
OCEAN
Model B: New mountain position. Major stream takes short route to ocean. Other streams are longer. One rerouted to secondary stream.
Mts.
Mts.
Mts.
Mts.
OCEAN

GIS Modeling

GIS helps to discover and to create spatial models. These illustrations show both spatial and process models. The upper illustration shows a generalization of spatial configurations. The "reality" of the landscape is in the top map, with mountain ranges having odd shapes and rivers flowing in their real paths. Also, a basic landscape description is given, explaining the hydrology, or river systems.

The bottom map of the upper illustration is a model—a spatial simplification (or "generalized reality" picture). The mountains and rivers have been given geometric shapes. Sometimes complex patterns, such as the river patterns here, are more clear and better understood.

The lower illustration shows use of the model in trying to understand what would happen under various conditions. This is a "what-if" scenario or condition. The top map is a repeat and explanation of the landscape model used in the upper illustration. Note that the long stream routes around a mountain range and the other streams feed directly into it.

The bottom map of the lower illustration shows what the landscape might look like *if* the Central Mountains were located farther east. The hydrology would be different, and apparently the resultant landscape effects would be different, such as vegetation distribution and topography.

These types of "what-if" GIS models are useful in trying to evaluate landscape changes under various conditions (though we seldom move entire mountain ranges!). The cultural landscape is highly dynamic (changing), and modeling can be an important tool in landuse planning. For example, this illustration could be a city with various landuses, and evaluating the effects of relocating one activity for transportation rerouting could be very beneficial. Environmental management is a strong application of GIS modeling.

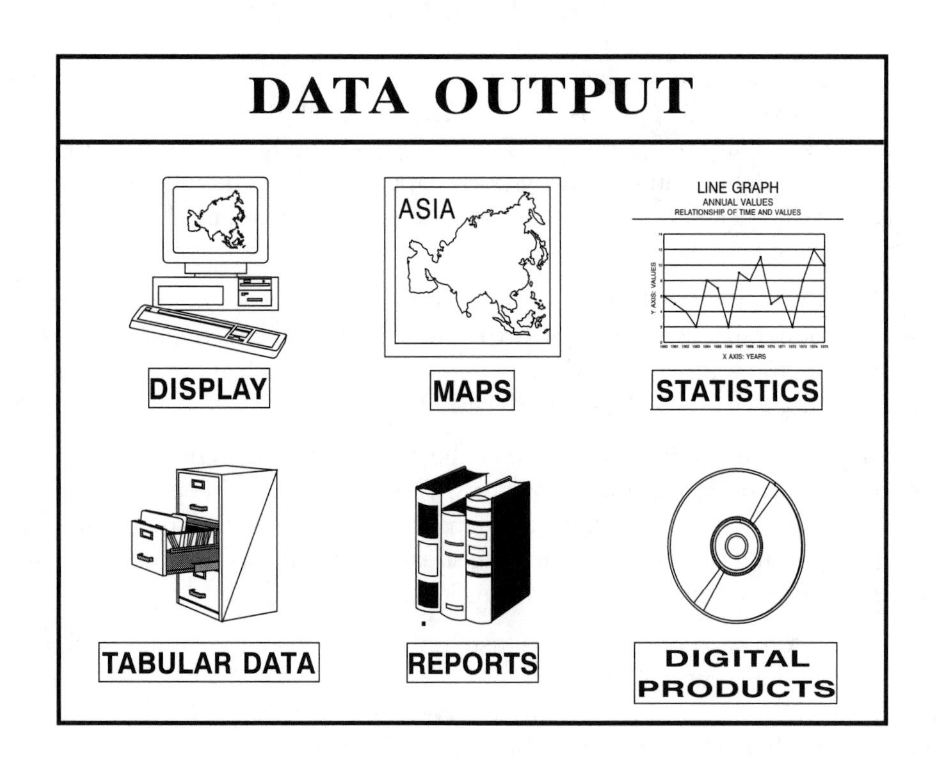

Data Output

GIS can provide a variety of products. Monitor display is the most common, obviously, because it is used constantly in GIS work. Maps are the most standard output format, but frequently are accompanied by statistics, perhaps tabular data listing records and other data, and reports to detail the project. Digital products are becoming more common due to efficient delivery and easy incorporation into the user's computer.

The Nature of GIS

A few general statements can be made that are important in understanding GIS technology and applications and their place in society. GIS is used in many fields and fulfills numerous roles in connection with its applications within those fields.

GIS Is Multidisciplinary

GIS is multidisciplinary; it is not just a geography tool, or just a piece of computer science equipment, but belongs to and is used by a variety of academic disciplines and applications. It is also equally useful and important to many businesses and government agencies. GIS is not confined to a single compartment of knowledge or to one branch of education, but is truly universal in its utility.

The world is holistic; few processes or activities in nature or culture are isolated and independent of all others. The natural and cultural worlds are connected in many ways; they are integrative (they tend to integrate). GIS is an ideal technology to use in dealing with diverse processes.

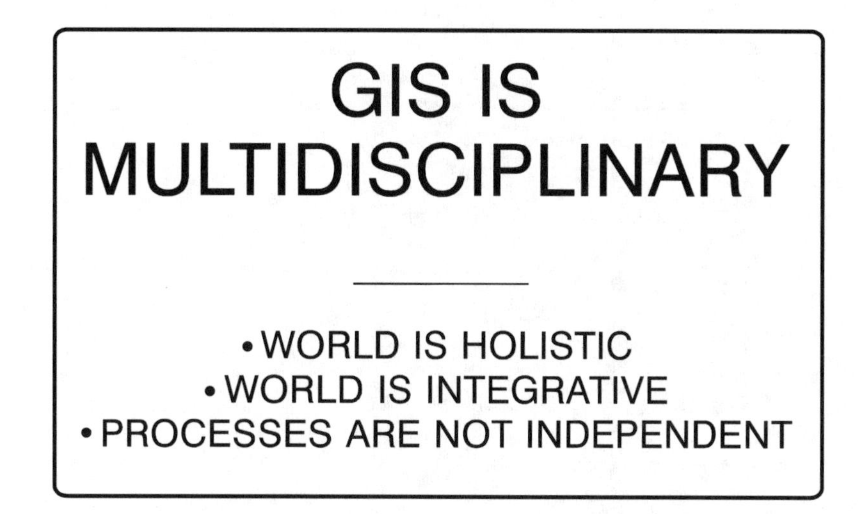

GIS IS
MULTIDISCIPLINARY
• WORLD IS HOLISTIC
• WORLD IS INTEGRATIVE
• PROCESSES ARE NOT INDEPENDENT

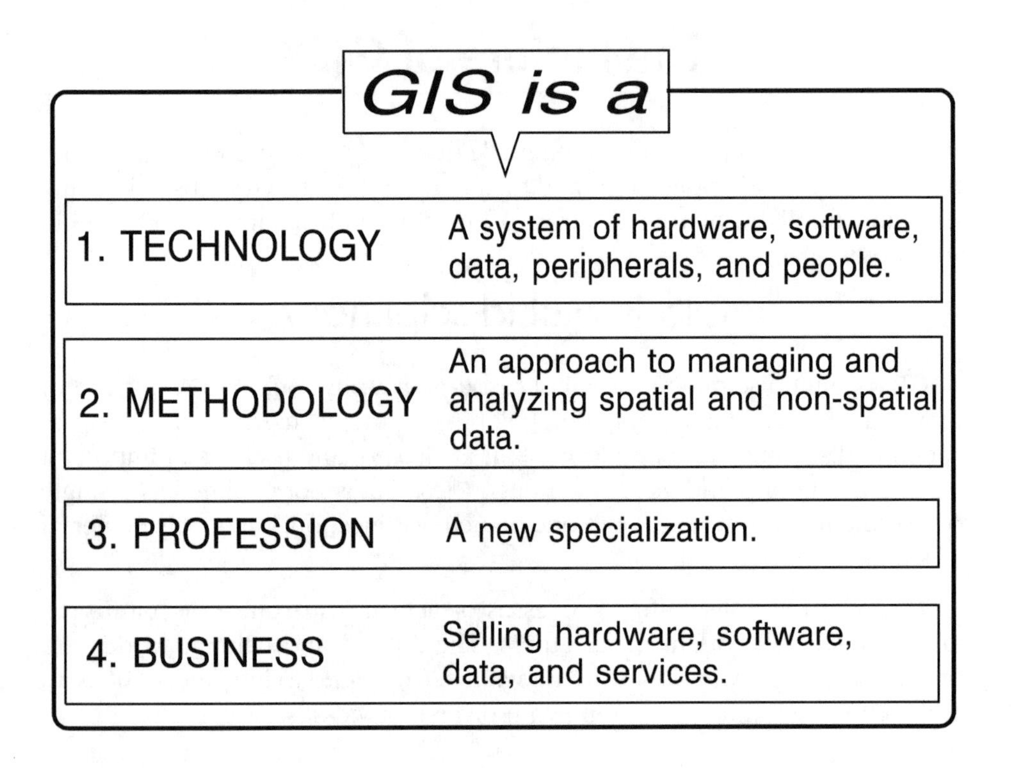
GIS is a
1. TECHNOLOGY
A system of hardware, software, data, peripherals, and people.
2. METHODOLOGY
An approach to managing and analyzing spatial and non-spatial data.
3. PROFESSION
A new specialization.
4. BUSINESS
Selling hardware, software, data, and services.

GIS Roles

As noted, GIS is much more than a computer system, and that it plays several roles. We can say that GIS is a

◆ *Technology:* GIS is a system of hardware, software, peripherals (associated equipment), and people.

◆ *Methodology:* GIS is an approach to managing and analyzing spatial and associated nonspatial data. GIS has led to new ways of accomplishing many traditional and innovative tasks, allowing us to take a different approach to solving problems.

◆ *Profession:* As a new field, there is a demand for GIS specialists, thereby making it a new profession or career. GIS operation is far more than a technical skill to be mastered by technicians and scientists. Because of the wide range of its applications and breadth of technology it encompasses, GIS is a field developing into a distinct profession. The first generation of GIS specialists is beginning to appear. The GIS professional helps to integrate a variety of needs and many types of data, and supplies the technical know-how to make GIS work.

◆ *Business:* Many new businesses have taken advantage of the rapidly growing commercial aspects of GIS to offer hardware, software, data, services, and other marketable items. GIS consulting and services are growing rapidly.

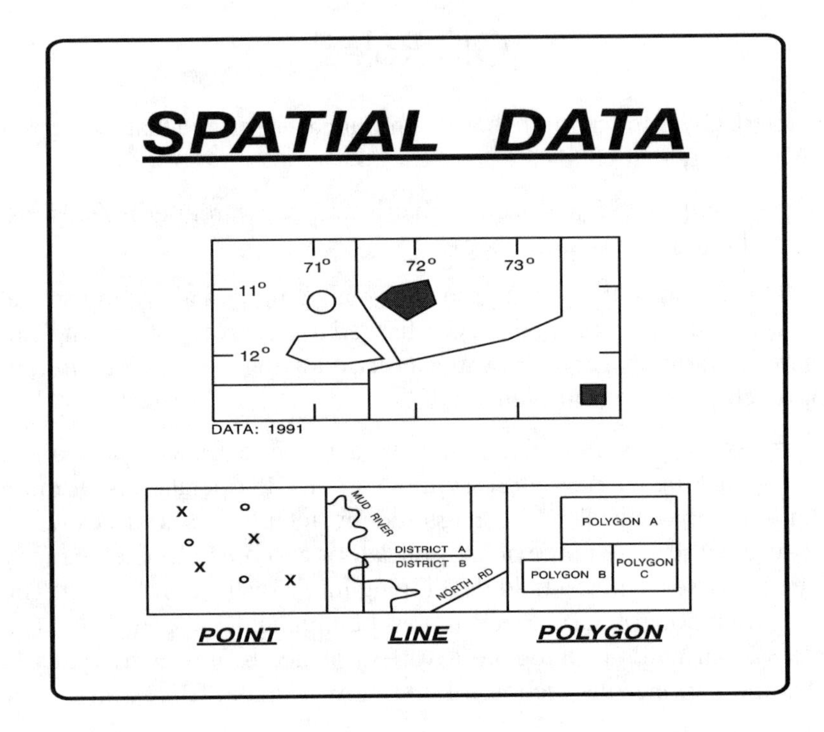
SPATIAL DATA
71°
72°
73°
11°
12°
DATA: 1991
X
O
X
O
X
O
X
MUD RIVER
DISTRICT A
DISTRICT B
NORTH RD
POLYGON A
POLYGON B
POLYGON C
POINT
LINE
POLYGON

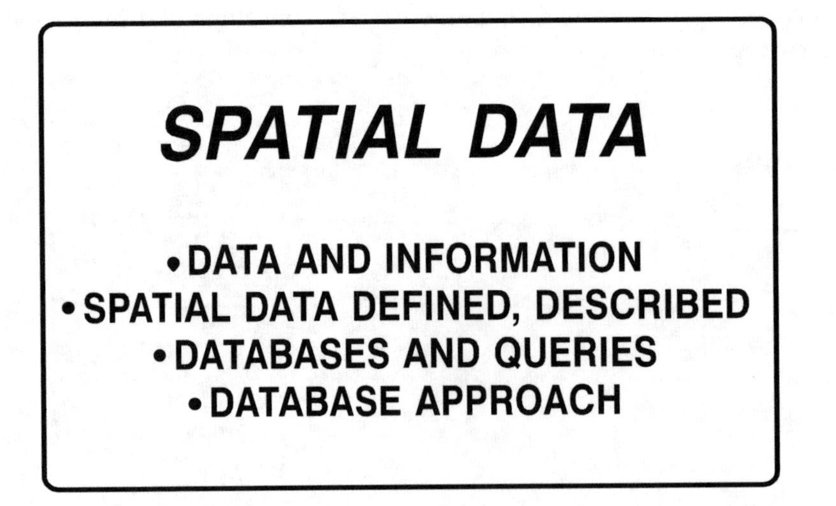
SPATIAL DATA
• DATA AND INFORMATION
• SPATIAL DATA DEFINED, DESCRIBED
• DATABASES AND QUERIES
• DATABASE APPROACH

3

Spatial Data

Introduction

To understand GIS, we must understand spatial data, also called geographic data. This is not a long chapter, but it is one of the most important. Topics include the following:

- *Data and Information:* Differences, value, and use
- *Spatial Data:* What it is, types, and characteristics
- *Databases:* What they are and how they work, particularly for GIS
- *Database Approach:* A wonderful way to manage and work with data

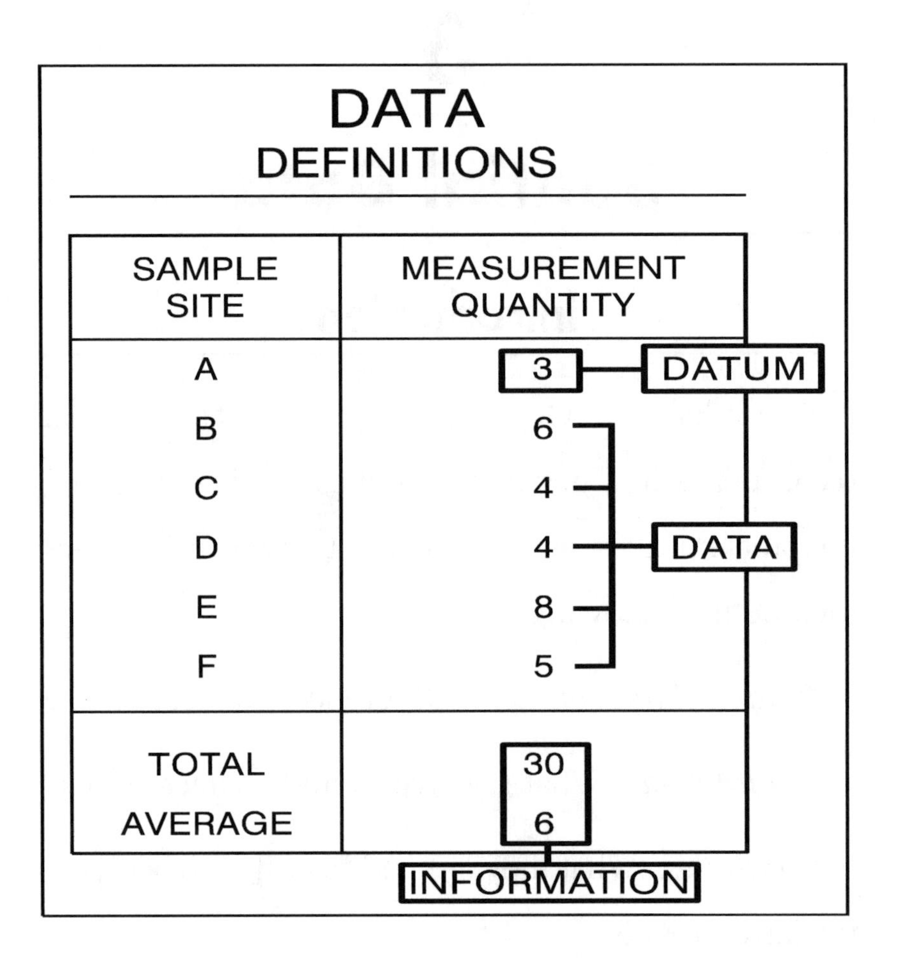
DATA
DEFINITIONS
SAMPLE SITE
MEASUREMENT QUANTITY
A
3
DATUM
B
6
C
4
D
4
DATA
E
8
F
5
TOTAL
30
AVERAGE
6
INFORMATION

Data Definitions

The terms *data* and *information* are often conveniently interchanged without real loss of meaning, but an important difference can exist. Working definitions are given in the following small database.

Data	Facts, or numbers representing facts. Data in its raw (uninterpreted) form does not offer meaning, only measures or facts.
Information	Meaning from multiple facts or numbers (although a single piece of data can represent information).

Technically we say "datum is" and "data are," although *data* is commonly used as both the singular and plural form.

Information implies meaning or significance of collective data. What is information to one person might be data to another. The total and average (bottom of the illustration) offer some meaning to the list of numbers (data), and are therefore the informational aspects of the small database in this illustration.

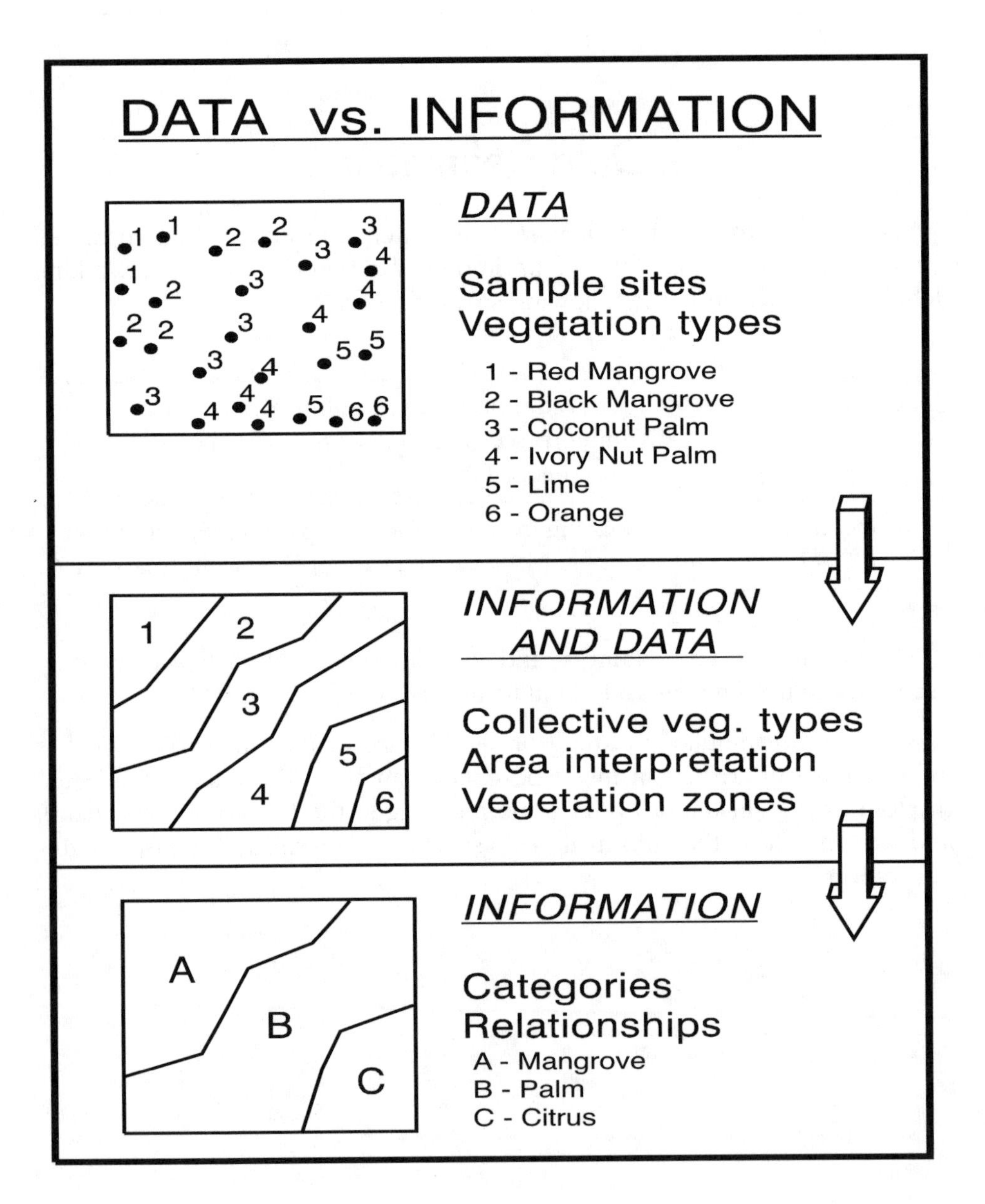
DATA vs. INFORMATION
DATA
Sample sites
Vegetation types
1 - Red Mangrove
2 - Black Mangrove
3 - Coconut Palm
4 - Ivory Nut Palm
5 - Lime
6 - Orange
INFORMATION AND DATA
Collective veg. types
Area interpretation
Vegetation zones
INFORMATION
Categories
Relationships
A - Mangrove
B - Palm
C - Citrus
A
B
C

Data Versus Information

As noted, information consists of multiple data. The illustration points out the difference in the terms.

The top map shows a study area containing sample sites (data points), each with a single measurement or observation. The numbers are codes identifying the type of vegetation found at each site. For example, 1 corresponds to Red Mangrove, and 2 to Black Mangrove. These are "data."

When data are combined and tell us something (in this case, about the area), they yield meaning; they build information. Placement of the lines on the middle map was determined by estimating vegetation areas according to the sample sites and code numbers (data). Each area has the vegetation type corresponding to the sample site numbers.

The middle map can be both information and data. It is information built from the top map's data. This information can also be used as data for a more generalized map (at bottom).

The Red (1) and Black (2) mangroves have been combined into the category Mangroves. Coconut (3) and Ivory Nut (4) palms have been combined into Palms. Lime (5) and Orange (6) have been merged into Citrus. Thus, collective data equals information, and information can be data for other information.

NOTE: *To the GIS user, all three maps may be useful, not just the one at bottom. Each map has unique data and information; therefore, all are saved as part of a GIS data set.*

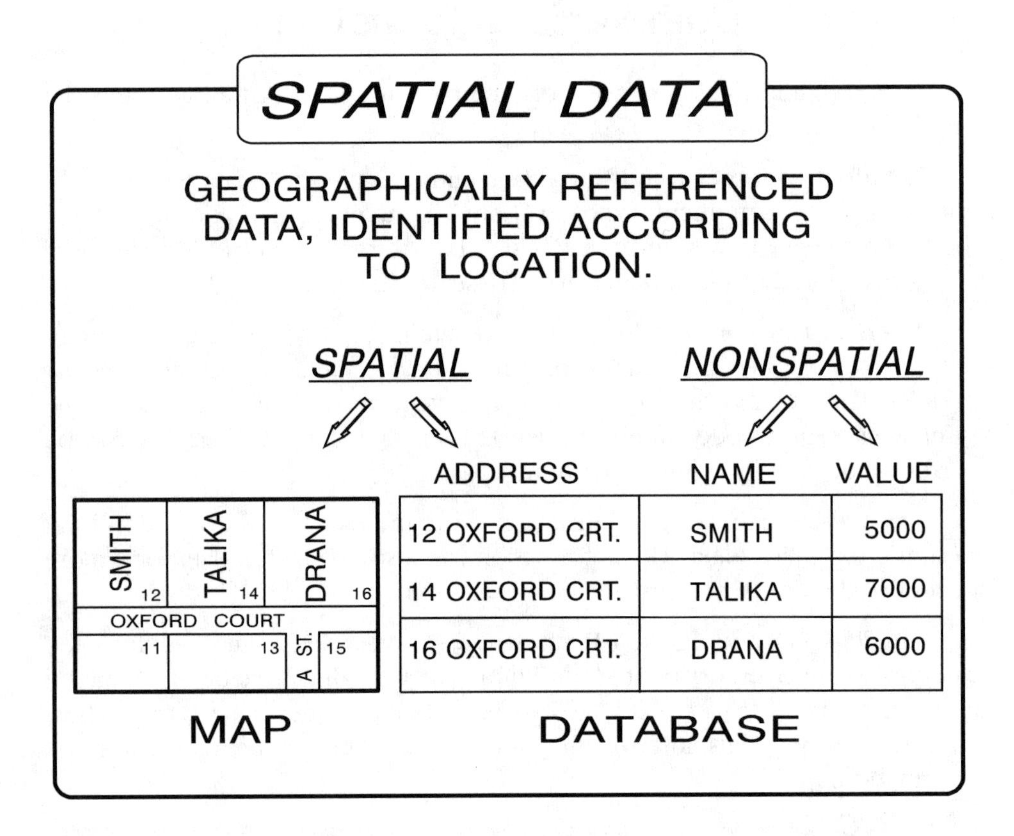
SPATIAL DATA
GEOGRAPHICALLY REFERENCED DATA, IDENTIFIED ACCORDING TO LOCATION.
SPATIAL
NONSPATIAL
ADDRESS
NAME
VALUE
SMITH
12
TALIKA
14
DRANA
16
OXFORD COURT
11
13
A ST.
15
12 OXFORD CRT.
SMITH
5000
14 OXFORD CRT.
TALIKA
7000
16 OXFORD CRT.
DRANA
6000
MAP
DATABASE

Spatial Data

What Is Spatial Data?

The glossary defines *spatial data* as data that occupies cartographic (mappable) space and that usually has specific location according to some geographic referencing system (e.g., Latitude-Longitude) or address. Spatial data are data defined by physical characteristics, usually including location or position. Spatial data may simply give an address (a specific location), or they can define size, such as a certain area of a forest (which is the same as defining the various positions of boundaries). The illustration shows a database containing addresses and a map (lower left) showing the area. Both the database and map represent spatial data.

The addresses in the database are locational, and are therefore considered spatial data. The map is based on location, so it, too, is spatial. The owner name and value of the property are nonspatial data. They are descriptive characteristics called "attributes" (see Chapter 2 under the heading Attributes). Many attributes are not location-specific but are meant to offer description about the spatial data or something that occurs at the location. For example, Talika is the owner of one address, but she could own, or move to, another address. The address doesn't change, but the attribute (owner) might. GIS data sets usually contain spatial data and associated nonspatial data.

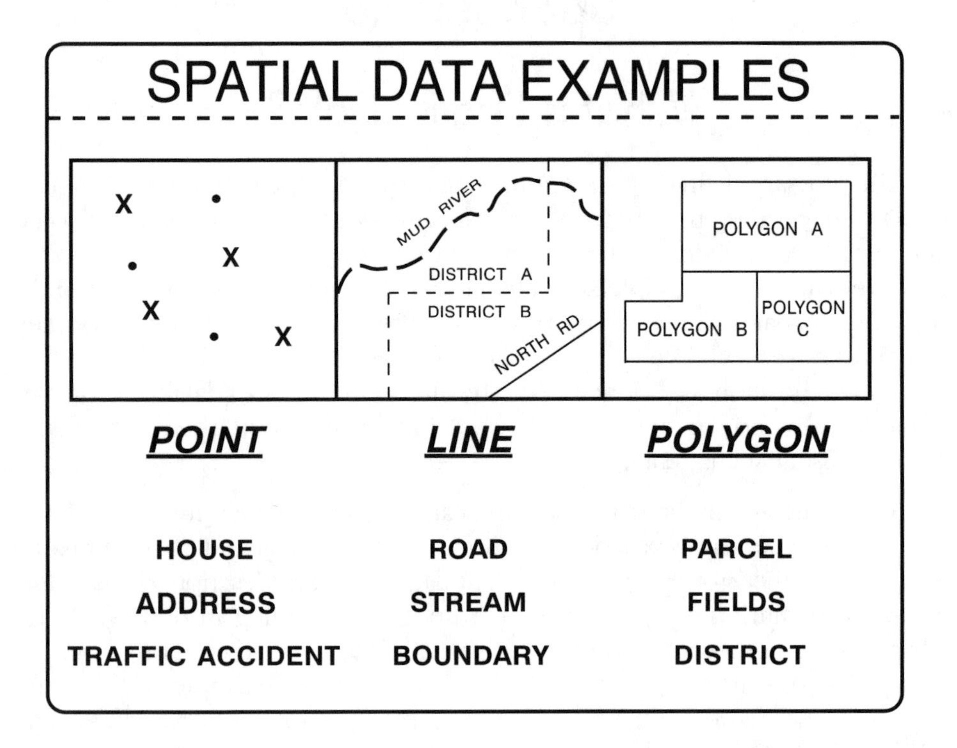
SPATIAL DATA EXAMPLES
X
X
X
X
MUD RIVER
DISTRICT A
DISTRICT B
NORTH RD
POLYGON A
POLYGON B
POLYGON C
POINT
LINE
POLYGON
HOUSE
ADDRESS
TRAFFIC ACCIDENT
ROAD
STREAM
BOUNDARY
PARCEL
FIELDS
DISTRICT

Types of GIS Spatial Data

In GIS, spatial data is classified as three main types: point, line, and polygon (area). Features, or landscape elements, are almost always depicted as a point, a line, or a polygon. (Spatial data can take other, advanced forms, such as trend surfaces, but those forms are beyond the scope of this book.) The illustration shows examples of these three data types.

A *point* is a spot (or location) that has no physical or actual spatial dimensions. That is, the point is shown as a convenient visual symbol (an X, dot, or meaningful graphic), but there is no real length or width to the feature. Points indicate specific locations of features, which are usually not shown in true size, especially for things that are too small to depict properly at a given scale, such as a house. Points also show locations of intangible (non physical entity) features, such as an address or the location of an occurrence, such as a traffic accident.

A *line* is a one-dimensional feature—meaning length only, no width. A line has a beginning and an end. Lines are linear features, either real (e.g., roads or streams) or administrative (e.g., boundaries). Sometimes the thickness of the line indicates a measure (such as amount of traffic on a road) or type of road (such as primary versus secondary).

A *polygon* is an enclosed area, a two-dimensional feature with at least three sides (and therefore with an inside area). A parcel of land, agricultural fields, or a political district are examples. Of course, polygons have area and perimeter, also, which are important characteristics.

We will see that sometimes in data entry the "feature type" must be declared so that the GIS can properly identify and manage the data. GIS coverages are usually stored as point, line, or polygon data files.

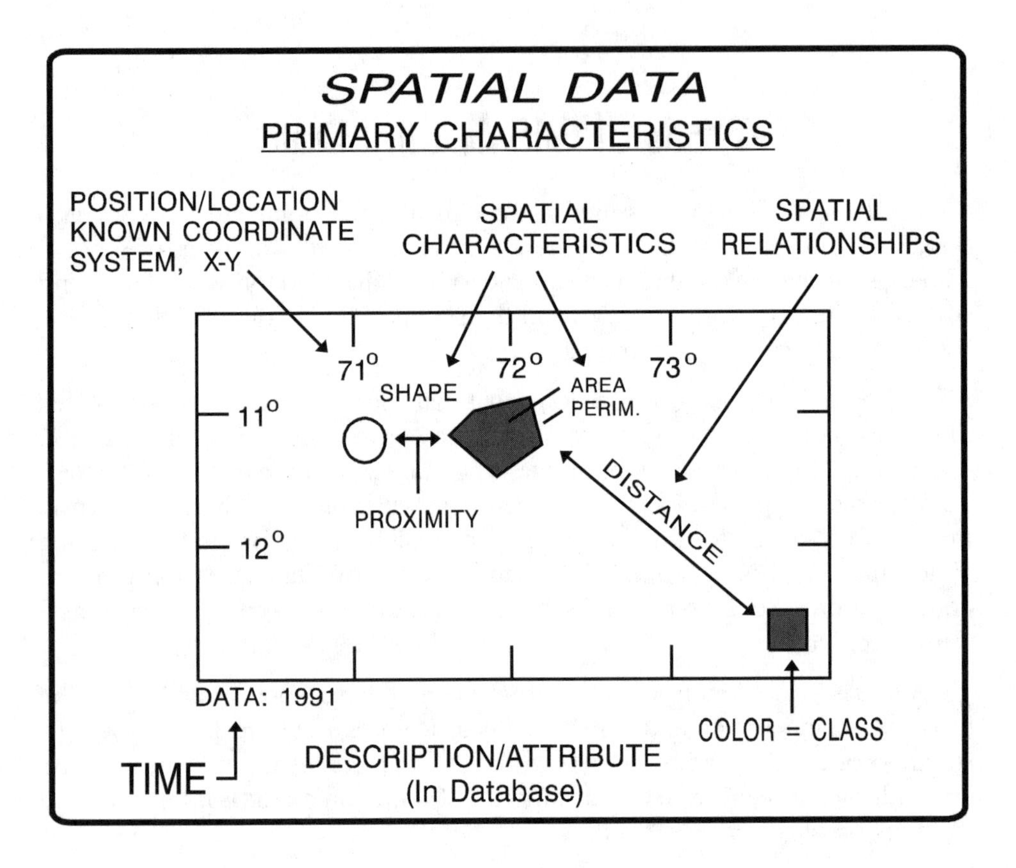
SPATIAL DATA
PRIMARY CHARACTERISTICS
POSITION/LOCATION
KNOWN COORDINATE
SYSTEM, X-Y
SPATIAL
CHARACTERISTICS
SPATIAL
RELATIONSHIPS
71°
72°
73°
SHAPE
AREA
PERIM.
11°
12°
PROXIMITY
DISTANCE
DATA: 1991
COLOR = CLASS
TIME
DESCRIPTION/ATTRIBUTE
(In Database)

Primary Characteristics

GIS spatial data typically has several primary characteristics that are seen (or sometimes not easily seen) on map. Although the eye can estimate many of these properties, GIS provides highly accurate measures. As indicated by the illustration, position (location) is the major starting point of measurement, normally using a known geographic coordinate system, such as Latitude-Longitude. This is the X-Y position (upper left in the illustration). The five-sided, shaded polygon (a feature) on the map in the illustration is at approximately 11 degrees South, 72 degrees East. Sometimes an elevation or depth is given, which is typically called the Z coordinate.

A map shows other spatial characteristics and relationships. Area and perimeter are also characteristics of features, and are easily calculated in most GISs. The shape of features is quickly appreciated by the eye, but computers need special programming to make spatial form descriptions.

On the other hand, spatial relationships between and among features can be important GIS project considerations and are not always apparent to the eye. Distance from one feature to another is available through simple measurement. Also, proximity of features, such as connections or nearness, may be determined through various measures to determine "neighborhood" characteristics. (See Chapter 5, under the heading Topology and Spatial Relationships, for more on proximity.)

Time can be an important part of GIS data. The date of the data is meaningful when determining trends or change. A population project needs up-to-date data, and old data may cause confusion. The lifetime of data can also be significant. Population numbers change constantly, and in many countries, demographic data can be expected to have an official ten-year lifetime, the interval between census taking.

Features on a map may have color or patterns to indicate a class or category of data, such as landuse identification or crop type. Technically, classification of data is nonspatial, but it is an important characteristic of a feature shown on a map. The GIS database includes map-related information; therefore, in GIS we normally consider the map and database necessary companions.

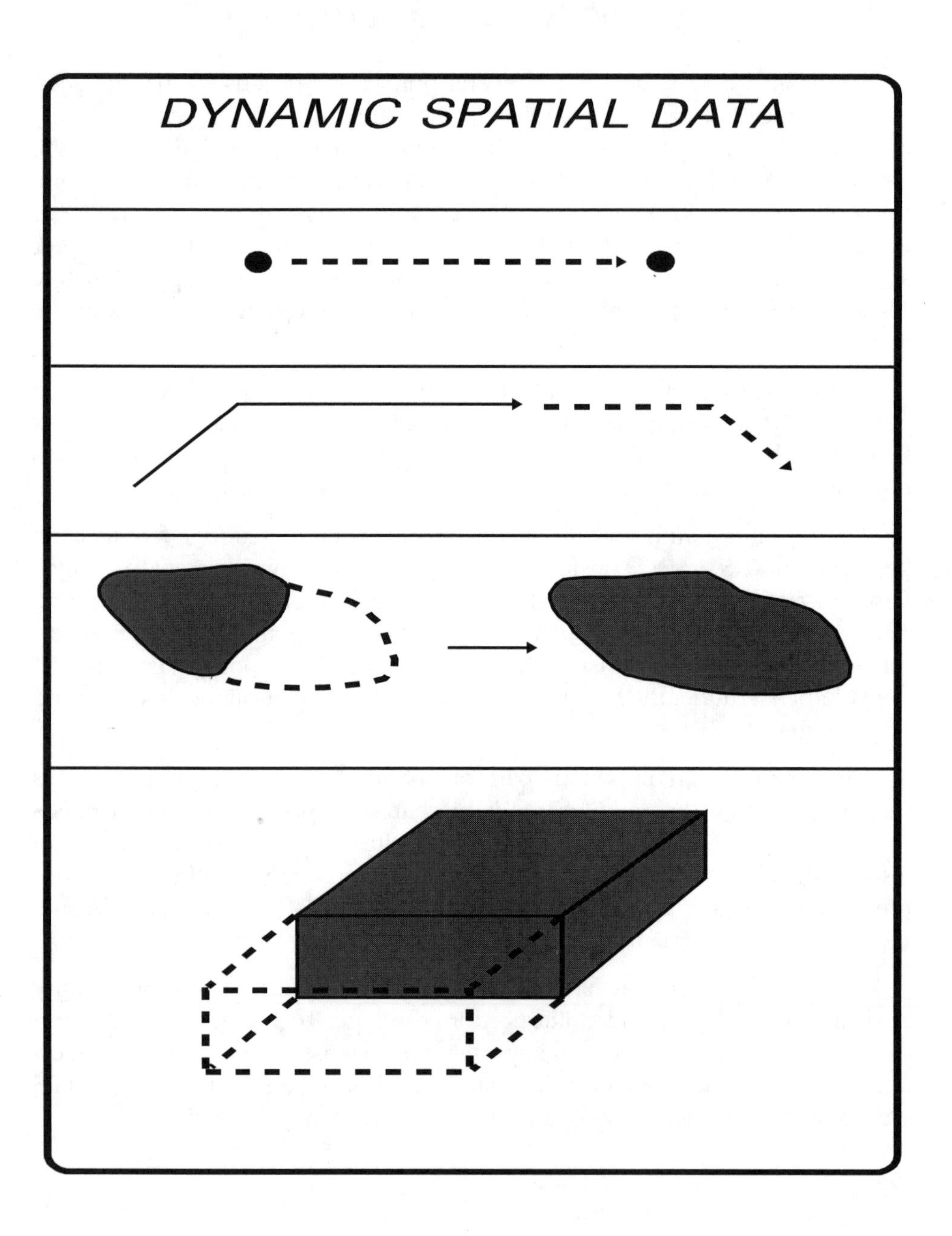
DYNAMIC SPATIAL DATA

Dynamic Data

We should think of spatial data as dynamic, or changing, entities rather than as simple, static features that have no "life." Time is an example of a dynamic component of a spatial data set. New technological developments are increasing computer hardware and software capabilities so that this dynamic aspect of data can be accounted for by GISs. Sometimes this is referred to as *hyperdata*. Viewing and analyzing change yields additional information, which lends greater value to data. A better understanding of land, for example, and the processes it undergoes can be gained.

A point can change position because many landscape features have a tendency to move, such as disaster events (earthquake epicenters), population centers, or new customer sites. A line might change length, either shorter or longer (as in road building). Its shape might also change—as with alternate routing around a damaged part of a road.

Polygons can also change shape. Over time, physical and cultural landscapes are highly dynamic. For example, vegetation patterns change (especially seasonal), and political districts may be altered on a periodic basis.

Cubic data, sometimes called volumetric (volume) data, uses X-Y-Z (depth) coordinate boundaries to give a realistic, 3D view. For example, marine scientists need data for the entire oceanic environment, not just surface data. Cubic features can change. Reservoir volume is a cubic feature that often varies seasonally, and planners need this information if they are to make accurate projections and effective policy.

Dynamic data may not be an immediate concern for beginning GIS students, but changing patterns and dimensions are becoming more important. However, this type of analysis demands a great deal of computing power, and therefore three-dimensional GIS is a frontier that is only now being explored.

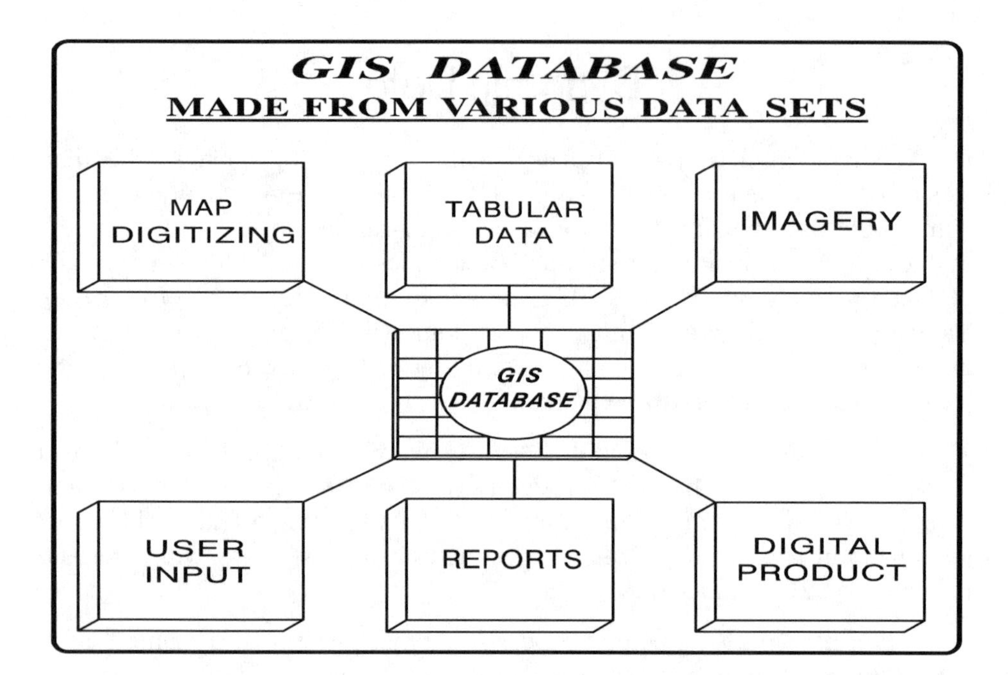

DATABASE

AREA	NAME	SIZE (SQ KM)	MAJOR CROP	POP (000)
A	DUBOP	11	RICE	1.1
B	JUMOM	22	RICE	1.7
C	TEROP	21	NONE	2.0
D	EERTO	17	RICE	0.7
E	BUROP	20	FRUIT	0.3

DATABASE: A list or table of data arranged by:
Columns: categories of data termed "Fields."
Rows: "Record" of each entry or observation.

The GIS Database

GIS data comes from a variety of sources: digitization of maps (electronic copying), tabular data such as census lists, remote sensing imagery, user input (typed-in data), existing reports (e.g., soil surveys), and digital products (e.g., digital line graphs sold in some countries). These diverse sets of data (upper illustration) are not naturally combined. It is difficult to hold up a satellite image, for instance, and compare it to a list of agricultural sites; they do not integrate easily.

All of these data sets must go somewhere for GIS use, and that place is the database—a program that stores and manages data. The database contains data and may point to other data sets, such as satellite imagery. A major strength of GIS is that it can accept and merge diverse data into a single database, giving the user a flexible and powerful set of data from which to work.

NOTE: *The main idea to grasp is that the database is the operation center of GIS, and normally much of the primary work is done here. Sometimes graphics (such as maps) are not even necessary, but rarely is the database unimportant.*

"Database" in GIS is a simple concept—a list or table of data arranged as columns (categories of data) and rows (each observation entry). Columns are called "fields," whereas rows are "records." The lower illustration shows a simple database, consisting of an initial identity of each feature (entered as an Area letter), the name of the area, and its size, major crop, and population. The first column is the landscape feature (Area) identity, and the other columns are the attribute fields (the descriptive characteristics of the areas).

Note that each area's attributes are read across, on a row. Reading vertically, the range of names or measures in the study area can be seen. Also note that only one field (Size) is spatial, whereas the others are nonspatial (but associated). GIS databases can be simple, such as this one, or very large, with many fields and hundreds of records. Later we will work with the database to produce a variety of useful GIS information.

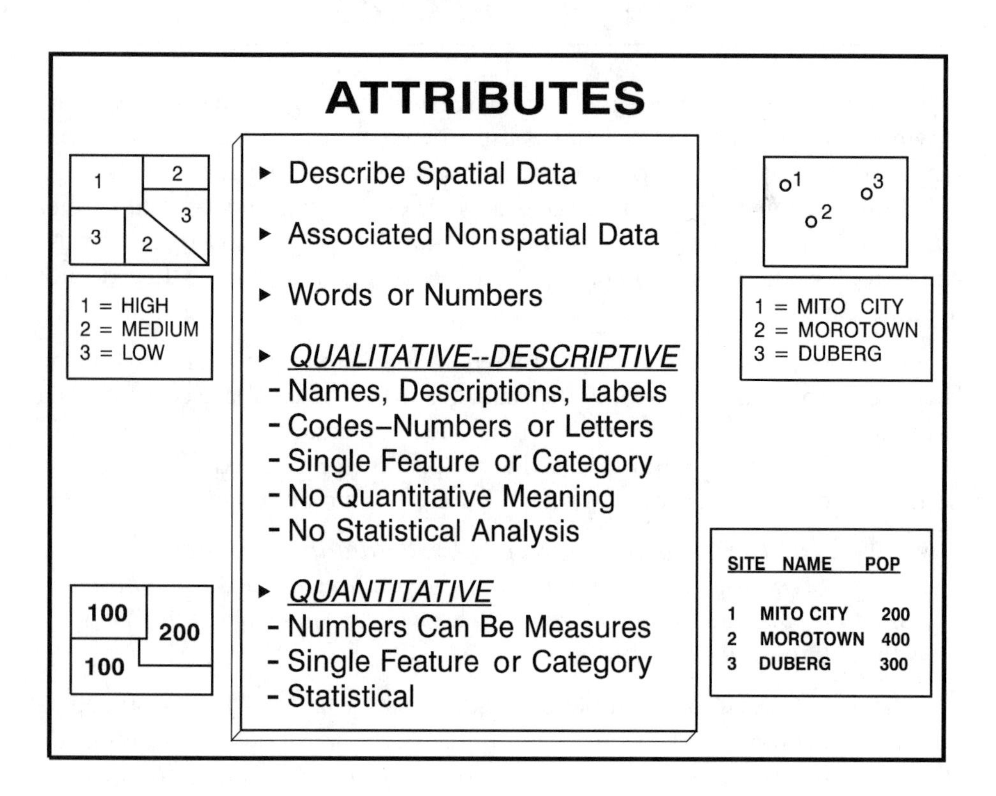
ATTRIBUTES
1
2
3
3
2
1 = HIGH
2 = MEDIUM
3 = LOW
100
200
100
▸ Describe Spatial Data
▸ Associated Nonspatial Data
▸ Words or Numbers
▸ QUALITATIVE--DESCRIPTIVE
- Names, Descriptions, Labels
- Codes–Numbers or Letters
- Single Feature or Category
- No Quantitative Meaning
- No Statistical Analysis
▸ QUANTITATIVE
- Numbers Can Be Measures
- Single Feature or Category
- Statistical
o1
o2
o3
1 = MITO CITY
2 = MOROTOWN
3 = DUBERG
SITE NAME POP
1 MITO CITY 200
2 MOROTOWN 400
3 DUBERG 300

Attributes

Attributes have been defined (see Chapter 2 under the heading Attributes) as descriptions of spatial data that are normally found in the database and map legend, such as color, ownership, magnitude, and classification. Attributes come in many forms, including text descriptions, numbers indicating quantities of some sort, and codes that are shorthand for something else (such as a name or classification). There can be confusion about the use of numbers; students sometimes mistake code numbers for actual measurements. To help reduce confusion, attributes are explained here in terms of qualitative and quantitative characteristics.

Qualitative Attributes

Qualitative attributes have no measurement or magnitude; they are non-numeric descriptions. Names, explanations, and labels on a map serve as descriptions. Numbers or letters may be used as codes for cumbersome words, but they have no actual quantitative (mathematical) meaning. Qualitative attributes play no role, for instance, in statistical analysis, such as calculating an average (the average of code numbers is meaningless).

The small map in the upper left of the illustration shows a simple High-Medium-Low classification (could be population densities). It uses numbers as convenient codes because words would be too long and visually messy. Colors, shading, or symbols could also be used. The map in the upper right of the illustration shows three cities, coded 1 through 3, with their respective names listed in the legend (lower right).

Quantitative Attributes

Unlike qualitative attributes, *quantitative attributes* do have mathematical meaning; the numbers serve as measurements or magnitudes of the feature to which they refer. The numbers might be a measurement for a single feature, such as a city's population, or a class (e.g., population level). Mathematical meaning allows for statistical measure. The small map in the lower left of the illustration shows three areas, with property values indicated. The box in the lower right illustrates a small GIS database, corresponding to the city map at upper right. This database contains a mix of qualitative (city name) and quantitative (population) attributes.

DATABASE

SELECTION OPTIONS

AREA	NAME	SIZE (Ha)	MAJOR CROP	POP (000)
A	DUBOP	11	RICE	1.1
B	JUMOM	22	RICE	1.7
C	TEROP	21	NONE	2.0
D	EERTO	17	RICE	0.7
E	BUROP	20	FRUIT	0.3

+ 0.3 = 1.4

1. List selected records. ***LIST B***
   List selected fields. ***LIST CROPS***

2. Sort any field alphabetically or numerically. ***SORT NAMES***
   Sort any field ascending or descending.

3. List by range of field data.
   SHOW POPULATION 1,100 - 1,800

4. Any combination of above.
   SHOW AREAS > 18 Ha, NOT RICE, POP > 1.3

5. Can be modified or appended.
   MODIFY: ADD 0.3 TO AREA A POP

Data Manipulation Options

GIS databases function as much more than mere data depositories. The real value of GIS databases is that they serve as systems for data management and analysis. Simple data selection, a common first step in many GIS applications, can be performed in a variety of ways. The following are some methods of data extraction from databases. (Numbers correspond to the list in the illustration.)

1. List selected records or selected fields.
   - ◆ Example 1: List all data for area B.
   - ◆ Example 2: List all major crops.
2. Sort fields alphabetically or numerically, in any order.
   - ◆ Example 1: Rearrange the database alphabetically according to name, in ascending (forward) order.
   - ◆ Example 2: Rearrange the database according to area size, from smallest to largest.
3. List by range of field data.
   - ◆ Example 1: Show only those records having populations between 1,100 and 1,800. (In this example, only areas A and B will result.)
   - ◆ Example 2: List only those areas having rice as the major crop.
4. Any combination of the foregoing. (A good database can provide a variety of operations.)
   - ◆ Example: Show all areas that do *not* have rice as a major crop and that are greater than 18 ha in size and have a population greater than 1.3.
5. Modifications, changes, and additions can easily be made to fields and records. (These are normal database operations.)
   - ◆ Example: Modify by adding 0.3 to area A population.

The GIS database is a powerful tool for data storage and extraction. We can see from these types of database tasks that both simple and complicated operations, often involving the coordination of multiple data sets, are easily handled by GIS.

RELATIONAL DATABASE

DATABASE CONSTRUCTED SO THAT EACH ITEM AND ITS ATTRIBUTES ARE LINKED AND RELATED TO EVERY OTHER ITEM AND ITS ATTRIBUTES.

EX: QUERY–SHOW ALL NATIONS HAVING
POPULATION >100,000
GROWTH RATE 2.0+
LAND AREA >15,000 KM^2
PRINT CAPITAL CITY

NATION	POP	GROWTH (%)	LAND (KM^2)	CAPITAL
FIJI	716,000	2.0	18,272	SUVA
NEW CALEDONIA	146,000	1.2	19,103	NOUMEA
SOLOMONS	286,000	3.3	29,785	HONIARA
TONGA	97,000	1.1	697	NUKU'ALOFA
VANUATU	140,000	3.1	12,189	PORT VILA
W. SAMOA	162,000	0.7	2,934	APIA

ANSWER

FIJI–SUVA
SOLOMONS–HONIARA

The GIS Relational Database

Extraction of Related Data

There are several types of databases, each with special advantages and disadvantages. Some offer only basic options. For example, a "flat-file" database system is like a simple data table. It is a rather elementary container of records that allows only uncomplicated extraction of data. It cannot perform some of the cross-referenced operations previously mentioned.

Debatably, the best type of database is the relational database. It is defined in the illustration as a "database constructed so that each item and its attributes are linked and related (cross-referenced) to every other item and its attributes [every other record's data]." Data extraction, called a "query," can be simple or complex. The illustration shows a three-part query: search for nations having the specified conditions (or data ranges) related to population, growth, and land area.

The program searches the first field for population requirements, selecting all records with populations over 100,000 (follow along: see the boxes inside each column of the database). From those selections, it moves to the second field and searches for growth rates of 2.0 and above. Then, from the three records it looks for those records containing land area data of over 15,000 square km, finding two, Fiji and the Solomon Islands. The query is finished when it prints the nations and their capitals, as instructed.

With a well-ordered and well-constructed relational database, the only limit to queries is the imagination. Try this by eye: Which nation(s) has/have a combination of population(s) under 200,000, growth rate(s) under 1.5%, and land area(s) less than 2,000 square km? Most powerful GISs employ the highly useful relational database to perform just this type of task.

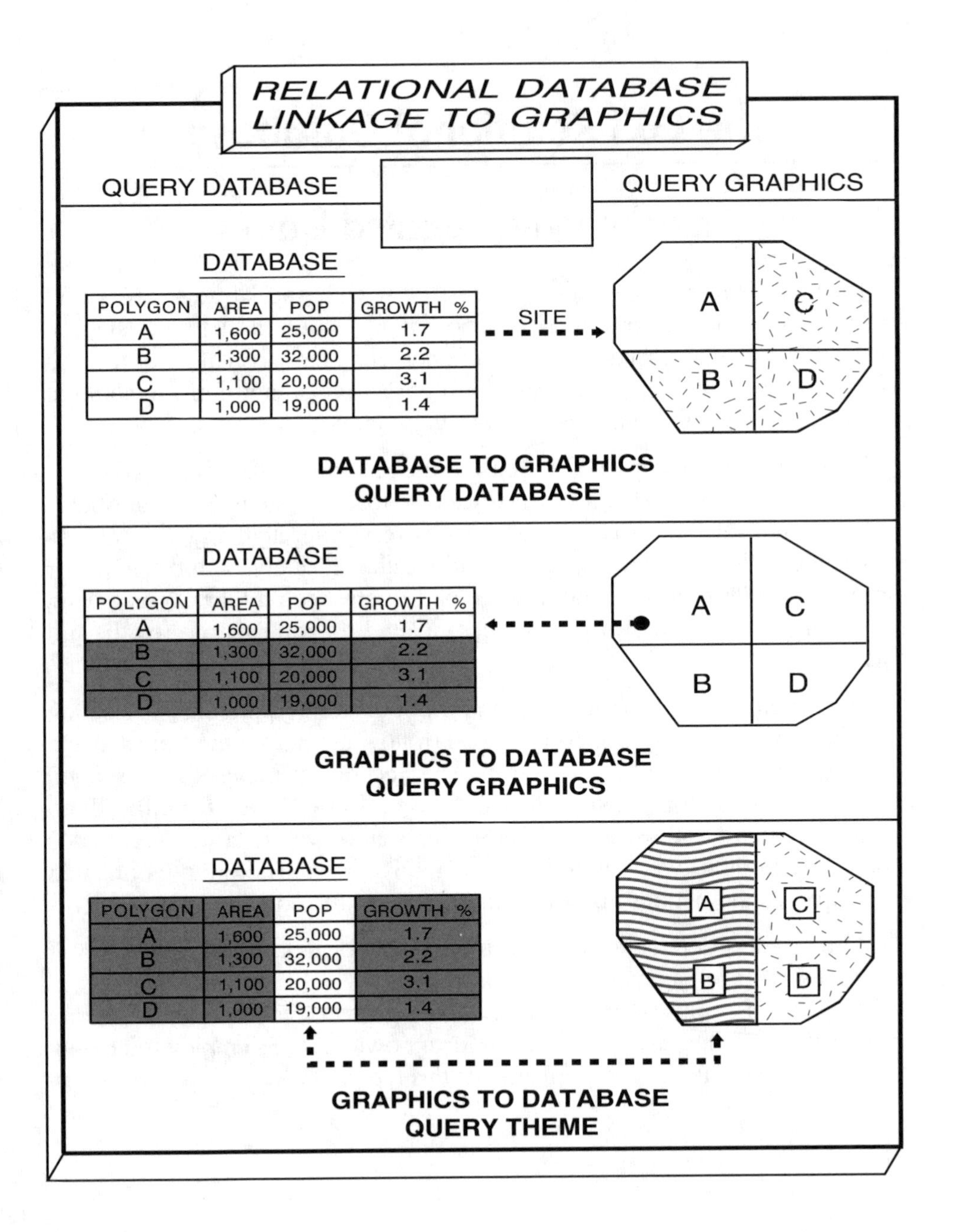

RELATIONAL DATABASE LINKAGE TO GRAPHICS
QUERY DATABASE
QUERY GRAPHICS
DATABASE
POLYGON | AREA | POP | GROWTH %
A | 1,600 | 25,000 | 1.7
B | 1,300 | 32,000 | 2.2
C | 1,100 | 20,000 | 3.1
D | 1,000 | 19,000 | 1.4
SITE
A
C
B
D
DATABASE TO GRAPHICS
QUERY DATABASE
DATABASE
POLYGON | AREA | POP | GROWTH %
A | 1,600 | 25,000 | 1.7
B | 1,300 | 32,000 | 2.2
C | 1,100 | 20,000 | 3.1
D | 1,000 | 19,000 | 1.4
A
C
B
D
GRAPHICS TO DATABASE
QUERY GRAPHICS
DATABASE
POLYGON | AREA | POP | GROWTH %
A | 1,600 | 25,000 | 1.7
B | 1,300 | 32,000 | 2.2
C | 1,100 | 20,000 | 3.1
D | 1,000 | 19,000 | 1.4
A
C
B
D
GRAPHICS TO DATABASE
QUERY THEME

Link to Graphics

The real strength of GIS comes in when a relational database is linked to graphics. A good GIS permits relational queries at the database, with results shown on the graphics, or the reverse: graphics selection to database display. For example, consider the database quiz question just posed: a map of the South Pacific would be shown, with Tonga highlighted because it is the only nation to meet the selected criteria.

The illustration shows a small database and an odd-shaped area of four districts (polygons A through D). In the top box of the illustration, district A has been selected in the database and highlighted on the graphics. The selection could simply be a matter of wanting to see polygon A, or A could be the result of a query: Show site(s) having area 1,200; population 24,000; and growth <2.0%.

Although the connection is simple in this illustration, consider a 300-polygon map of soils and wanting to find those that meet selected criteria. A database-graphics link would be essential. Also, selection on graphics can show records at the database. For example, by pointing (middle box of the illustration) to a specific area on the displayed coverage, the database is presented, showing the desired data.

It is also useful to select a field as a mapping theme and then have the graphics present results. The operator chooses a type of classification, which reduces the field measurements into generalized categories, which are then mapped according to some coloring or shading scheme. In the bottom box of the illustration, population has been given two classes (<24,000 and 24,000+) and shaded accordingly.

We can see from the foregoing discussion and examples that we can work in GIS from graphics to database or database to graphics. This is just one of the many "flexibilities" of GIS that make it such a powerful and useful tool.

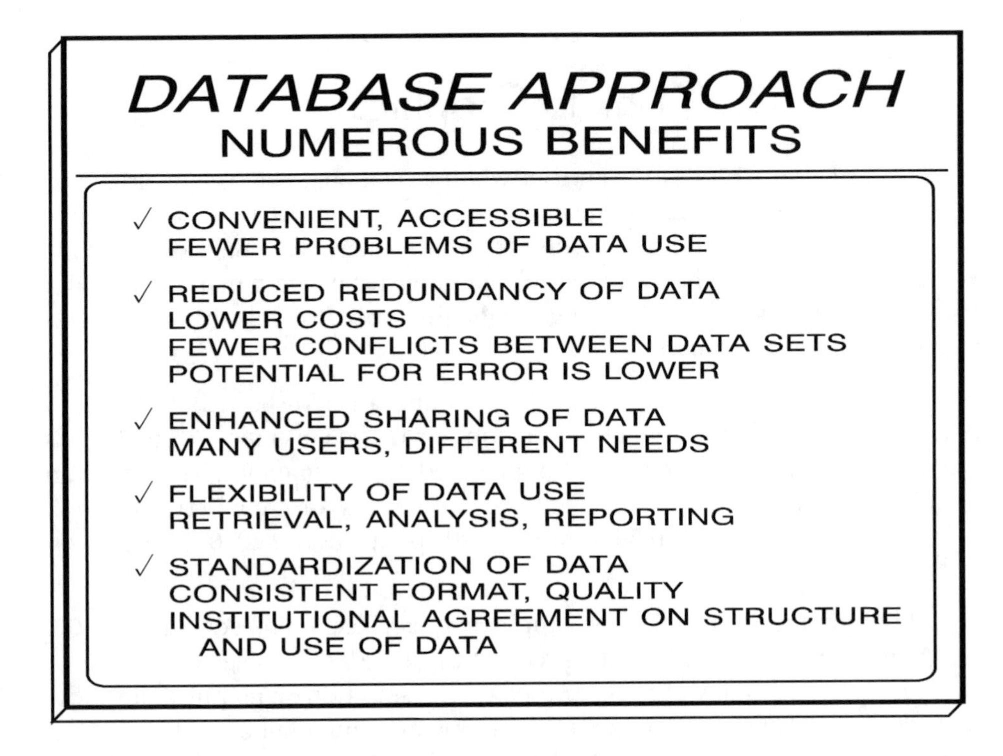

The Database Approach

This chapter has covered the difference between, and uses of, data and information.

Datum → Data → Information → More Information

We have looked at spatial data: definition, types, characteristics, and uses. The benefits of databases, especially relational databases, have been seen, as well as the value of the database approach. GIS is a spatial data technology that integrates graphics and databases to offer powerful and flexible capabilities.

There are other options, operations, and capabilities offered by good GIS databases, some of which will be discussed in upcoming sections. The important aspect here is that the "database approach" is highly valuable in GIS; it offers much more than a simple table or elementary mapping. The following are among the numerous benefits to using the database approach.

- *Convenience:* Data are very accessible, which results in fewer problems than searching map cabinets, reports, and statistics in various locations. A quick computer operation produces the desired data.

- *Reduced Redundancy:* This means that identical data are not needed, thereby decreasing costs related to duplication and data management. Also, there are fewer conflicts between data sets that contain identical data that could be changed in one set but not in all. The potential for error is lower.

- *Enhanced Sharing:* Several users in different offices or agencies can use the same database, thereby maximizing the efficiency of management and quality control. One database can serve multiple needs.

- *Flexibility:* A good database offers a great deal of flexibility in the use of data, from a variety of retrieval options to diverse analysis and various reporting formats. It can "grow" to accommodate new users and new applications, while remaining stable for existing users.

- *Standardization:* Because a typical setup includes many data contributors and users, control of data format (form) and quality is important. Usually there must be agreement on how the database is structured (such as naming conventions), what types of data go into it, how the data are structured, and how modifications are to be made. In short, GIS promotes "standardization" of data so that every user understands what is included. Often, the database approach introduces this type of agreement and standardization as a new idea. This also helps to commit management to a specific direction, which in turn makes arbitrary changes more difficult.

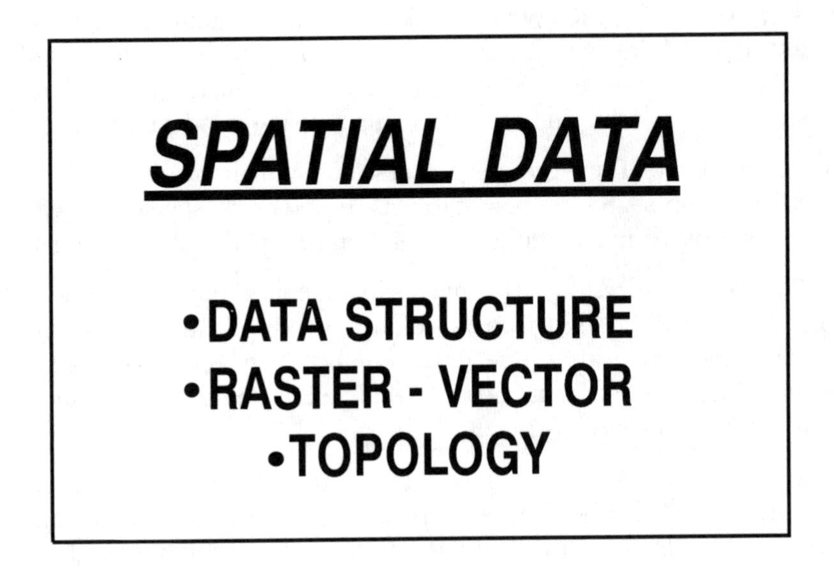
SPATIAL DATA
•DATA STRUCTURE
•RASTER - VECTOR
•TOPOLOGY

1
2
3
4
5
6
7

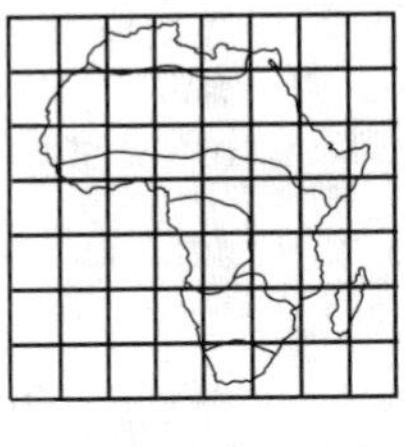

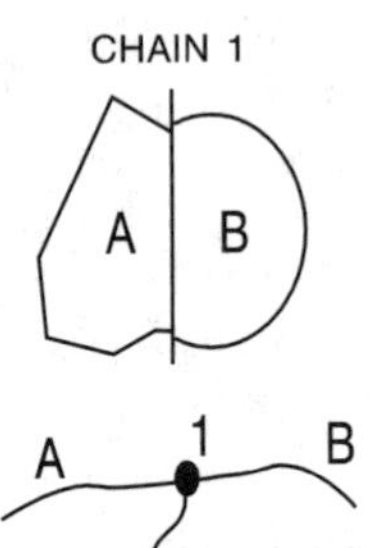
CHAIN 1
A
B
A
1
B
C

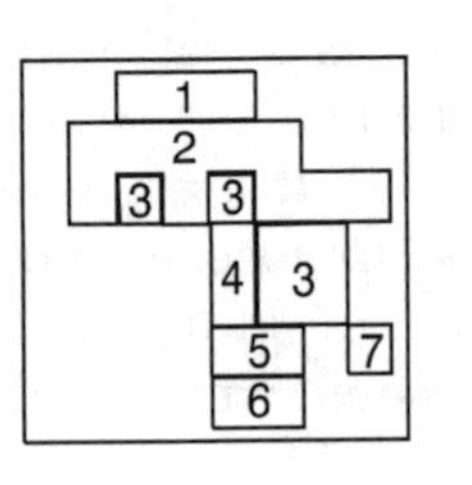
1
2
3
3
4
3
5
7
6

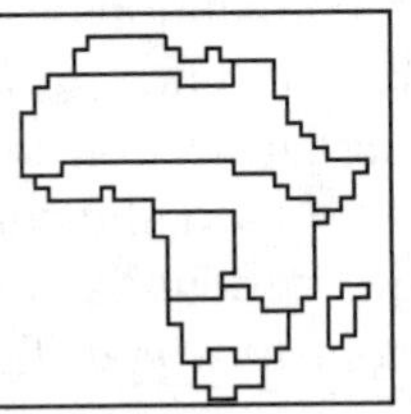

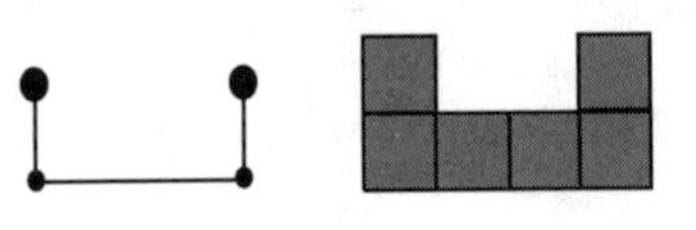

4

Raster and Vector Data

Introduction

This chapter discusses an aspect of spatial data structure important to an understanding of how GIS operates: raster and vector formats. These, along with topology (see Chapter 5), form the basic foundation of data construction and function. Raster and vector formats are a way of defining spatial data in the computer. Topology is a special characteristic of spatial data that establishes powerful relationships among features. Before we explore raster and vector formats, we need to look at some definitions.

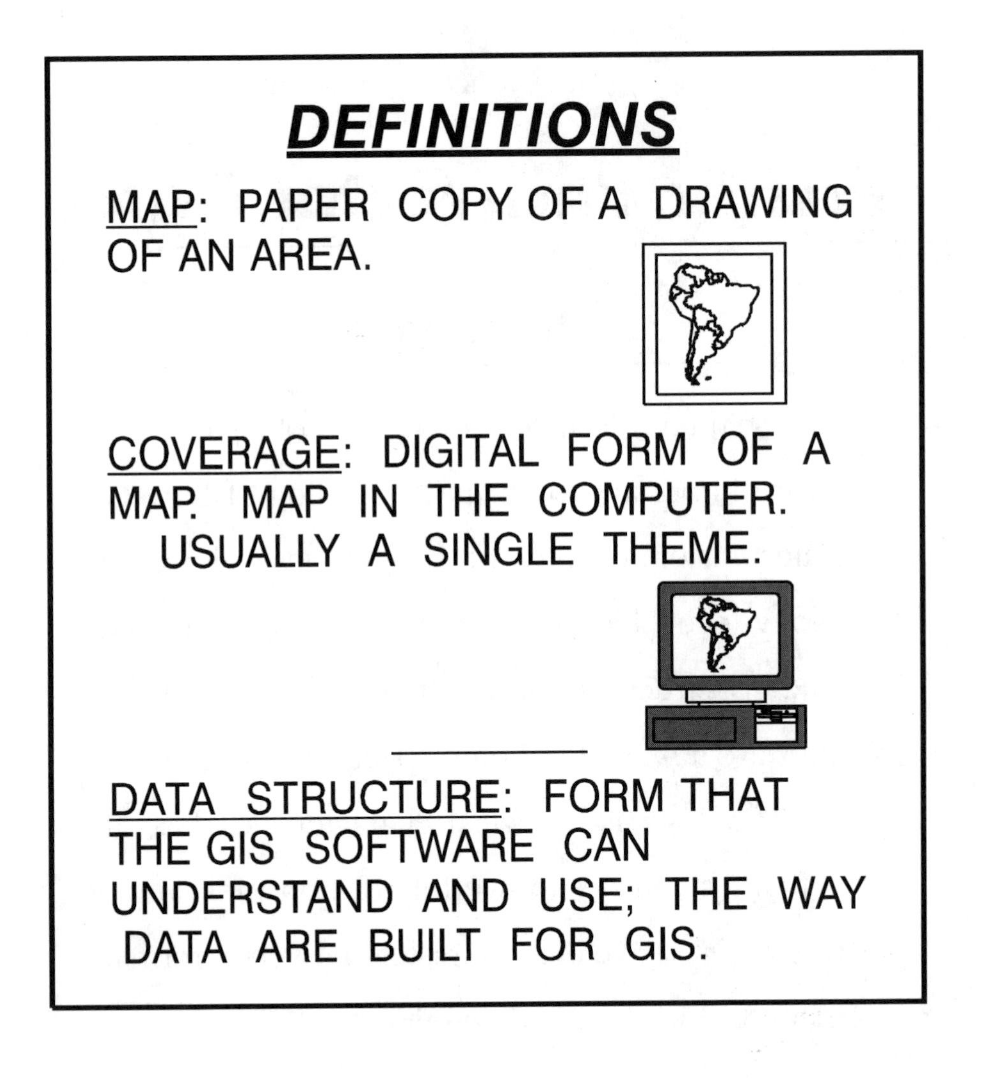
DEFINITIONS
MAP: PAPER COPY OF A DRAWING OF AN AREA.
COVERAGE: DIGITAL FORM OF A MAP. MAP IN THE COMPUTER. USUALLY A SINGLE THEME.
DATA STRUCTURE: FORM THAT THE GIS SOFTWARE CAN UNDERSTAND AND USE; THE WAY DATA ARE BUILT FOR GIS.

Definitions

It is important to understand several terms as defined in the world of GIS.

Map	A map is the paper version of a landscape drawing. It is a chart that can be held. Also called hard copy in GIS.
Coverage	The digital form or version of a map; the map in the computer. Usually, GIS coverages have a single, major theme, such as landuse or vegetation. They are normally *not* "general reference" (multiple themes and different types of data), as many maps are.
Data Structure	The form of the data in the computer. It is the format (type of construction) of data the GIS program understands and uses; the way GIS data are built, stored, and displayed.

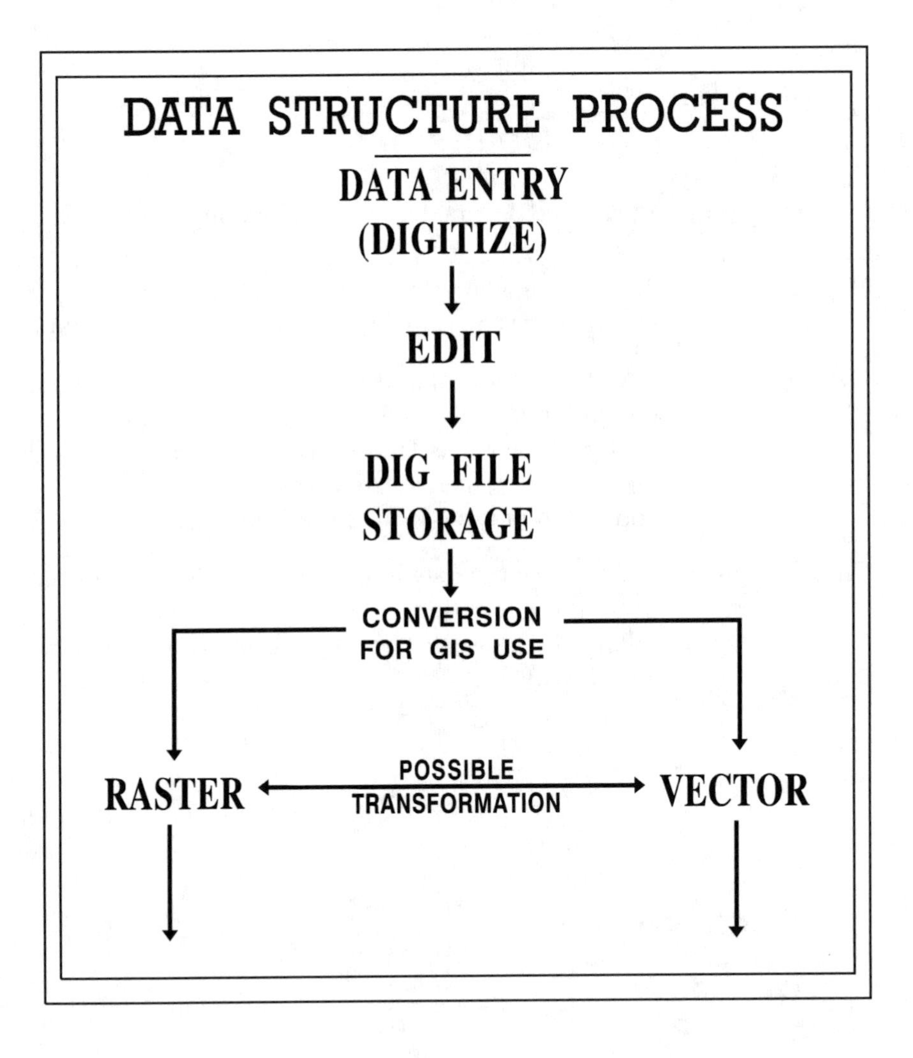
DATA STRUCTURE PROCESS
DATA ENTRY
(DIGITIZE)
EDIT
DIG FILE
STORAGE
CONVERSION
FOR GIS USE
RASTER
POSSIBLE
TRANSFORMATION
VECTOR

Data Structure Process

Data structure is a bit difficult to understand by simply defining it, but it should become clearer as a concept as we move along. However, before discussing data structure, we need to know where in the GIS process it is determined or developed. Follow the illustration.

Data entry is the first step. Maps are turned into coverages through digitizing—tracing the map electronically and changing it into digital form. This transfers the map data into the computer. (The digitizing process is discussed in Chapter 6).

After maps are digitized and edited, the data are temporarily stored, waiting to be entered into the GIS. Data seldom go directly into the GIS from digitizing because there are preparations to be made. In the meantime, digitized data are stored in a "dig" (short for "digital" and pronounced "dij") file.

This file must be converted to fit into either a raster or a vector GIS. These are two different constructions (or format structures) for the data, and both have advantages and disadvantages (see this chapter's heading Raster and Vector Pros and Cons). As will be seen, raster can be transformed (changed) to vector, and vector to raster. Therefore, data structure is one of the first major decisions to be made in the GIS process. Most GISs are either raster-based or vector-based, but many can employ both formats.

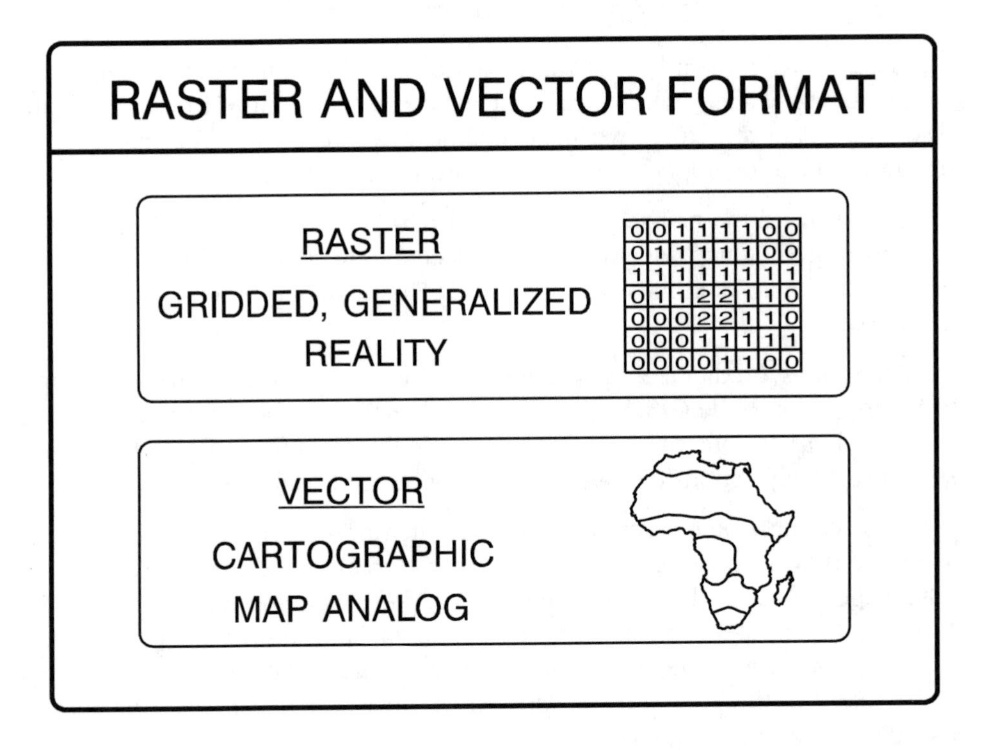
RASTER AND VECTOR FORMAT
RASTER
GRIDDED, GENERALIZED
REALITY
0 0 1 1 1 1 0 0
0 1 1 1 1 1 0 0
1 1 1 1 1 1 1 1
0 1 1 2 2 1 1 0
0 0 0 2 2 1 1 0
0 0 0 1 1 1 1 1
0 0 0 0 1 1 0 0
VECTOR
CARTOGRAPHIC
MAP ANALOG

Descriptions: Raster and Vector

Let's take a closer look at the two basic data structures used to store and display data in GIS: raster and vector.

Raster

In the raster format, a landscape scene is gridded, and each cell in the grid is given a single landscape identity, usually a code number that refers to a specific attribute measure (e.g., a particular soil type or landuse). The number might also be an actual measurement value, such as an amount of rainfall. (Recall the discussion on qualitative and quantitative attributes in Chapter 4.)

The grid in the illustration shows Africa, with code 0 standing for "outside," code 1 for "African land," and code 2 for "central tropical jungle area." The continent of Africa is impossible to recognize in grid form because the raster format generalizes data. This illustration is actually an exaggeration of the effect of generalization; normally, a coverage of a land mass the size of Africa would use several thousand cells and would make the shape of the land mass more recognizable.

> **NOTE:** *The main idea to grasp is that all of the land in the cell is reduced to a single code (a number or letter). This is "generalized reality," meaning that the number or letter represents all of the land and its attributes.*

Vector

Vector implies the maplike drawing of features, with the generalizing effect of a raster grid minimized. Shape is better retained, much like an actual map. The lines are continuous and are not broken into a grid structure.

Before going on, try to think of the advantages and disadvantages of each type of data structure. Which one is probably more spatially accurate?

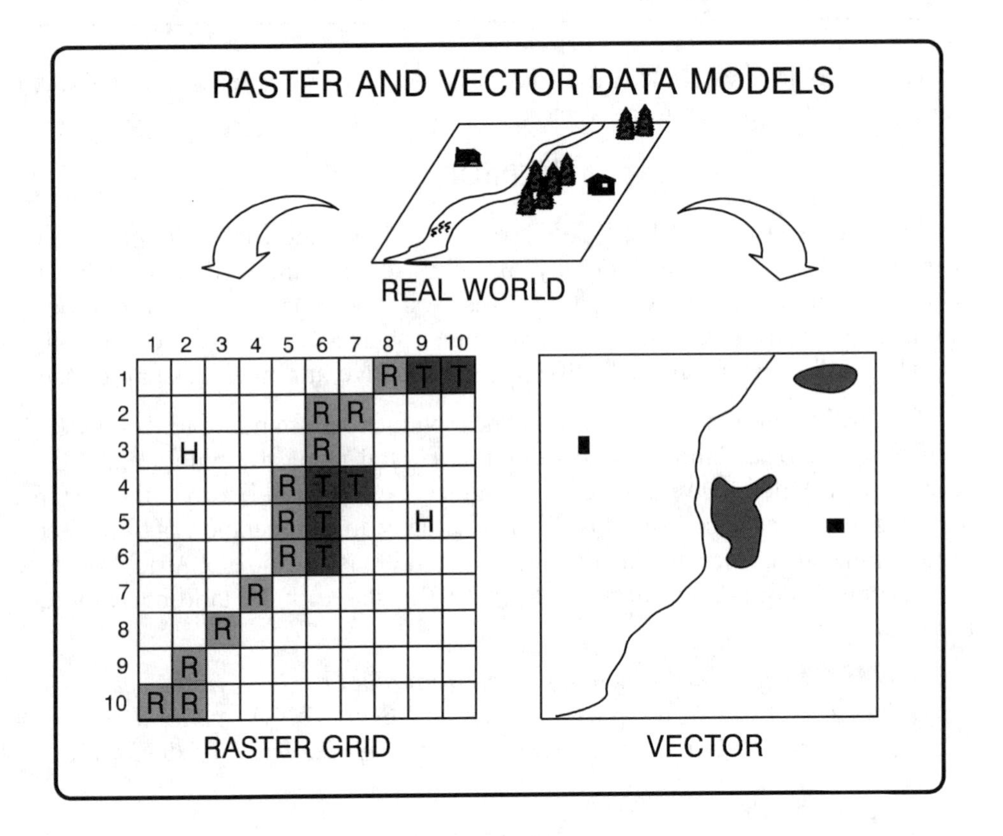
RASTER AND VECTOR DATA MODELS
REAL WORLD
1 2 3 4 5 6 7 8 9 10
1
2
3
4
5
6
7
8
9
10
R T T
R R
H R
R T T
R T H
R T
R
R
R
R R
RASTER GRID
VECTOR

Raster and Vector Data Models

To illustrate the differences between raster and vector, the "real-world" scene at top is depicted in both data structures. The left-hand illustration shows how the data are reduced to a grid, with letter codes indicating the river (R), trees (T), and houses (H). Normally, the background would also have a code, which has been excluded here for clarity. Each cell has only a single code; it cannot have two codes.

In a raster format, the cell is the "minimum mapping unit," meaning that it is the smallest size any landscape feature can be represented and drawn. Real sizes and shapes cannot be kept by the raster cell. The river, for example, might actually be more narrow than a cell, but only the entire cell can be coded as river. Also, note its shape change—more geometric than curvy—from vector to raster.

The vector format (right-hand illustration), however, uses symbols that depict size and shape. The river maintains its curves, and the forest areas have their realistic odd shape. The only limitation would be the thickness of the line used to draw the features. Also, houses may be shown in their actual shape if the map scale permits; otherwise, a square symbol of a standard size and shape might represent all individual houses.

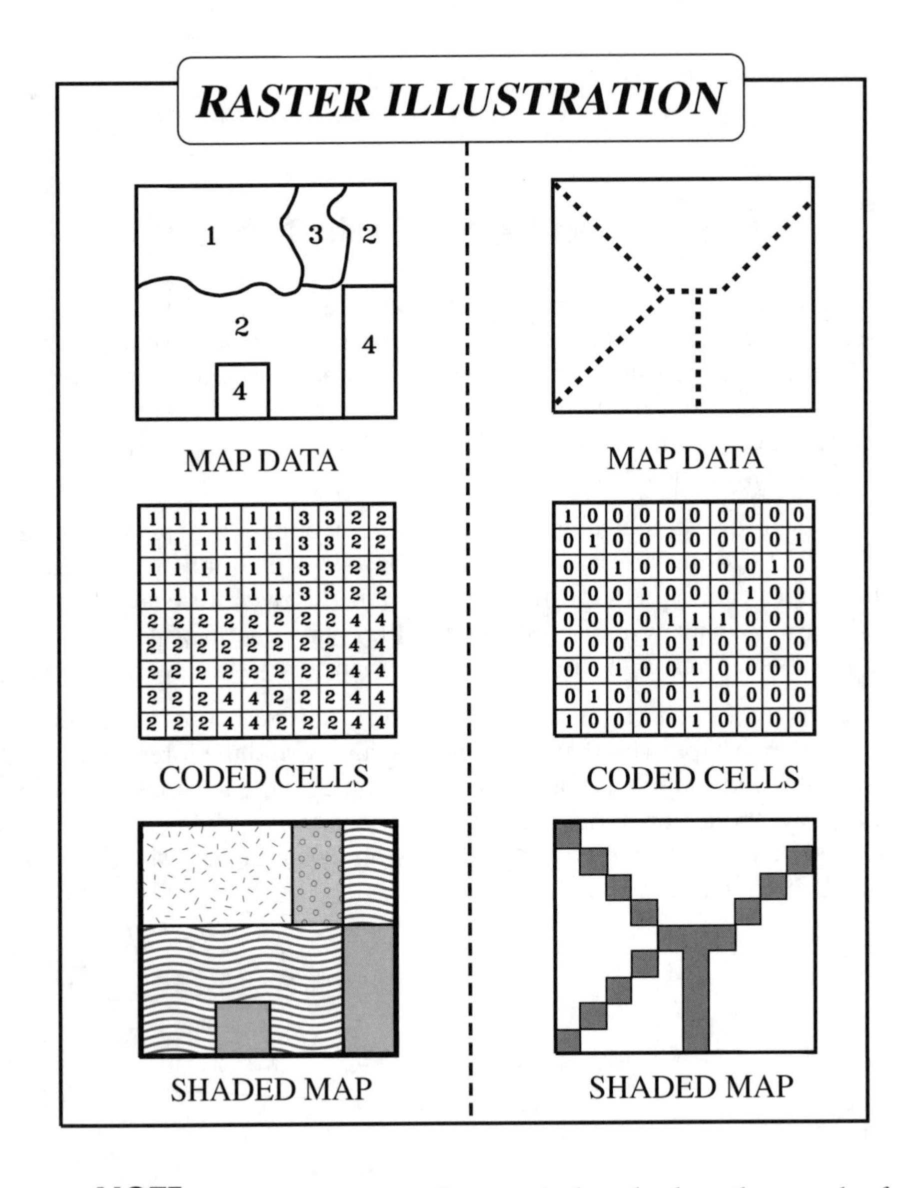

➾ **NOTE:** *Raster maps normally contain hundreds or thousands of cells, rather than a few, as seen in these illustrations. Roads actually look better than shown here. However, when magnified, those same roads would show the same stair-step effect (it's just that the eye can't see the same effect without magnification). Sometimes the changed shapes are important, and sometimes they are not.*

Raster Data

Illustration

A major problem with the raster structure is that the shape of features is forced into an artificial grid cell format. For right-angled features, such as square agricultural fields or rectangular political districts, this may not present a problem. However, for many features, size and shape can become undesirably distorted.

Illustrated are two sets of map data. The left-hand set shows areas coded 1 through 4 (e.g., four different landuses). The boundaries between 1, 2, and 3 are uneven and slightly odd-shaped. The Coded Cells box (middle of the left-hand set) represents raster cells, which force uneven lines into geometric shapes. Thus, landuses 1 through 3 have been generalized into right-angled, straight-line features. This results in errors of spatial accuracy in terms of size and exact location.

Landuse 4 features have square and rectangular shapes in the original Map Data display (top of left-hand set), so their raster versions are not significantly affected. The only change imposed on them might be slight adjustment of borders at the edges of cells, which would result in small spatial errors.

The shaded map (bottom box of the left-hand set) shows how codes are shaded or colored for better visualization (called a choropleth map). Obviously, a grid full of numbers is unreadable. Changing it into an eye-pleasing and readable map with colors, shades, and symbols is more effective.

Linear features may be greatly affected by rasterization. The Map Data box of the right-hand set of images shows roads, three of which are diagonals. Two important changes are made in the raster-cell shaded map (right-hand set): (1) shape change of the diagonal features and (2) width increase of the linear features because of minimum mapping cell sizes.

The diagonals are turned into stair-steps because of the influence of diagonal linking cells. The width of all linear features is "generalized" to the cell size because there is no smaller datum size than the single cell.

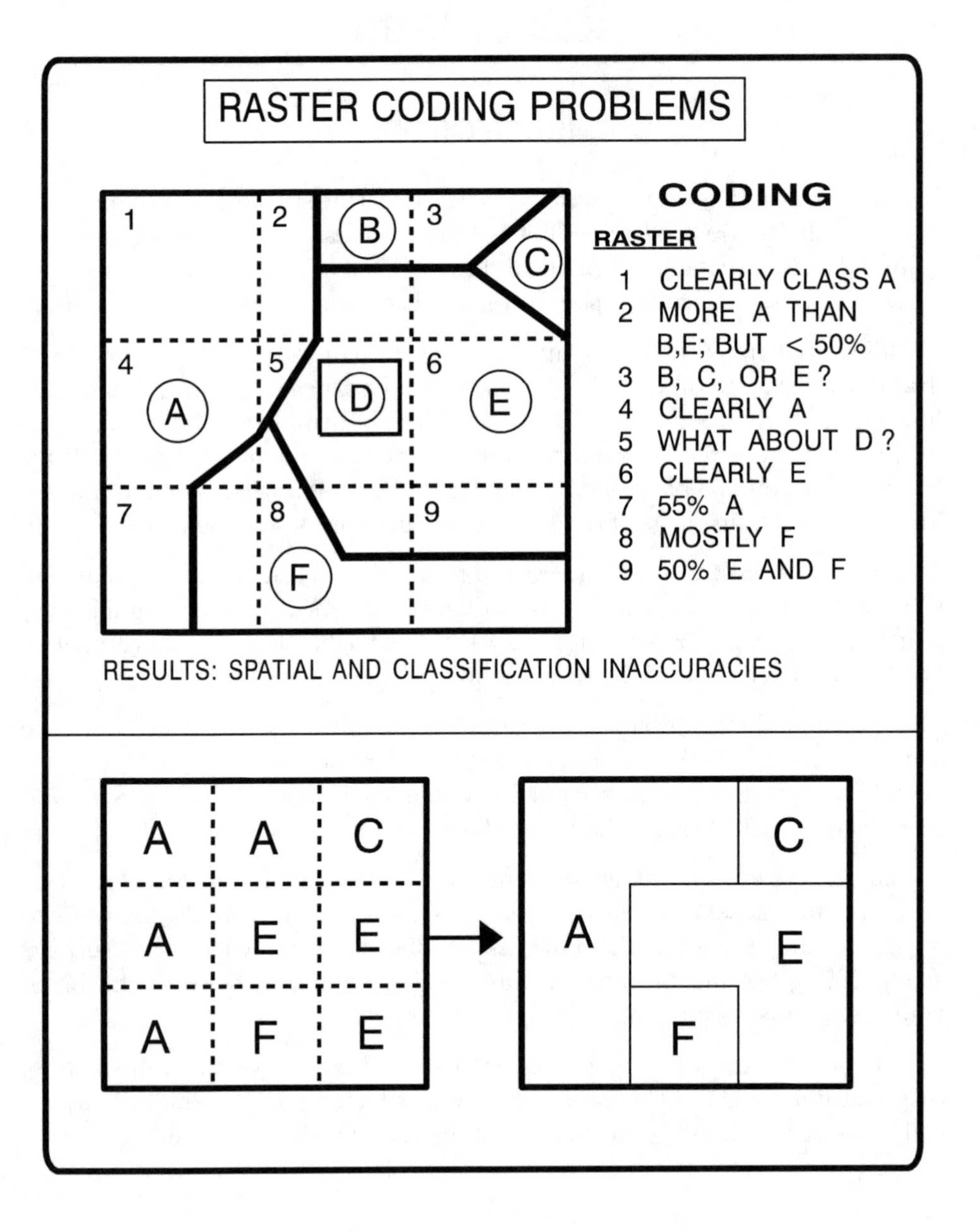
RASTER CODING PROBLEMS
1
2
3
4
5
6
7
8
9
A
B
C
D
E
F
CODING
RASTER
1 CLEARLY CLASS A
2 MORE A THAN B,E; BUT < 50%
3 B, C, OR E ?
4 CLEARLY A
5 WHAT ABOUT D ?
6 CLEARLY E
7 55% A
8 MOSTLY F
9 50% E AND F
RESULTS: SPATIAL AND CLASSIFICATION INACCURACIES
A A C
A E E
A F E
A
C
E
F

Coding Problems

The upper illustration shows landuse categories (divided by thick lines) of A through F. Cell 1 is clearly part of class A, but cell 2 is divided into classes A, B, and E. Because a cell can have only one code, the decision to assign the dominant class gives cell 2 a code of A, despite the fact that A constitutes less than 50% of the cell. This decision may be a significant loss of the desirable data effects of categories B and E.

NOTE: *This way of showing coding problems might be an exaggeration, but the concepts are important.*

Cell 3 has equal proportions (B, C, and E), which makes it difficult to classify this cell. It is coded C because that corresponding landuse appears to be at the center of the cell. Cell 4 is almost all A—no problem. Cell 5 is mostly E, but the small area of D will be completely lost. This is another special problem: D clearly exists in the area, and may be important, but rasterization might eliminate it as a data category. (On the other hand, if this is a large magnification, perhaps D is too small to be significant.)

Cell 6 is clearly E. Cell 7 is classed A because of the 55% coverage, but that means that the remaining 45%, F, will be lost, giving inaccurate final area statistics. Cell 8 will be classed as F. Maybe the E part gained (converted to F) will help to compensate for the lost F in cell 7. Cell 9 is 50% E and 50% F (no obvious center point), and so is given the first letter (E).

Some raster code assignment programs use the dominant class, whereas others read the very center of the cell. In the center method, landuse D would get the code in cell 5, thereby eliminating significant portions of surrounding E, and some of A. A potential problem with this method is that the centers of cells 3 and 9 might be difficult to determine correctly.

Assigning attribute codes to each cell can be a problem when the cells cover several features or classes. Problems of coding are made more apparent in the greatly magnified views (lower illustration) of these nine cells, where the results are shown. Note that all landuse types (A through F) are in the original data set, but only A, C, E, and F survive in the end. This could pose a problem in terms of the accuracy of final output. For example, class B is much larger than C, but its unfortunate placement in the cells effectively "killed" it as an area feature.

Raster systems generalize a landscape and yield spatial and classification inaccuracies. This might not be important for some purposes, but it could be critical for others. One possible solution is to increase the number of cells, making each one smaller and therefore more sensitive to accurate classification. This parameter, resolution, is examined next.

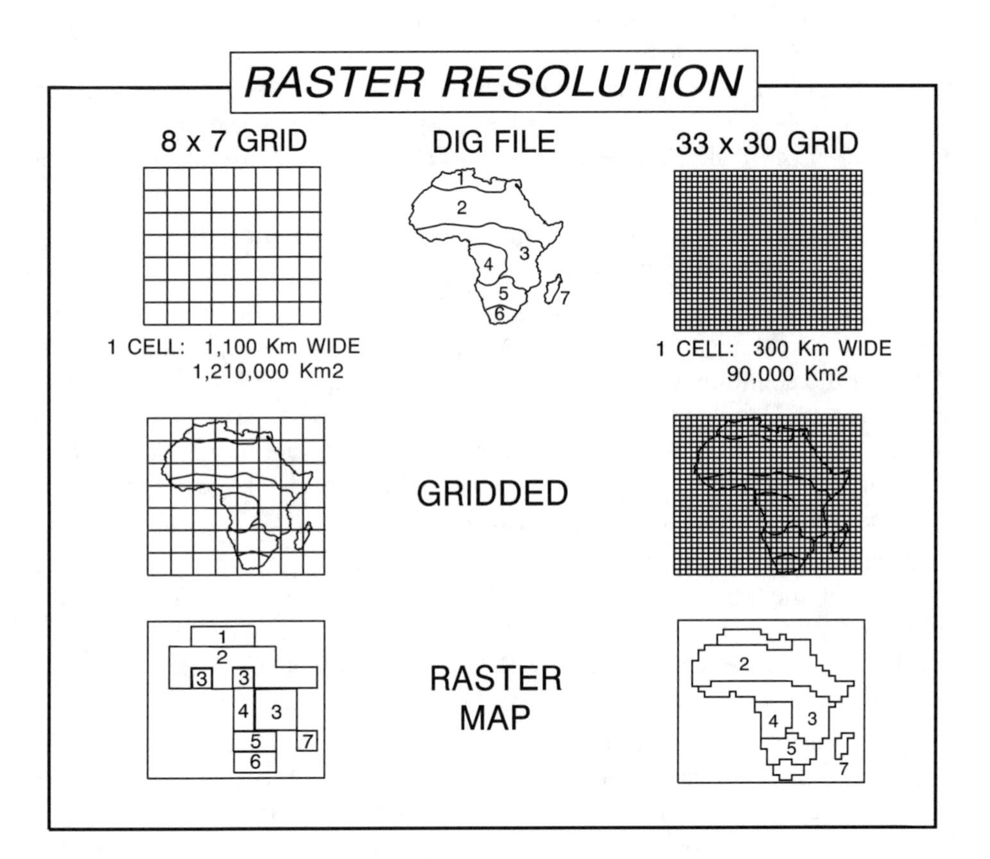
RASTER RESOLUTION
8 x 7 GRID
DIG FILE
33 x 30 GRID
1 CELL: 1,100 Km WIDE
1,210,000 Km2
1 CELL: 300 Km WIDE
90,000 Km2
GRIDDED
RASTER
MAP

Resolution

Increasing the number of cells on a coverage increases spatial resolution, which helps to increase spatial accuracy. Consider the ecosystem "dig-file" vector map in the top middle of the illustration and the two grids on each side used for rasterization. The left-hand grid is rather coarse, with few (56) cells. When used to represent the African continent, the results are a nearly unrecognizable shape (bottom left). One advantage to using relatively few cells is the associated short processing time and ease of analysis. The true shape might be unimportant to a project in which the emphasis is fast analysis.

The much more dense grid on the right (990 cells) makes a recognizable shape, which might be preferred for mapping and analysis. However, this grid takes more storage space and has a slower computer processing time. Compare the cell resolution of each grid (1,210,000 versus 90,000 square km). Depending on the application, the 300-km cell may be too detailed, too general, or just right. Most GIS raster files contain many more cells and much better spatial definition than these examples.

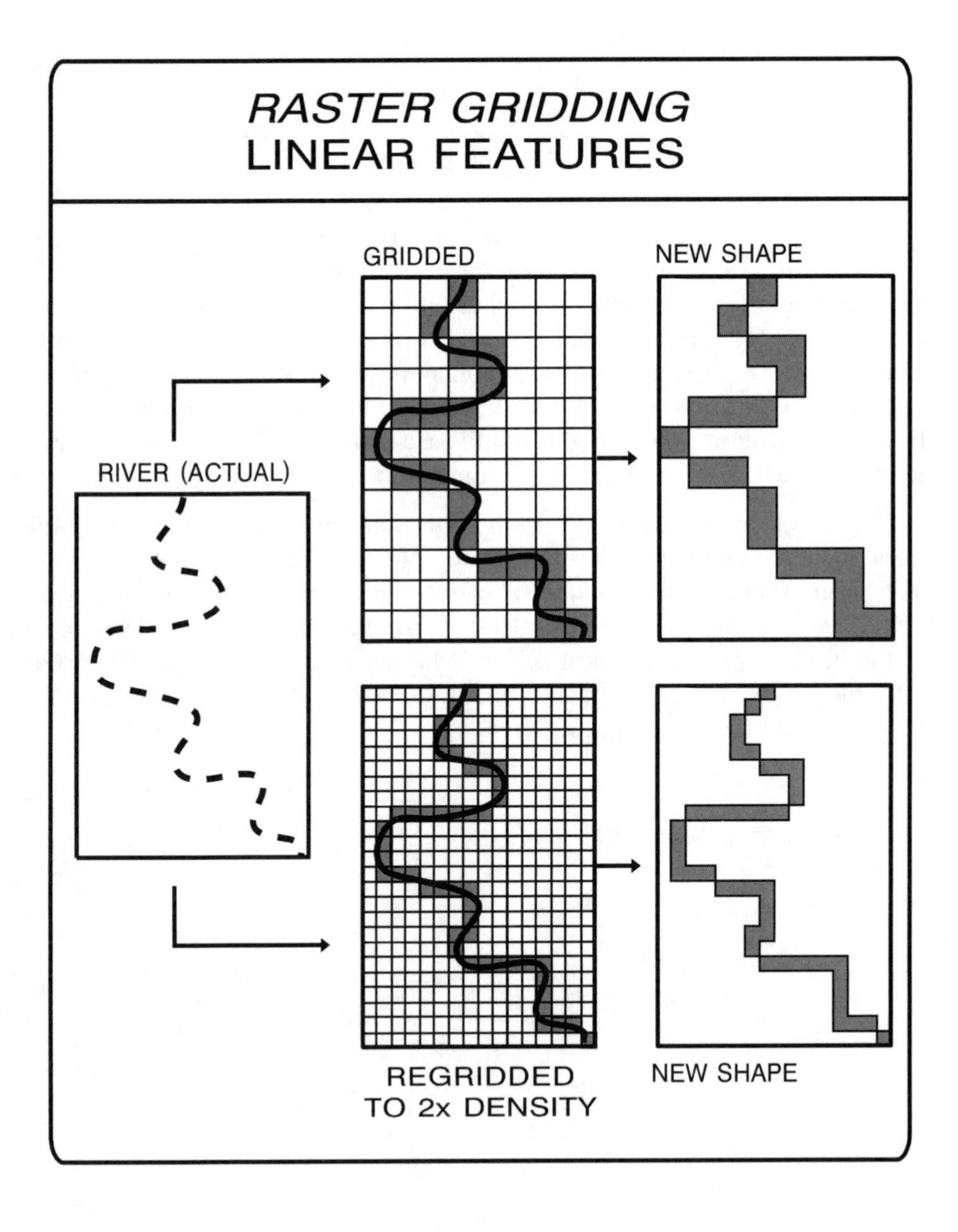
RASTER GRIDDING
LINEAR FEATURES
GRIDDED
NEW SHAPE
RIVER (ACTUAL)
REGRIDDED
TO 2x DENSITY
NEW SHAPE

Gridding and Linear Features

To see how the raster format can create spatial inaccuracies, note the shape of the river in the left-hand box of the illustration, and then compare it to the two gridded versions in the middle. Each cell containing a significant part of the river has been coded (thereby shaded) as "river." The top example uses an 8 x 12 grid, with the result a rather generalized and crude shape (top New Shape box). The bottom grid uses twice as many cells (16 x 24 grid), whereby the New Shape (lower right) appears more realistic, though still a long way from the vector shape.

For spatial measurement, suppose the side of each cell in the top grid represents 1 km. A small measuring device indicates a river length of 24 km, but there are 19 river cells in the top New Shape version, thereby resulting in a 5-km error. However, if we count every cell the river actually *touches,* even slightly, there are 24 cells. "Touch" is a matter of luck in this case; a very slight shift in initial placement of the grid might have resulted in fewer cells.

Cells on the doubled grid at bottom measure one-half km, and the river count is 46 cells, yielding a length of 23 km—a better measure, but not perfect. A higher density of cells in a raster system usually implies more accurate measurements (but not as good as vector). The size of the raster cells is therefore important. However, the purpose of a project usually determines data accuracy needs.

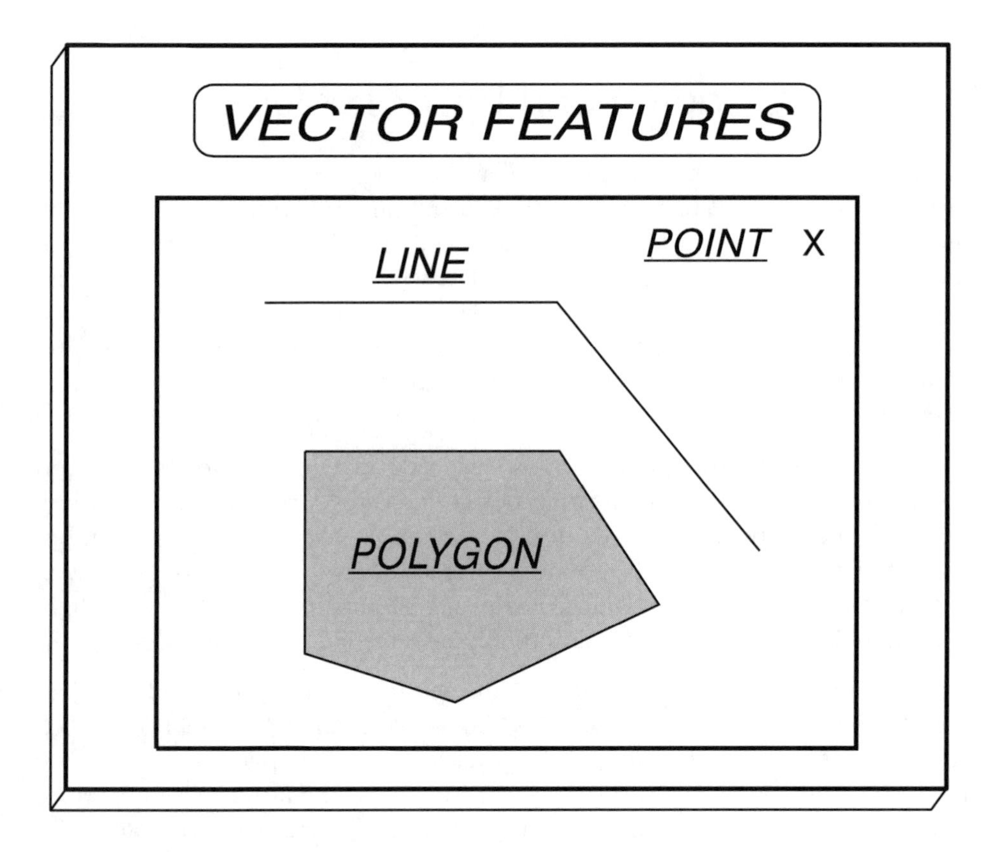
VECTOR FEATURES
LINE
POINT X
POLYGON

Vector Data

Features

You've seen how the raster format works. Here we examine the vector format. A reminder about the vector data structure: it is maplike, usually very similar to the original data. Illustrated are the three features types: point, line, and polygon. A point has no dimensions and is represented as a symbol that indicates its position. A line is a one-dimensional feature that has length and direction, but no width. Here is a complex line consisting of two parts. A polygon is an enclosed area, the shaded feature. The next illustration shows how the vector format builds these feature types.

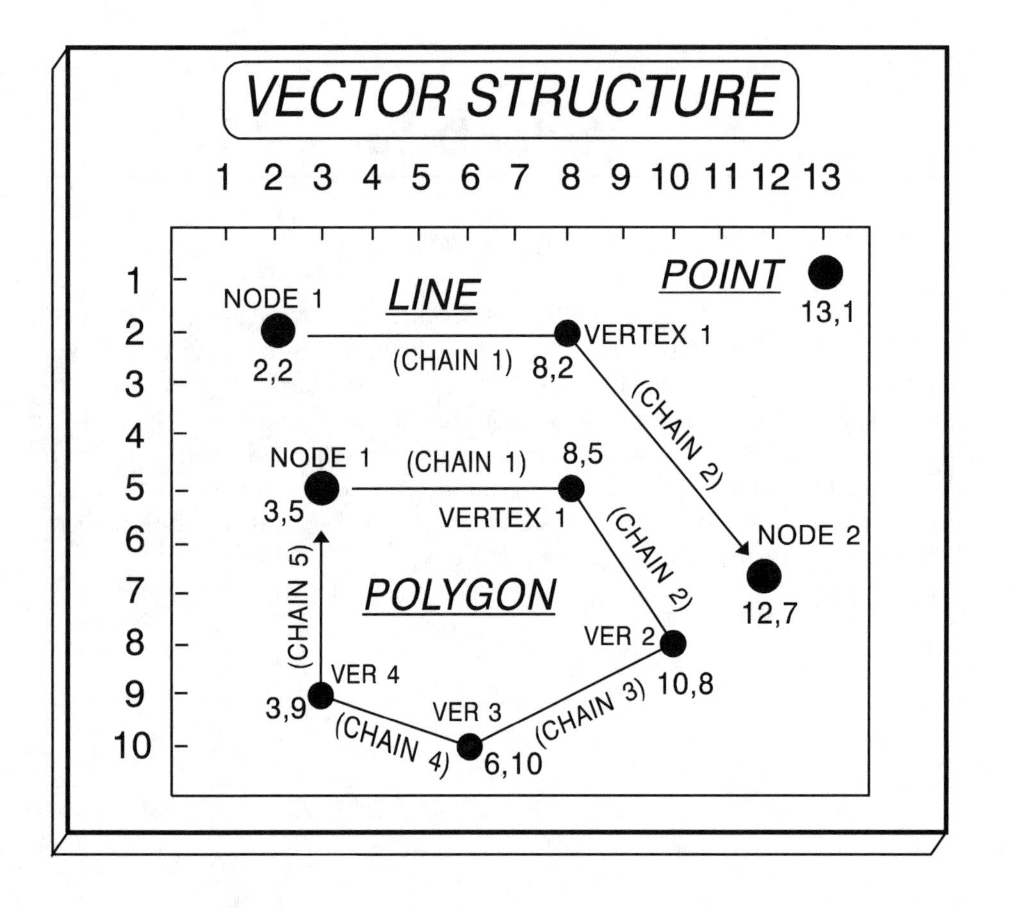

VECTOR STRUCTURE
1 2 3 4 5 6 7 8 9 10 11 12 13
1
2
3
4
5
6
7
8
9
10
LINE
NODE 1
2,2
(CHAIN 1)
8,2
VERTEX 1
(CHAIN 2)
NODE 2
12,7
POINT
13,1
POLYGON
NODE 1
3,5
(CHAIN 1)
8,5
VERTEX 1
(CHAIN 2)
VER 2
10,8
(CHAIN 3)
VER 3
6,10
(CHAIN 4)
VER 4
3,9
(CHAIN 5)

Structure of Vector Data

Instead of being built by raster cells, vector features are actually defined by "coordinate points"—spots located precisely by the X-Y coordinates. Then "chains" (special lines) connect the points to draw the feature.

At the beginning and end of every line or polygon feature is a node—a special point that defines the ends. At each "bend" (or change of direction) in a line is a vertex (plural: vertices). The illustration shows a vector structure, indicating the X-Y coordinate of each node and vertex. The connecting lines are called chains (sometimes termed *arcs*).

NOTE: *Nodes are end points and vertices are in between; chains are the connecting line segments.*

A vector point is a location of a feature—a coordinate location that has no spatial dimension; that is, no width or length because in reality they are either too small to be seen at the mapped scale or are featureless, designated locations. A point is expressed as a single X-Y coordinate (position 13,1 in the illustration). The program fills in a dot or some other symbol for visual convenience.

A vector line feature has only one dimension: length (no depth). There are two nodes on a line, and if the line is straight, it has no vertices. A more complex line has vertices, one at each change of direction. The line in the illustration begins at node 1 and ends at node 2, with vertex 1 between them. A node in the middle of a chain is actually the separation of two vector features. Remember: nodes are end points.

A polygon is an enclosure; therefore, the beginning node is also the ending node (no opening or gap is permitted once the chains are placed). For a single polygon (no others attached), there is only one node (start and stop), but there must be at least two vertices to make an area.

In a coverage display, normally only the chains are seen, defining the line or polygon feature, but under special editing views, nodes and vertices can be inspected. Actually, vector system data files have only the coordinates of each node and vertex; the connecting chains are drawn by the program or hardware. This is an efficient data storage format.

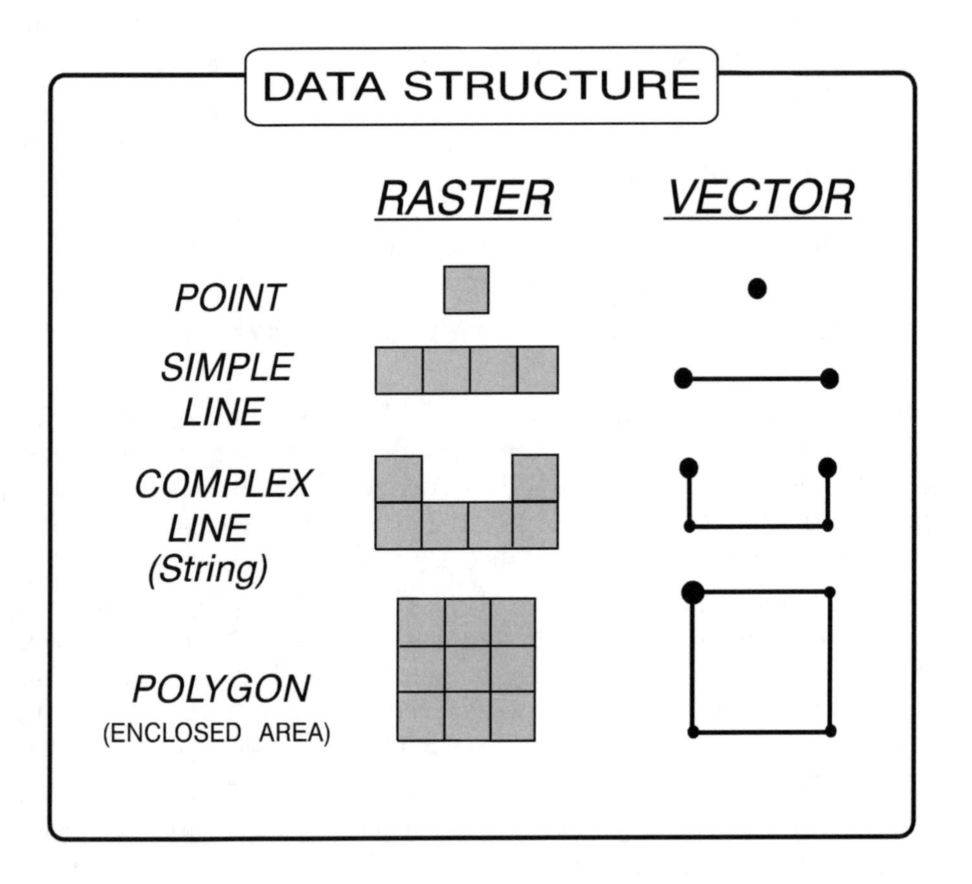
DATA STRUCTURE
RASTER
VECTOR
POINT
SIMPLE LINE
COMPLEX LINE (String)
POLYGON
(ENCLOSED AREA)

Raster Versus Vector

Raster and vector structures have different methods of storing and displaying spatial data. The illustration compares raster and vector points, lines, and polygons.

A *point* in a raster system is a single cell, but in a vector system it is only a symbol representing a precise coordinate position.

A simple *line* in a raster system consists of a sequence of cells, which mistakenly gives the impression of width. In a vector system, a line consists of two nodes (at each end) and a chain that connects them. A more complex line (sometimes called a string) in raster includes cells representing lines that run in more than one direction. In a vector system, in addition to the two nodes and chain, vertices mark changes in direction.

Raster *polygons* fill the area within their borders with cells. The vector format uses a single node (as a beginning and end) and several vertices to mark the boundary direction changes.

For programming reasons, usually only one data type is in a GIS file (point file, line file, or polygon file). One reason is to avoid confusion between features that may appear alike but are different, such as four connecting streets that could be interpreted as a square polygon. The Line file defines them as lines. For presentation and mapping, the final coverage can have a mix of feature types—point, line, and polygon.

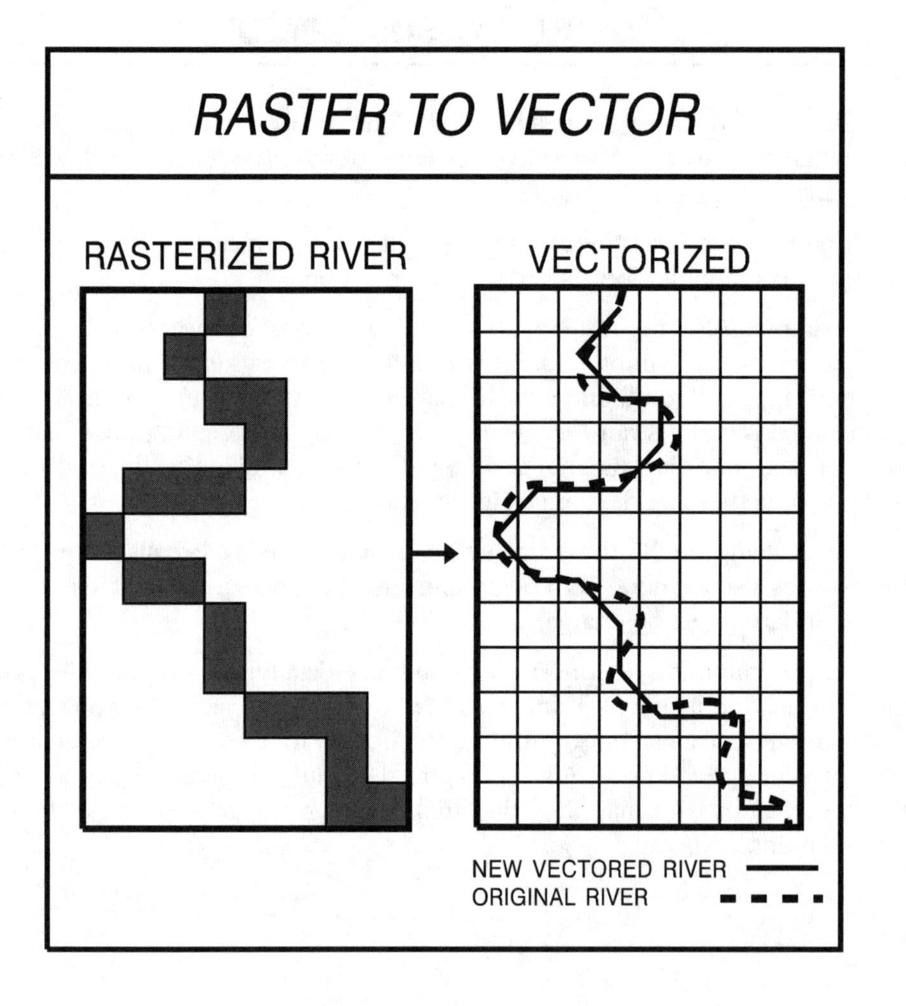
RASTER TO VECTOR
RASTERIZED RIVER
VECTORIZED
NEW VECTORED RIVER
ORIGINAL RIVER

Raster to Vector

Feature Conversion

It is possible to change (convert) features from raster to vector format. However, in doing this, even though the vector version looks more accurate, it is not. Remember, we changed accurate vector dig file data to generalized raster format. It is not possible to regain the accuracy simply by vectorizing (converting to vector format) raster data. However, there may be reasons for making new vector files from the raster data (discussed under the next topic).

The process of changing raster features to vector features normally involves connecting cell centers. There is no way to know precisely where the feature actually existed within the cell area originally, so the center is the logical vertex location. This uncertainty causes problems in accuracy. The final product *appears* very accurate because it is vector, but really the new river is nothing more than chains connected with "guessed" vertices.

The raster version of the river is presented on the left. When cell centers are linked in the vector version, the resulting river shape is different from the original. Remember: the original digitized river had the most accurate shape and location, but it was rasterized, which generalized the data and created spatial inaccuracy. This process of revectorizing the data does not return the river back to its original accuracy; we get only another generation of data. The user must be careful to separate appearance from true spatial accuracy.

NOTE: *Some GISs can make the new vector version look smoother by "softening" the sharp vertices, but again, the user can be fooled into believing the nice appearance is spatially accurate, when in reality it is not.*

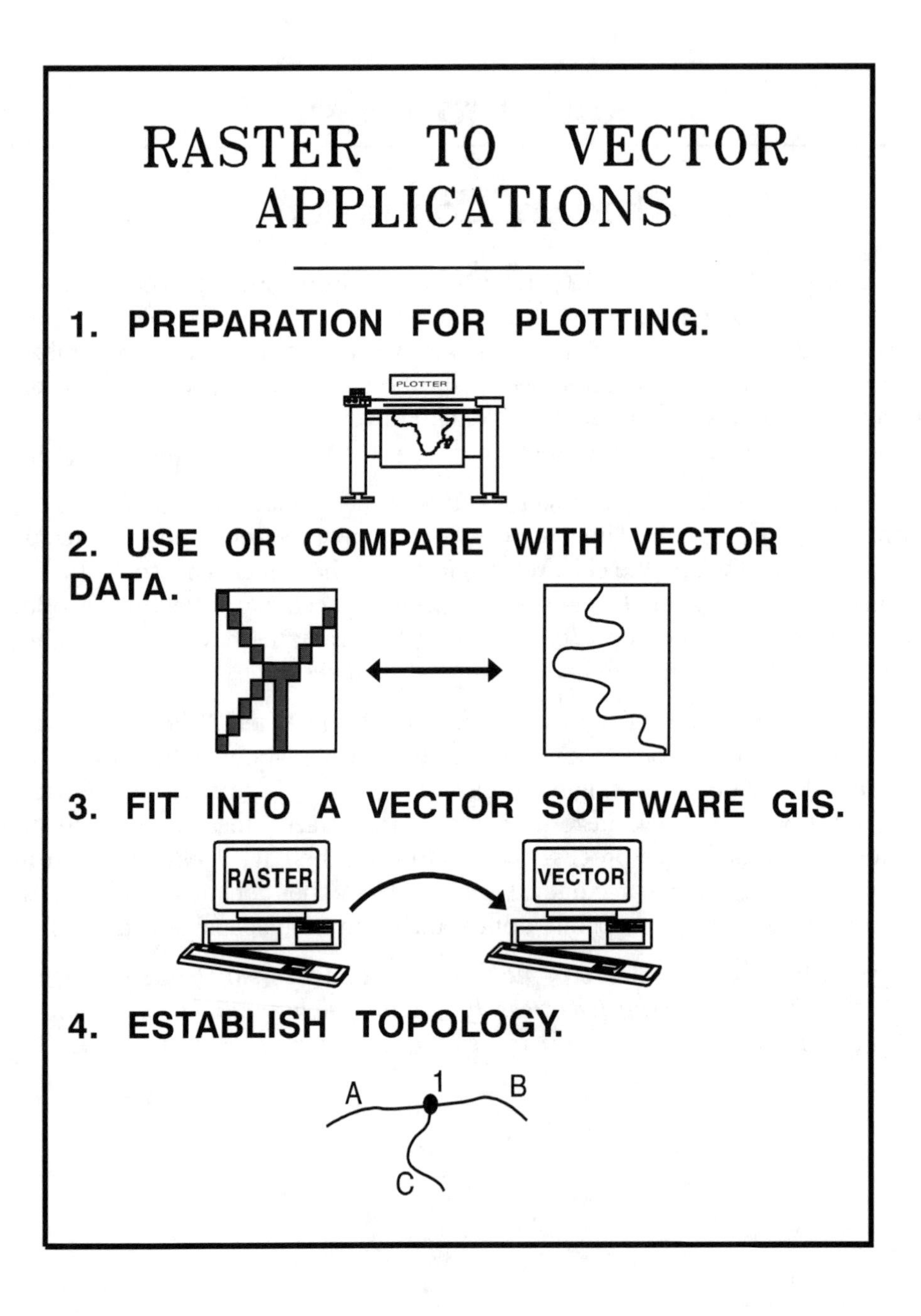
RASTER TO VECTOR APPLICATIONS
1. PREPARATION FOR PLOTTING.
PLOTTER
2. USE OR COMPARE WITH VECTOR DATA.
3. FIT INTO A VECTOR SOFTWARE GIS.
RASTER
VECTOR
4. ESTABLISH TOPOLOGY.
A
1
B
C

Applications of Raster to Vector

There may be several reasons to convert raster data to vector. The following are a few.

- *Preparation for Plotting:* Plotters prefer vector formats. It is easier and faster to draw lines than to print squares.
- *Visual Form of Features:* Most features look better when displayed as vector data.
- *Comparison with Other Vector Data:* Coverage or data comparisons work best when the formats are identical.
- *Fit with Another GIS:* GISs usually have either raster or vector as the central operating structure, and data formats typically must be identical to facilitate data management. However, many modern systems permit a mix of formats, so this requirement is becoming outdated.
- *Topology:* A special, highly useful aspect of many GISs, topology uses vector formats (see Chapter 5).

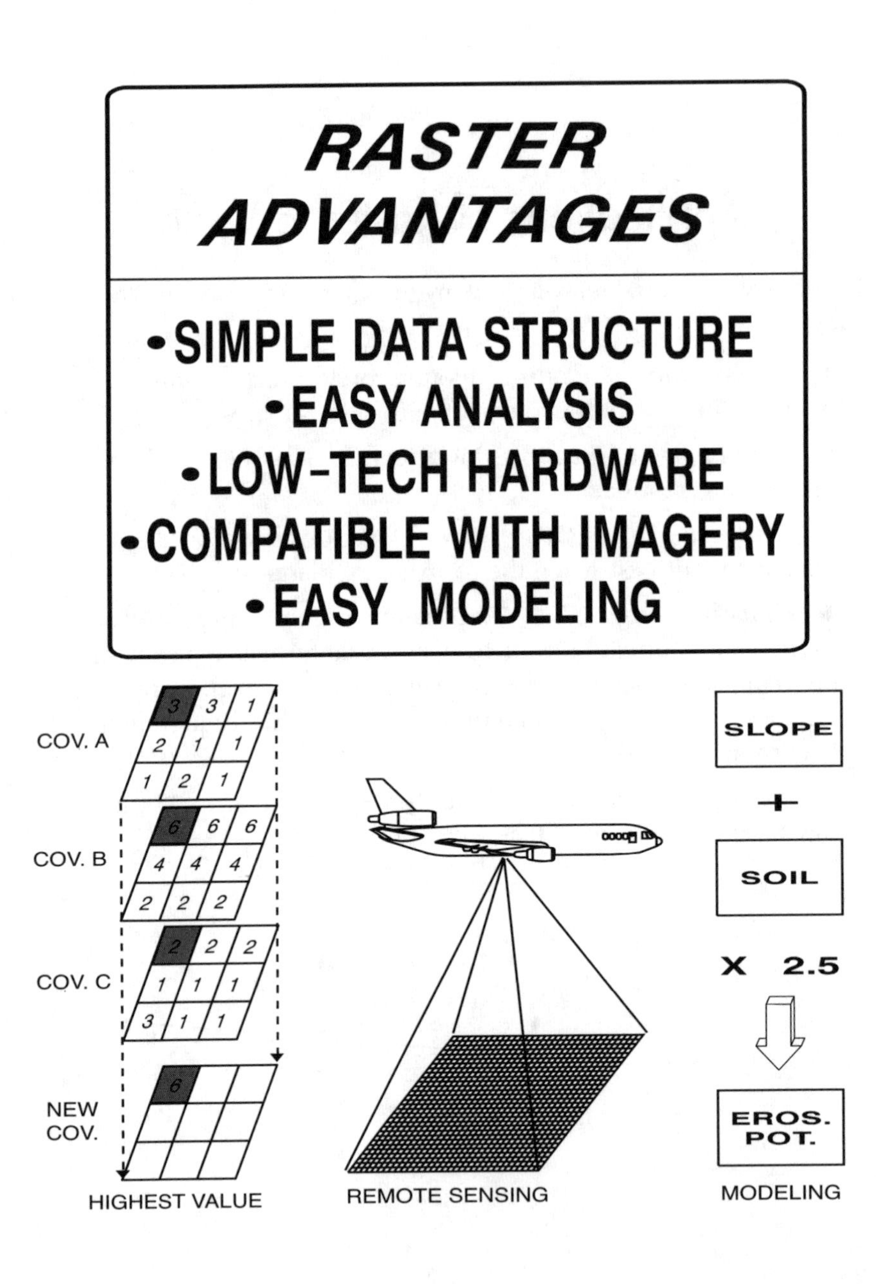
RASTER ADVANTAGES
•SIMPLE DATA STRUCTURE
•EASY ANALYSIS
•LOW-TECH HARDWARE
•COMPATIBLE WITH IMAGERY
•EASY MODELING
COV. A
3 3 1
2 1 1
1 2 1
COV. B
6 6 6
4 4 4
2 2 2
COV. C
2 2 2
1 1 1
3 1 1
NEW COV.
6
HIGHEST VALUE
REMOTE SENSING
SLOPE
+
SOIL
X 2.5
EROS. POT.
MODELING

Raster and Vector Pros and Cons

Raster Advantages

The advantages of raster format include the following:

- It is a relatively simple data structure: a grid with a single number (representing code) in each cell. It is easy to understand and use, even by beginners.
- The simple, coded grid structure makes analysis easier; computers are good at comparing numbers. Even if there is a "stack" of data files to be manipulated for some complex analysis, the computer reads each grid cell position one by one and does the analysis on that cell for each data file. For example, determining which data file has the highest-value number in each grid cell position is a simple matter of comparing numbers.

The illustration shows three coverages (A through C), with the first grid cell in each one shaded (cell position 1). The program compares raster values and enters the highest in New Cov. This type of operation can be accomplished very rapidly over the entire raster grid.

- Because of the relative simplicity of raster formats, the required computer system can be "low tech" and inexpensive.
- Remote sensing imagery—say, from aircraft or satellite—is obtained in raster (because imaging equipment uses this format). This allows comparison between imagery and GIS coverages—two different types of data that combine easily. The illustration shows a remote sensing aircraft and a diagram of a raster landscape image obtained by it. The image is easily integrated into a raster-format GIS.
- Modeling (see Chapter 2 under the heading Modeling), which uses raster numbers, is the creation of a generalized data file or a set of universal procedures to accomplish a certain GIS task. Raster cell values are easy to use in a formula. If, for example, the predicted soil erosion in a given area is expressed as a formula (using slope and soil values), raster numbers are more appropriate to use than vector formats. The illustration shows a simple example of adding the slope value of a given cell to its corresponding soil value cell in a second coverage, and then multiplying the sum by 2.5 to arrive at a number meaningful for evaluating the erosion potential. Performed for each cell in the coverage, the potential for erosion over the entire area can be mapped and analyzed.

- SPATIAL INACCURACIES
- IMPLICIT STRUCTURE
- (WYSINWYG)
- LOW RESOLUTION
- LARGE DATA SET

0	0	0	0	0	0	0	0
0	0	0	0	0	0	0	0
0	0	0	2	2	0	0	0
0	0	0	2	2	0	0	0
0	0	0	0	0	0	0	0
0	0	0	0	0	0	0	0
0	0	0	0	0	0	0	0

Raster Disadvantages

Disadvantages of the raster format include the following:

- Spatial inaccuracies are common with raster systems. It is usually hoped that losses are compensated by gains and that, overall, inaccuracies are therefore canceled out. This may be wishful thinking, however, and projects that need high accuracy either have to use more cells (greater resolution), or convert to vector format.
- Because of spatial inaccuracies caused by data generalization, a raster structure can not tell precisely what exists at a given location and therefore "implies" truth (an implicit structure). In computer graphics the acronym WYSIWYG means "What You See Is What You Get." We can say that raster formats result in views that WYSINWYG: "What You See Isn't Necessarily What You Get."
- Because each cell tends to generalize a landscape, the result is relatively low resolution compared to the vector format. Even the use of a very high number of cells (which makes files much larger and slows down computation and display) can only guarantee better resolution, not necessarily satisfactory accuracy.
- Each cell must have a code, even where nothing exists. That is, even "nothing" must be coded (usually 0). Therefore, every cell is coded, making computer storage needs high, especially for high-resolution grid cell formats (many cells), used when higher accuracy is desired. Thus, raster systems can have very large data sets. The grid at the bottom of the illustration shows an example of a raster grid with one feature, coded 2, in the center. All other cells must be given a "nothing" code (value 0). In actuality, the feature is circular, but has been generalized into four raster cell squares.
- The general public does not usually understand raster imagery.

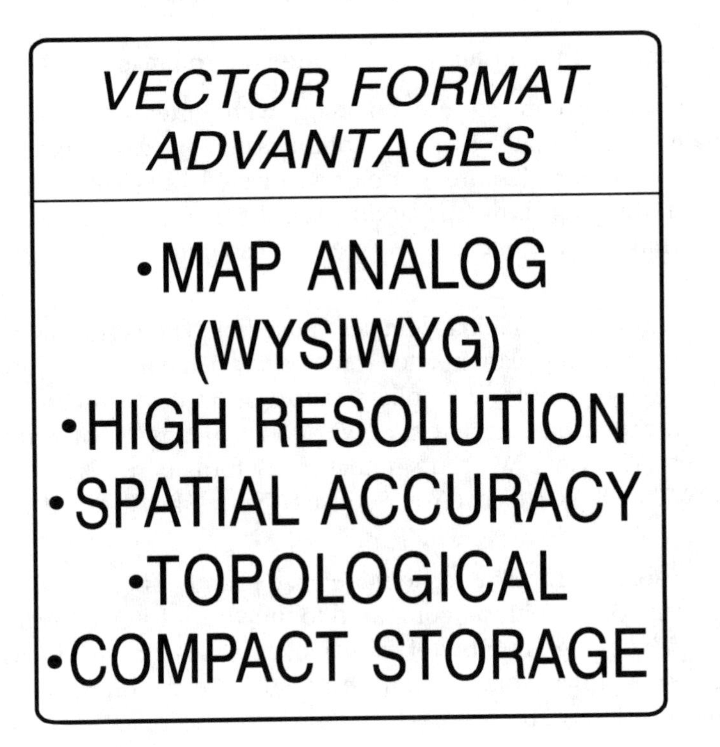
VECTOR FORMAT ADVANTAGES
•MAP ANALOG (WYSIWYG)
•HIGH RESOLUTION
•SPATIAL ACCURACY
•TOPOLOGICAL
•COMPACT STORAGE

Vector Advantages

The advantages of using the vector format include the following:

- It is more maplike. The acronym WYSIWYG applies to the view depicted by a vector display: it shows what will be mapped. Vector displays are also more pleasing to the eye.
- Because of the nature of vector data, high resolution is the norm. In fact, vector data can be more detailed than products from a computer monitor or mapping device. In fact, vector data detail and accuracy may be higher than the display resolution of the computer monitor or mapping device; limitations are due to hardware restrictions rather than the data.
- The high resolution supports high spatial accuracy.
- Vector data can be topological. A topological data structure has distinctive benefits that raster data formats do not offer (see Chapter 5).
- Systems and data managers are always concerned with computer use and data storage. Vector formats take less storage space and usually offer better storage capabilities than raster formats. This is because where vector features are defined and stored only as nodes and vertices, whereas raster coverages have every cell coded. This means that vector data files can be smaller and faster than raster files.
- The general public usually understands what is shown on vector maps.

Vector data structure seems to be the system of choice for many GIS users, particularly among those whose projects require control over data accuracy and outcome.

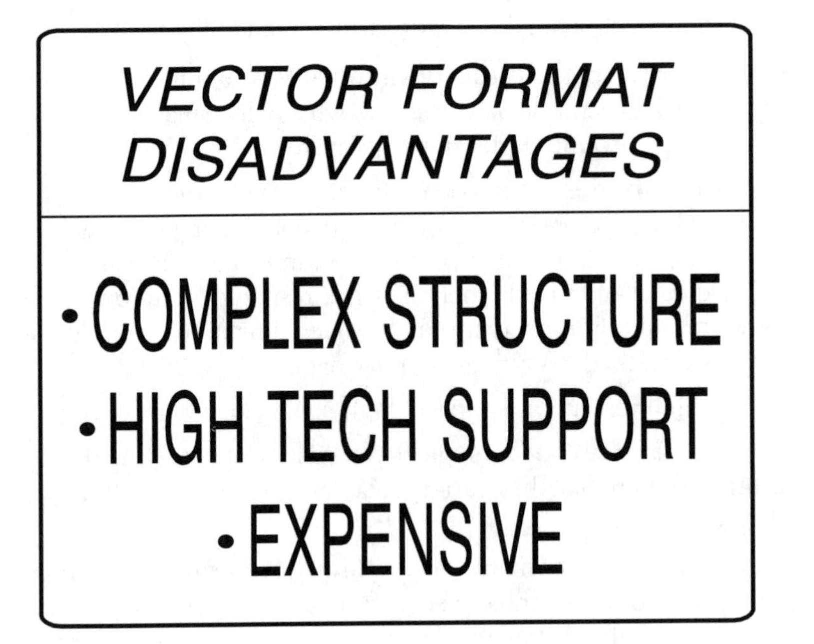
VECTOR FORMAT DISADVANTAGES
• COMPLEX STRUCTURE
• HIGH TECH SUPPORT
• EXPENSIVE

Vector Disadvantages

The following are some of the disadvantages of the vector format:

- Vector data formats may be more difficult to manage than raster formats. The are usually stored in a long list of coordinates for nodes and vertices—easy for the computer to understand but difficult for editing by the user. Knowing how to read and work a data file can be demanding.
- Whereas very simple, "low-end" computers can operate many raster-based GISs, vector formats require more powerful, high-tech machines. Management of computer equipment becomes more of a problem.
- The use of better computers, increased management needs, and other considerations often make the vector format more expensive.

Because of these considerations, some users prefer the more simple and less expensive raster format. For example, raster systems might be easier to use for introductory training.

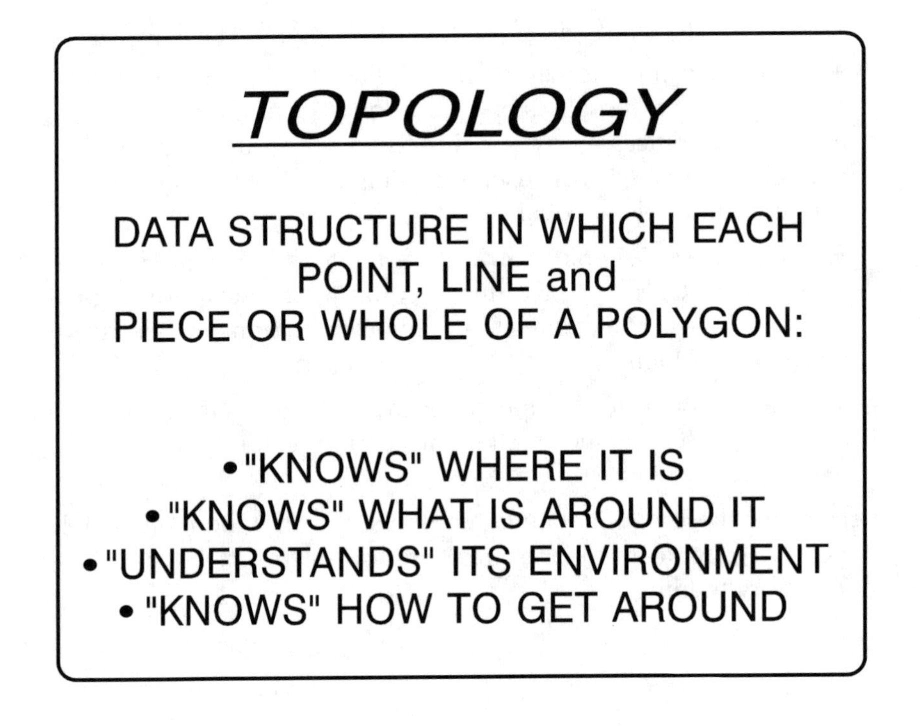
TOPOLOGY
DATA STRUCTURE IN WHICH EACH
POINT, LINE and
PIECE OR WHOLE OF A POLYGON:
• "KNOWS" WHERE IT IS
• "KNOWS" WHAT IS AROUND IT
• "UNDERSTANDS" ITS ENVIRONMENT
• "KNOWS" HOW TO GET AROUND

5

Topology

Introduction

Topology is one of the most useful data structure concepts in GIS. It allows great flexibility and powerful "connections" to be made among data. It is defined as an addition to a vector data structure in which each point, line, and partial or whole polygon has the following characteristics:

- "Knows" where it is: its position is part of the data knowledge.
- "Knows" what is around it: the attached and surrounding features are recognized.
- "Understands" its environment: by virtue of recognizing its surroundings, topology identifies features and uses their attributes to accomplish tasks.
- "Knows" how to get around: gets from point A to point B via the shortest (or least cost) path.

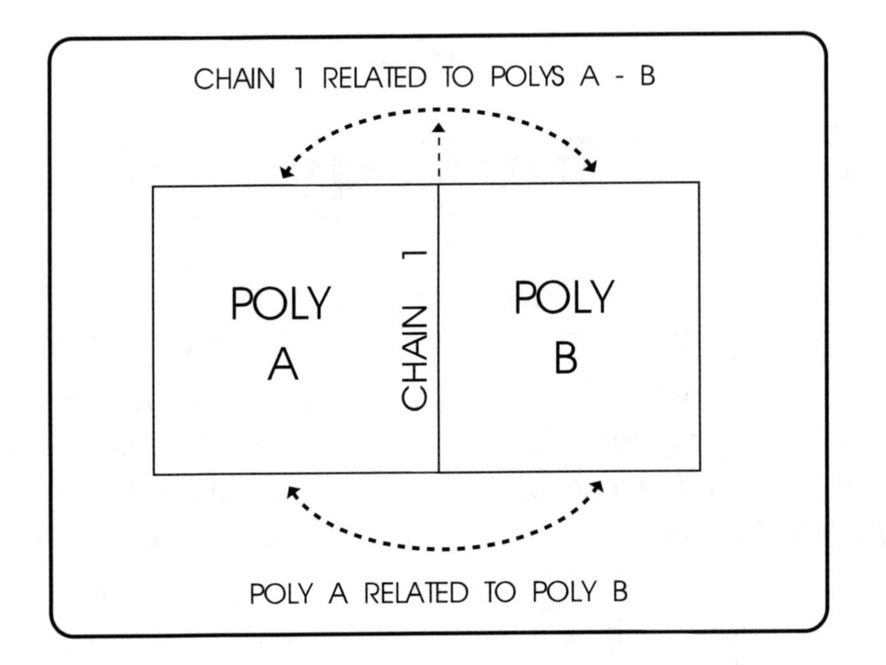
CHAIN 1 RELATED TO POLYS A - B
POLY A
CHAIN 1
POLY B
POLY A RELATED TO POLY B

Topological functions are possible because the database contains several data properties that are effectively and efficiently linked, such as identification of the polygons to the right and left of each chain or node connection. Topology is an advanced concept that gives the GIS user special options for inventory and analysis operations. Let's see how topology works, and a few of the operations that can be performed with it.

The illustration serves as a brief preview of topology. Shown are polygons A and B, with chain 1 separating them. Chain 1 actually belongs to both polygons; it defines A's right side and B's left side. Topology recognizes this defining property. It "knows" that chain 1 is part of both polygons and thus relates one to the other. Polygon A is directly related to polygon B because chain 1 is their common topological feature.

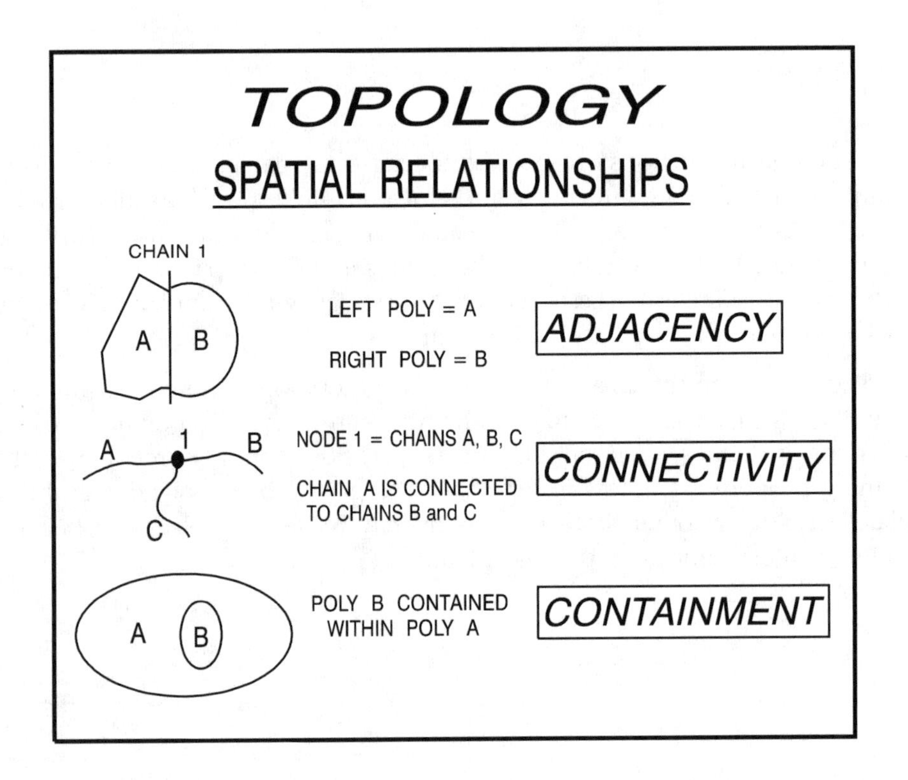
TOPOLOGY
SPATIAL RELATIONSHIPS
CHAIN 1
A
B
LEFT POLY = A
RIGHT POLY = B
ADJACENCY
A
1
B
C
NODE 1 = CHAINS A, B, C
CHAIN A IS CONNECTED TO CHAINS B and C
CONNECTIVITY
A
B
POLY B CONTAINED WITHIN POLY A
CONTAINMENT

Topology and Spatial Relationships

Topology uses several major data structure characteristics to perform its "magic." First, each piece of vector data structure—nodes, vertices, coordinates, chains, and so on—is stored in a database. A relational database can then functionally link every part of the data in some way.

The important aspect of topology is its capability to recognize and use surrounding features. Topology automatically unites (links) all parts of the data structure in order to function as it is intended. The following are three of the most important properties that come with topology.

◆ *Adjacency:* For any given feature, topology links adjacent features, usually in terms of what is to the left and what is to the right. In the illustration, the left and right polygons would be attributes listed in the database for chain 1.

◆ *Connectivity:* Topology keeps track of all connected features. For node 1, chains A, B, and C would be listed in the database. This also helps chain A to be connected to the other chains by way of node 1.

◆ *Containment:* This refers to what is *within* a polygon. Polygon B is within polygon A (or polygon A has polygon B within it). Containment is a particularly difficult data-structure programming problem, but is necessary in representing such items as islands, holes, and features that surround something else.

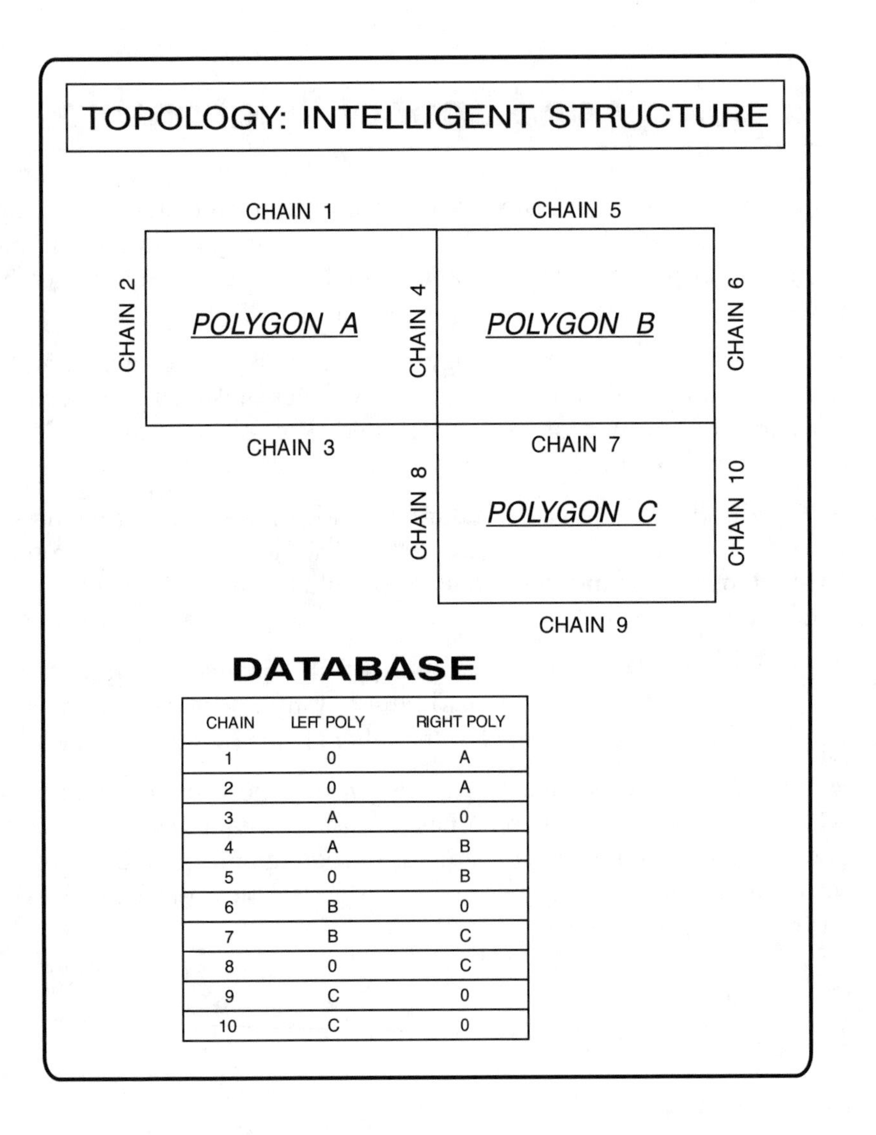

CHAIN	LEFT POLY	RIGHT POLY
1	0	A
2	0	A
3	A	0
4	A	B
5	0	B
6	B	0
7	B	C
8	0	C
9	C	0
10	C	0

Intelligent Structure

The term *intelligent structure* is used for topology because there is some sense of "recognition" in the way it builds, connects, and uses standard data features such as chains, nodes, and vertices. Because the database links, by way of what is to the right and left of a feature, topology has the ability to make functional connections.

Topology is "established" from the data structure by a special computer program that creates the links and their connections, then enters them into the database. Normally, the program is run after digitizing and editing, typically with a simple command, such as "Build Topology." The database has the ability to search for connections and make the data structure relationships.

Consider the three polygons and the database in the illustrated feature. Polygon A has four chains, with number 4 serving as its eastern (right) side. Polygon B also has four chains, with number 4 serving as its western (left) side. Topologically, chain 4 has polygon A to its left, and polygon B to its right. Similarly, chain 7 has polygon B to its left, and polygon C to its right. Actually, a sophisticated GIS also keeps a polygon database to identify chains.

These topological characteristics are very useful for spatial analysis. For example, it is easy to see how polygon A is connected to polygon B, and B to C. Topology uses adjacent chains to make connections. The next illustration shows the process.

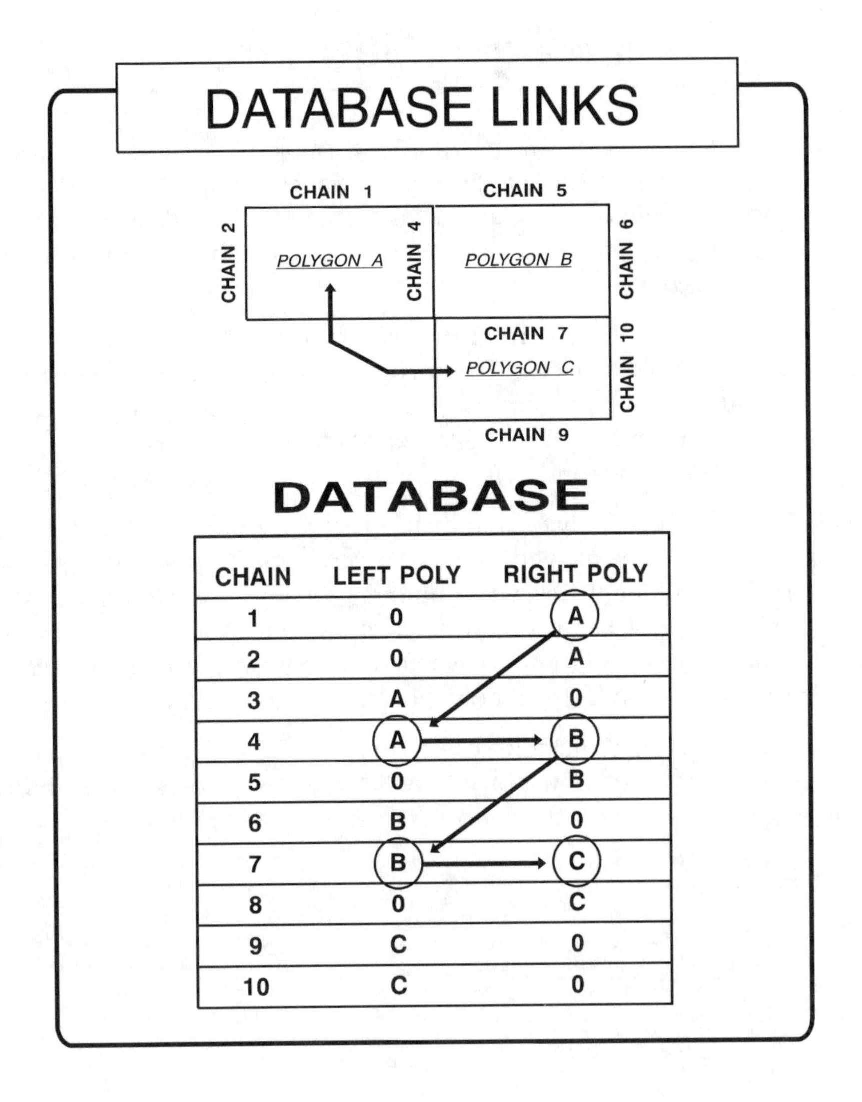

CHAIN	LEFT POLY	RIGHT POLY
1	0	A
2	0	A
3	A	0
4	A	B
5	0	B
6	B	0
7	B	C
8	0	C
9	C	0
10	C	0

Database Links

As long as there are connections, topology can build "bridges" across a coverage to link features. For example, polygon A is linked to polygon C using the logic

Polygon A → chain 4 → polygon B → chain 7 → polygon C.

The illustration shows the database links, where the real work of topology actually occurs. This database lists Chain as the main record, with the Left Poly(gon) and Right Poly(gon) for each chain registered as attributes. The path from chain 1 and polygon A to polygon C is easily followed. (Polygon 0 is the area outside the polygons and is not considered a normal link unless needed.)

The path works as follows. First, the query finds A, which in the illustration is a right poly (right polygon) of chain 1. The search goes to the next mention of polygon A that includes another polygon (not 0, the outside), which is found as a left polygon of chain 4. Chain 4 also has polygon B as its right polygon. The search moves to the next mention of B that has a polygon, which is the left polygon of chain 7. Chain 7, in turn, has C as its right polygon, the final destination. As long as there are topological links (bridges), GIS can relate features that are not apparently or directly connected.

NOTE: *This is a simplified version of the topological database and search process. Sophisticated GISs have several connected databases, and the search jumps from one to another as needed. A polygon table (database) may be used, with each polygon's chains listed. The program then goes to the chain table (see illustration) and searches for connecting polygons, reporting back to the polygon table to continue the process. This makes a very powerful GIS tool; however, as inferred in the Vector Disadvantages section (see Chapter 4), these formats can be complicated.*

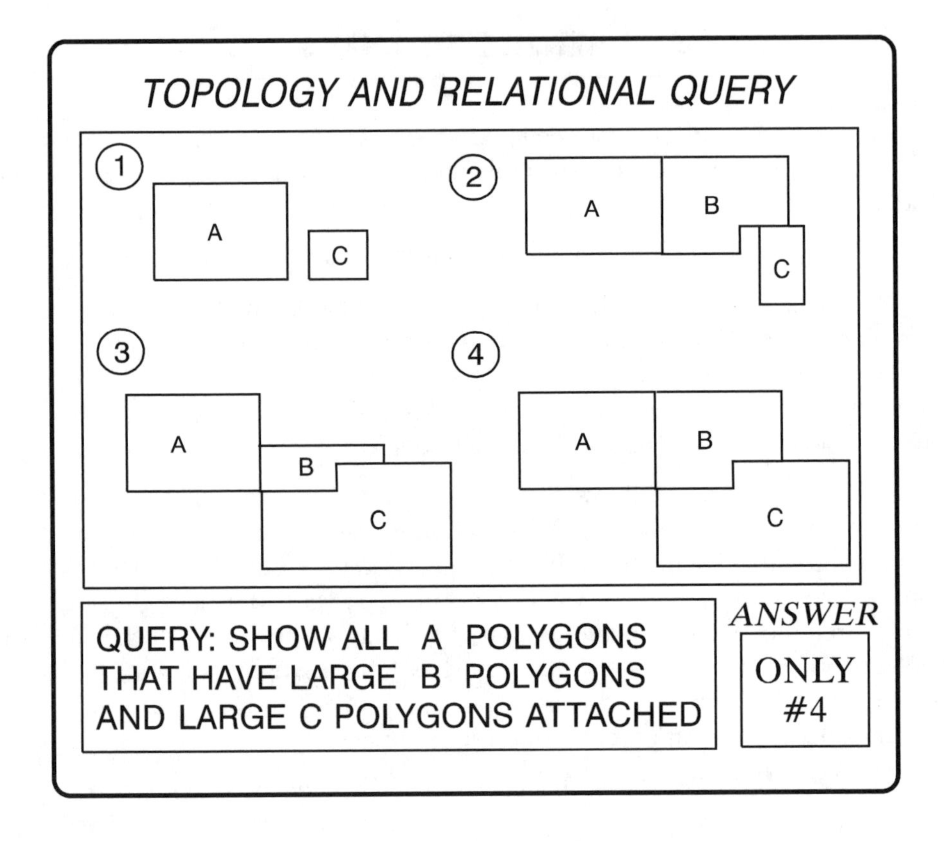
TOPOLOGY AND RELATIONAL QUERY
1
A
C
2
A
B
C
3
A
B
C
4
A
B
C
QUERY: SHOW ALL A POLYGONS THAT HAVE LARGE B POLYGONS AND LARGE C POLYGONS ATTACHED
ANSWER
ONLY #4

Topology and Relational Query

Combining the functions of a relational database and topology, we can see a powerful GIS capability. The illustration shows a mapped area with four sets of features (polygons A through C). Let's say we need to know if there are any sets that have polygon A with a large B and large C attached ("large" and "small" are used here for simplicity; normally, a quantitative figure is given, such as 10 ha). Topology will provide the connections, and the relational database will sort through the size attributes.

- *Set 1:* No attached B and no connections.
- *Set 2:* Large B attached, but C is small.
- *Set 3:* Small B attached, though C is large.
- *Set 4:* Large B and large C attached.

Set 4 is the only one meeting all of the necessary conditions (criteria). Without topology and a relational database, this search would have to be done largely by sight, and by reading lists of features and sizes. A good GIS can perform this simple search almost instantly. A large coverage with many features would be virtually impossible without topological functions.

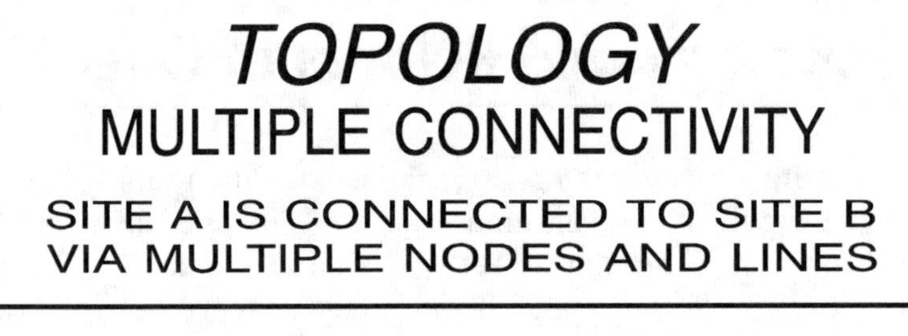
TOPOLOGY
MULTIPLE CONNECTIVITY
SITE A IS CONNECTED TO SITE B
VIA MULTIPLE NODES AND LINES

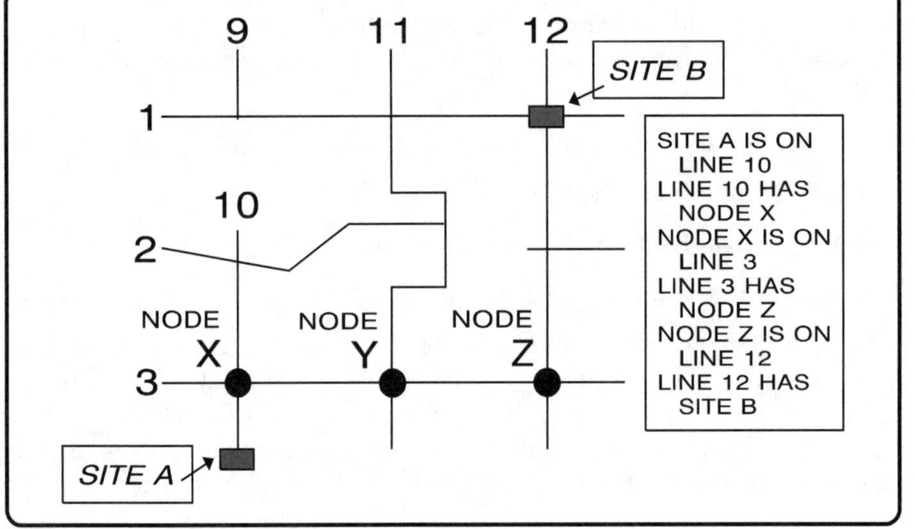
9
11
12
SITE B
1
10
2
NODE
X
NODE
Y
NODE
Z
3
SITE A
SITE A IS ON LINE 10
LINE 10 HAS NODE X
NODE X IS ON LINE 3
LINE 3 HAS NODE Z
NODE Z IS ON LINE 12
LINE 12 HAS SITE B

Multiple Connectivity

Topological connections can work on a set of lines, called a "network." Networks have nodes at each intersection, with chains in between. It is relatively easy for topology to find a path from, say, site A to site B. It uses the line (chain) to which site A is attached to search the nodes to find other lines (or chains) that may eventually lead to site B.

The boxed text within the illustration lists the set of connections used in this application. There are several other routes that could be used, most of which are more indirect than the one illustrated. In fact, the shortest path between points is easily determined by the database, which defines all possible routes and simply compares their respective total lengths and chooses the shortest.

Other attributes, not shown here, could be used in the search, such as one-way streets, or barriers that prevent continued movement. There are numerous applications for this type of topological operation, such as emergency response routing and transportation planning.

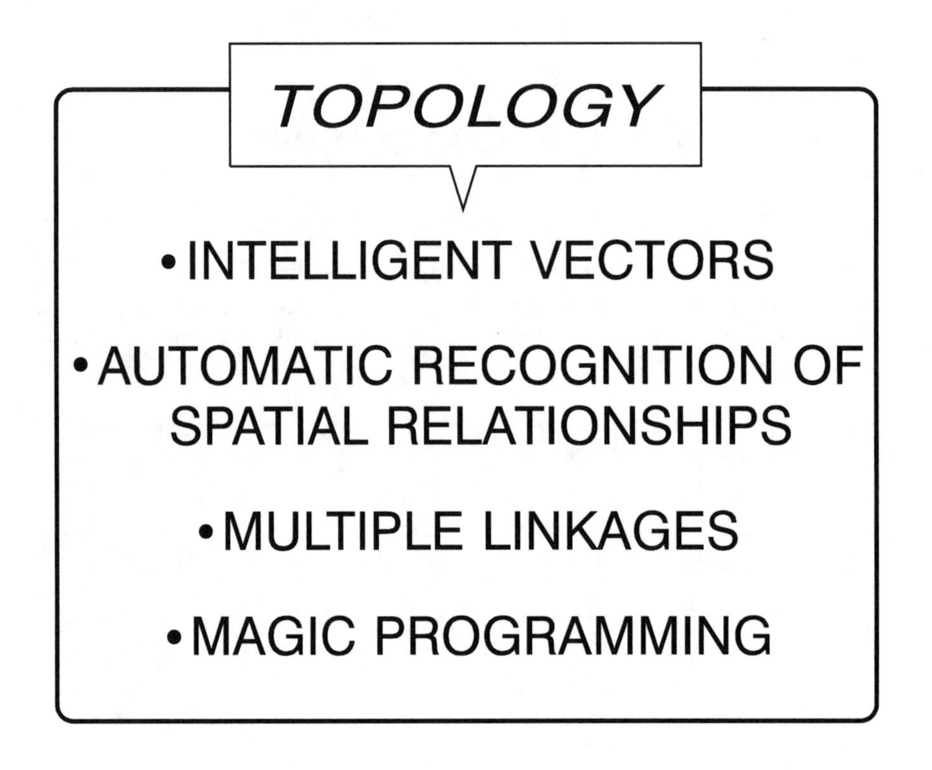
TOPOLOGY
•INTELLIGENT VECTORS
•AUTOMATIC RECOGNITION OF SPATIAL RELATIONSHIPS
•MULTIPLE LINKAGES
•MAGIC PROGRAMMING

Topology's Advantages

Topology offers special functions for spatial analysis, and most high-level GISs use it. In summary, we can say that topology offers several advantages, the following among them.

- *Intelligent Vectors:* There is "functionality" attached to topology's data structure. Nodes and chains "know" much about their surroundings, and this knowledge can be used for a variety of powerful analytical tricks.

- *Automatic Recognition of Spatial Relationships:* Relation-ships such as connectivity, adjacency, and containment are included in topology. There are other spatial relationships that can be constructed.

- *Multiple Linkages:* Each feature is linked to other features, providing multiple connections (linkages) that join (unite) them.

- *Magic Programming:* Because topology possesses the previously mentioned powerful characteristics, the programming seems almost magical to many users.

NOTE: *There are several data structures for GIS, only two of which are used in this book (in order to illustrate contrasts): a relatively simple raster system and the topological vector system with relational database. Low-cost GISs normally use raster structures, whereas more powerful and expensive ones employ a vector relational database format.*

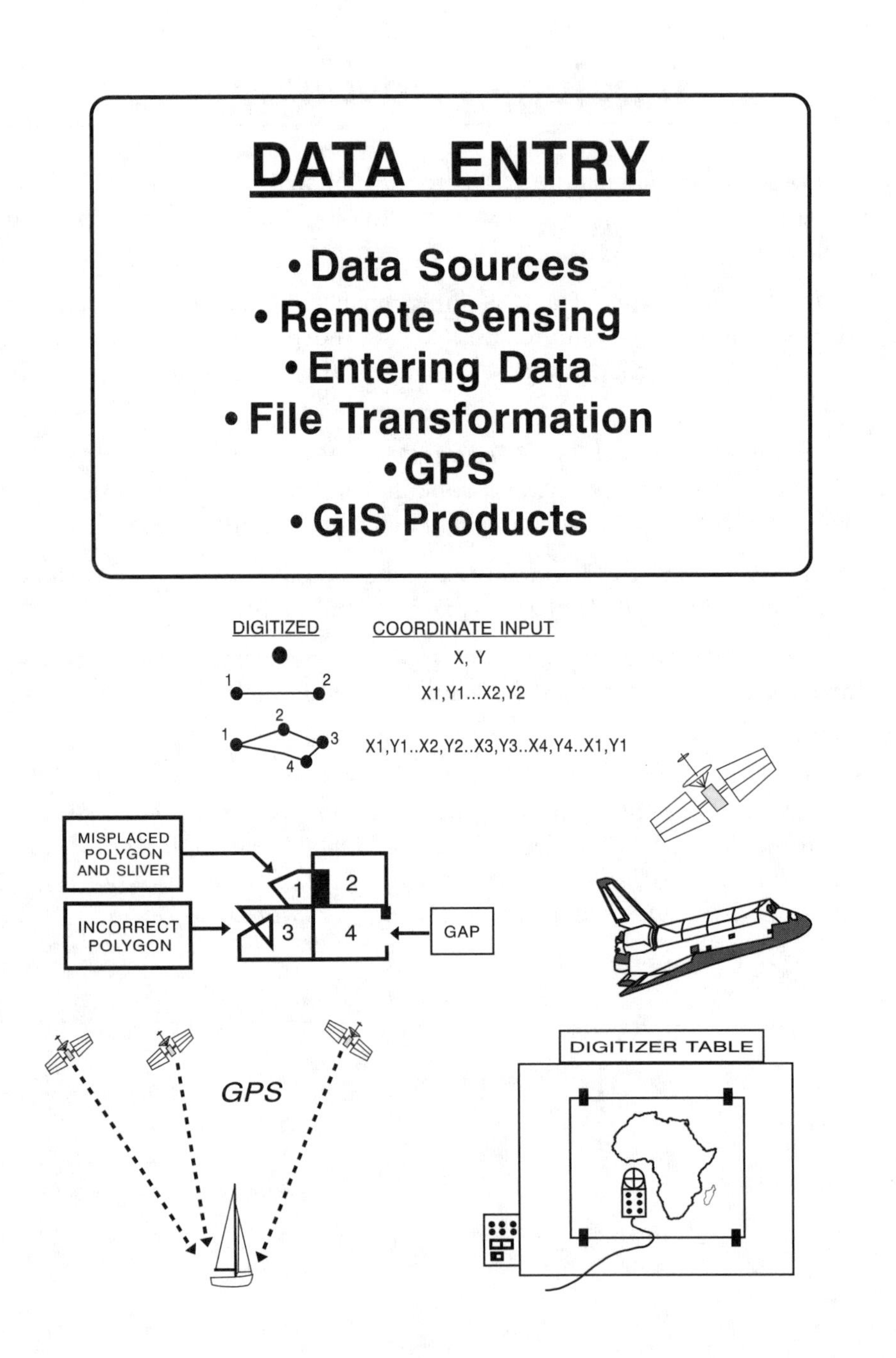
DATA ENTRY
• Data Sources
• Remote Sensing
• Entering Data
• File Transformation
• GPS
• GIS Products
DIGITIZED
COORDINATE INPUT
X, Y
X1,Y1...X2,Y2
X1,Y1..X2,Y2..X3,Y3..X4,Y4..X1,Y1
MISPLACED POLYGON AND SLIVER
INCORRECT POLYGON
GAP
GPS
DIGITIZER TABLE

6

Data Entry

Introduction

Finding appropriate data for a GIS project and getting it into the system can take a great deal of time and effort. Briefly discussed in this chapter are the primary sources of GIS data, including remote sensing. Also discussed is the process of data entry, including georeferencing and projections. A relatively new technology to GIS, called GPS (Global Positioning System), is presented at the end.

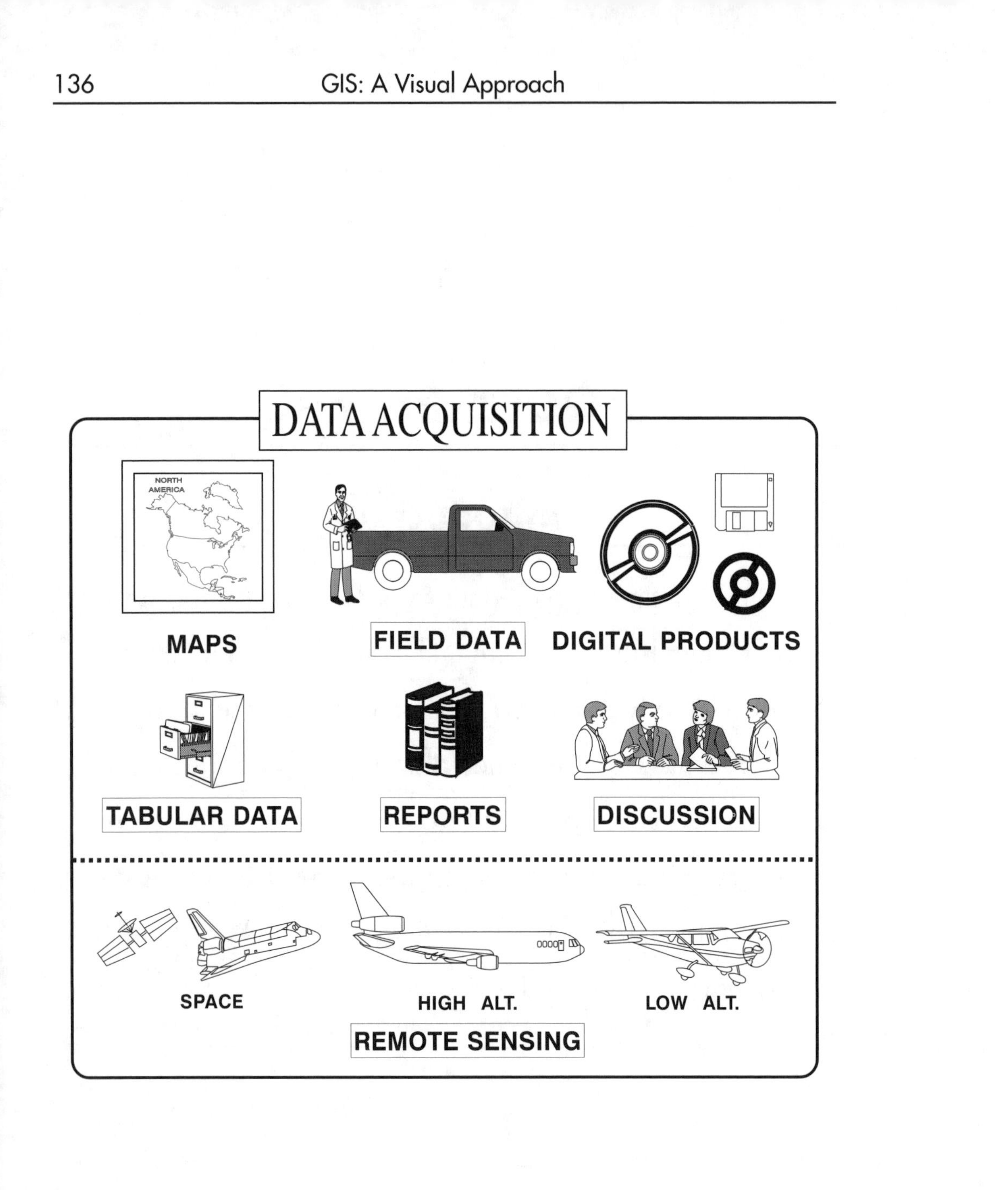
DATA ACQUISITION
NORTH AMERICA
MAPS
FIELD DATA
DIGITAL PRODUCTS
TABULAR DATA
REPORTS
DISCUSSION
SPACE
HIGH ALT.
LOW ALT.
REMOTE SENSING

GIS Data Acquisition

There are many sources of GIS data. The illustration shows a few of the more common types.

Maps are a primary source. Many organizations produce maps, and GIS accepts just about every type.

Data collected in the field (study area) is considered "primary" data—the first "raw" input into the database. This usually involves researchers or survey teams taking measurements and photographs, and making interpretations, of the places under investigation. These field personnel return their data to a central computing station for initial GIS data input and processing.

The types of data processed include digital products, which are processed data sets—sometimes complete GIS databases and coverages compiled by another organization. Digital products are becoming a major medium (vehicle) for GIS data acquisition because of their efficient storage, ease of transfer into computers, and convenience of update.

Tabular data may be standard lists, such as census reports or business marketing information, typically in printed form. The data might be typed into the GIS or copied by a scanner, an instrument that transfers copied information directly into the computer.

Reports contain a variety of data that can be copied into the GIS, such as maps and tabular data. Text discussions may not be easily reduced to GIS format, but the information can be important, thereby requiring translation by a user. Sometimes, copies of the text may be scanned and stored as an associated part of the database. Digital products are increasingly a support item associated with reports.

Human input, such as decisions on classification derived from discussion, can also be included in GIS. This is an important flexibility of GIS because judgments, interpretations, and even new data based on personal knowledge can be as valuable as field data.

Remote sensing involves the collection of landscape and other data from above, such as gathered by low-altitude aircraft carrying cameras, high-altitude aircraft carrying sophisticated electronic sensors, and space vehicles (shuttle and satellites) equipped with special imaging devices.

REMOTE SENSING

DATA GATHERING
FROM ABOVE

Remote Sensing

What Is Remote Sensing?

Geographic remote sensing is gathering data from above, usually by aircraft or space sensors (though other means also exist). It is a major source of GIS data, and although expensive, the large area covered makes the cost per ground unit lower than time-consuming and labor-intensive ground crew collection.

The illustration shows a plane (too low for reality, of course) imaging a portion of land. Many square kilometers can be covered rapidly and with relatively low cost per image. A field team takes much longer—a disadvantage for some applications, such as disaster evaluation. Also, images give a permanent record and are a more consistent view of the land than can be provided by various team members over a long period of time. Space imagery covers entire countries sometimes, giving nations poor in information resources a particularly useful set of data.

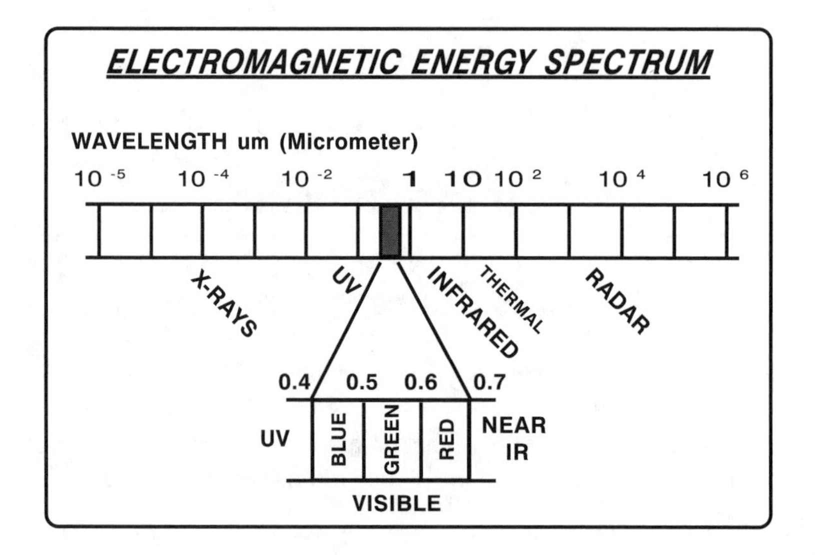
ELECTROMAGNETIC ENERGY SPECTRUM
WAVELENGTH um (Micrometer)
10 -5
10 -4
10 -2
1
10
10 2
10 4
10 6
X-RAYS
UV
INFRARED
THERMAL
RADAR
0.4
0.5
0.6
0.7
UV
BLUE
GREEN
RED
NEAR IR
VISIBLE

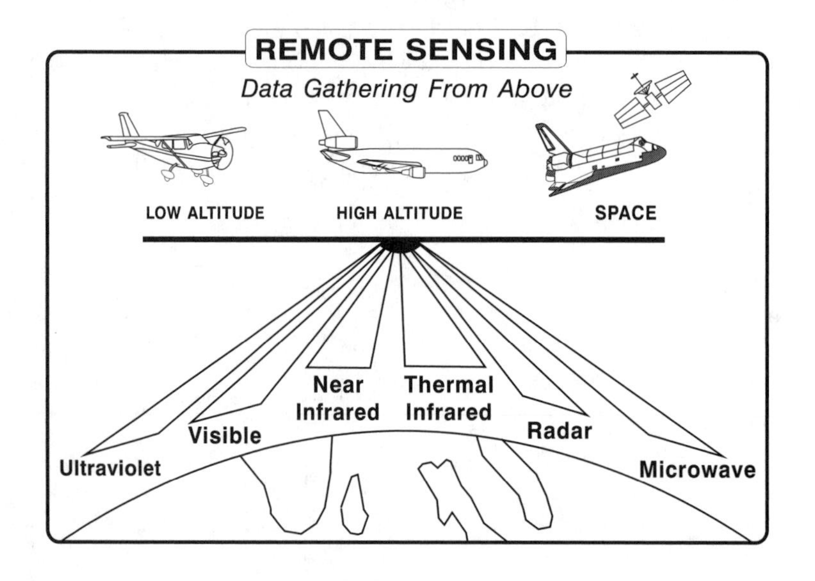
REMOTE SENSING
Data Gathering From Above
LOW ALTITUDE
HIGH ALTITUDE
SPACE
Ultraviolet
Visible
Near Infrared
Thermal Infrared
Radar
Microwave

Electromagnetic Energy Spectrum

Aerial photography was the first remote sensing data medium, providing visual imagery of landscape. In recent decades, sophisticated electronics have led to high-tech sensors that gather data from parts of the electromagnetic (EM) energy spectrum that are invisible to the human eye. The upper illustration shows that portions of the EM spectrum are expressed as wavelengths of energy, ranging from very short (on the left), through the visible middle, to very long (on the right).

Note the major remote sensing portions of the EM spectrum. X rays, for example, used in medicine and engineering, are not an important GIS data source.

The short wavelengths of ultraviolet (UV) are affected by atmospheric scattering and therefore are not an effective source of landscape data. UV remote sensing has been used successfully from the moon, however.

The middle range contains wavelengths ranging from blue to red, visible to the human eye. Ordinary camera film is sensitive to this spectrum, producing standard air photography. Some films capture enough of the next longer wavelength, the near-infrared, to reveal information not available in the visible spectrum, such as plant stress. Film can show damaged vegetation long before it is apparent to the observer on the ground. This is very useful for many environmental GIS projects.

Thermal infrared sensors detect very small temperature differences and display them on film or electronic media. This is useful in thermal pollution monitoring, for example, where industrial effluence can be analyzed in terms of heat characteristics.

Radar and microwave data are long-wave, producing land and water information much different from that of the visible spectrum. Radar is useful in penetrating (seeing through) cloud cover, for instance, and can map topography in humid tropical regions where aerial photography has been unsatisfactory. Radar also detects subtle geologic features, such as fault lines.

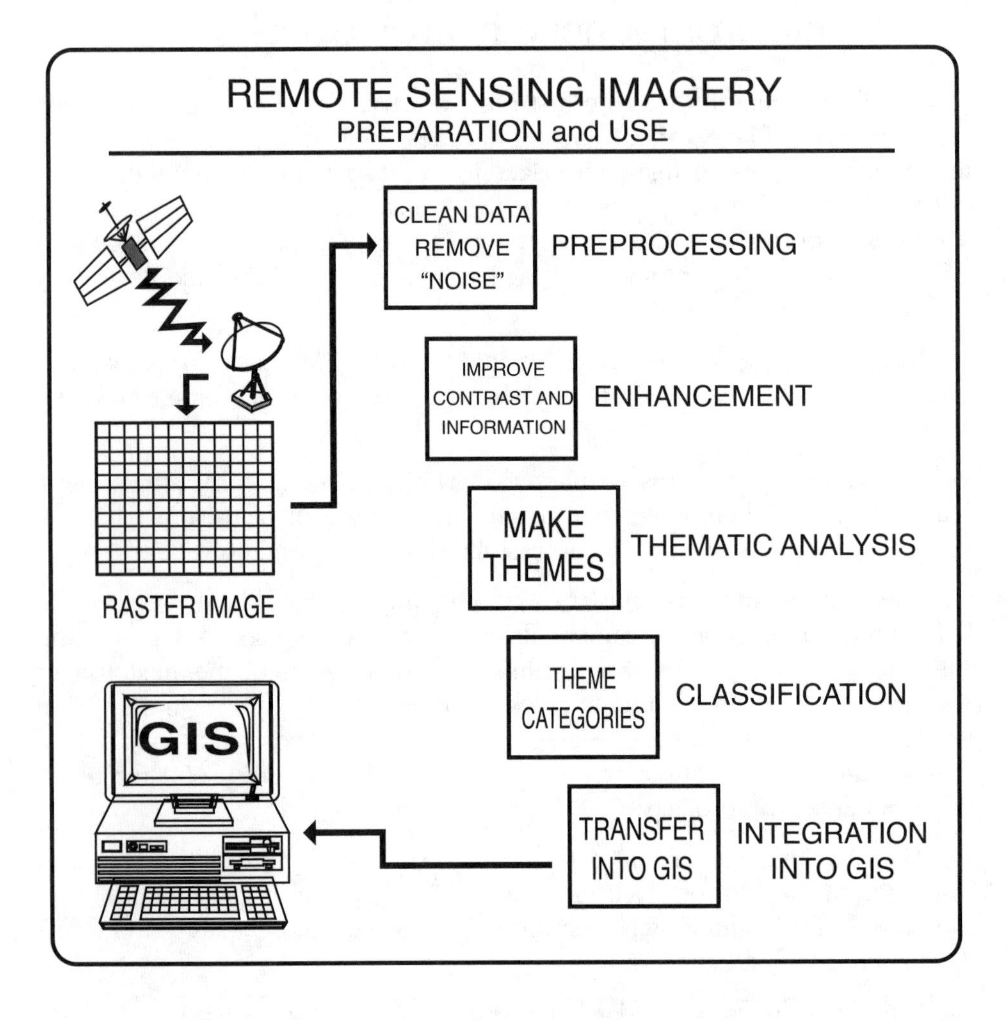
REMOTE SENSING IMAGERY
PREPARATION and USE
CLEAN DATA REMOVE "NOISE"
PREPROCESSING
IMPROVE CONTRAST AND INFORMATION
ENHANCEMENT
MAKE THEMES
THEMATIC ANALYSIS
RASTER IMAGE
THEME CATEGORIES
CLASSIFICATION
GIS
TRANSFER INTO GIS
INTEGRATION INTO GIS

Imagery

Remote sensing data, usually in the form of imagery, can be incorporated into GIS, but this requires preparation. Organizations using remote sensing data with GIS have to be proficient in handling this special type of data. The process (corresponding to the illustration) follows.

- The first step is "pre-processing," which "cleans" the data by removing electronic "noise" and correcting mistakes. Remote sensing data is in raster format, consisting of cells with numbers that are determined by the sensors. When first received, the "raw" imagery is usually not useful and must be processed by the computer to produce satisfactory data quality.

- The next step is to enhance the data so that better information can be obtained. This often includes improving visual contrast, such as changing subtle differences in gray tones into more distinctive shades.

- Thematic analysis involves turning enhanced data into selected themes. For example, a landscape image may contain several types of data (such as ground cover and landuse; themes common to many GIS projects) that will be useful as separate coverages.

- Classification of themes into distinct categories is a logical next step, though it is not always necessary. A vegetation theme, for example, might include a hierarchy of classifications, starting at overall land cover (forest, grassland, agricultural land, barren areas), to the community level (deciduous forest, coniferous forest, mixed forest), and down to areas of particular plant species.

The transfer of remote sensing data into a GIS requires several technical steps, with the basic aim of making sure that information quality remains and is expressed in the database and its product (e.g., graphics). Special software may be necessary in achieving this goal, though many GISs now include remote sensing data-integration capabilities. Many vector GISs, for example, might include raster-reading programs that handle imagery.

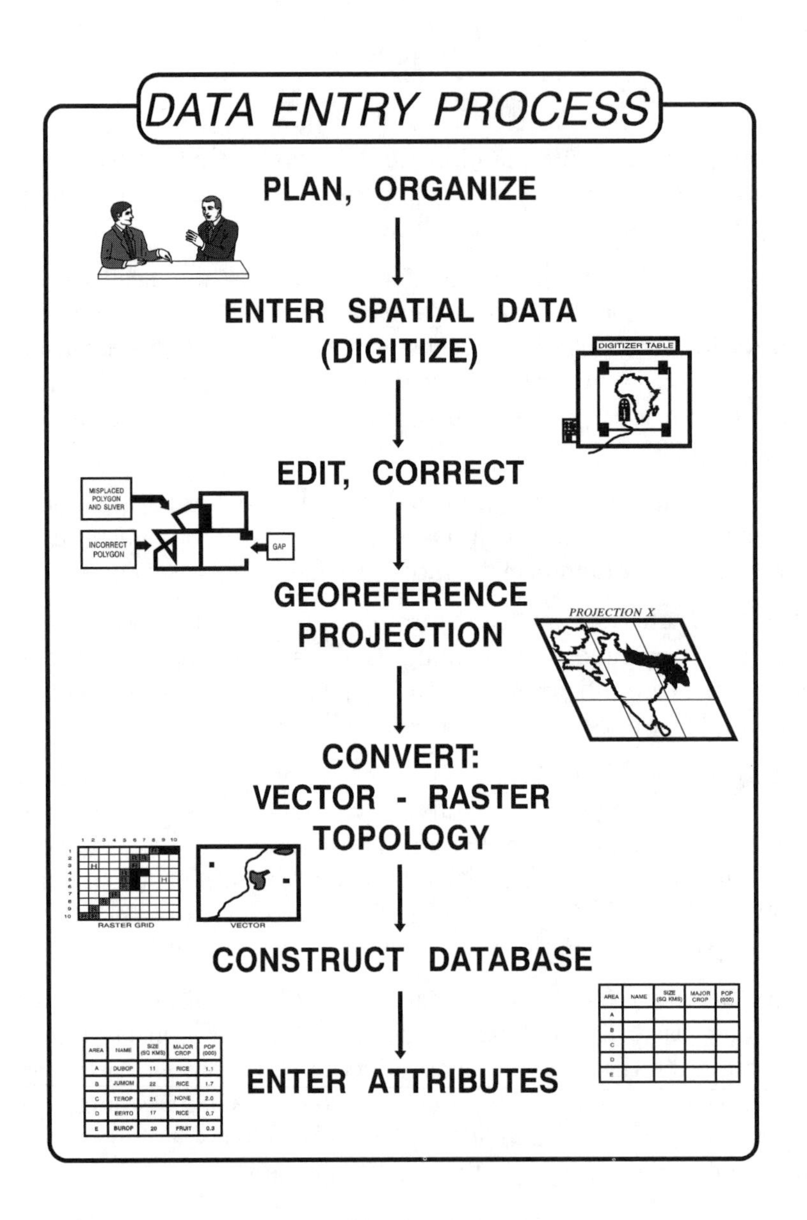

Data Entry Process

Data entry is a simple process that can take a great deal of time and effort, sometimes involving over half a project's life span. Because data entry is perhaps the most critical stage, patience and care are needed. The following list (corresponding to the illustration) describes the data process.

◆ *Plan and Organize:* Easy to say, difficult to achieve properly. The end product must be known, and all steps in the procedure must lead to it efficiently. Obviously, this is a critical foundation step that must be given patience and focus, yet too often it is hurried or ignored.

◆ *Enter Spatial Data (Digitize):* Entering spatial data can be time-consuming. Manual digitizing is the most common process, and it can be tedious. Other techniques can be used, such as typing, reading tape or CD-ROM data, and use of remote sensing imagery.

◆ *Edit and Correct:* Regardless of entry procedures, the data must be checked for accuracy. Mistakes are normal and must be corrected. As will be discussed, this process is also laborious, but necessary.

◆ *Georeference and Project:* Spatial data normally must be associated with a real-world coordinate system, such as Latitude-Longitude. Also, a desired world projection must be related to the digitized data so that the program knows how to lay out (or project) the points. Depending on the GIS used, the proper georeference system and projection may be given before digitizing.

◆ *Convert:* Digitized data must be given a specific data structure—either *vector* or *raster*. If vector is selected, *topology* may be assigned for those GIS programs that use it. These steps are fairly simple, normally involving a few commands to select the desired operation.

◆ *Construct Database:* Once spatial data have been secured, attributes must be attached. The first step is to develop the database—constructing fields, record formats, and so on. Some databases require considerable work to develop, whereas others are very easy and flexible.

◆ *Enter Attributes:* Some attribute data can be entered at the digitizing phase, but other data must be entered at this step. There are various ways to enter attributes, from typing to loading tables. Again, some databases make the task easy; others make it difficult.

NOTE: *In GIS, map files are called* coverages, *the digitized form of the map.*

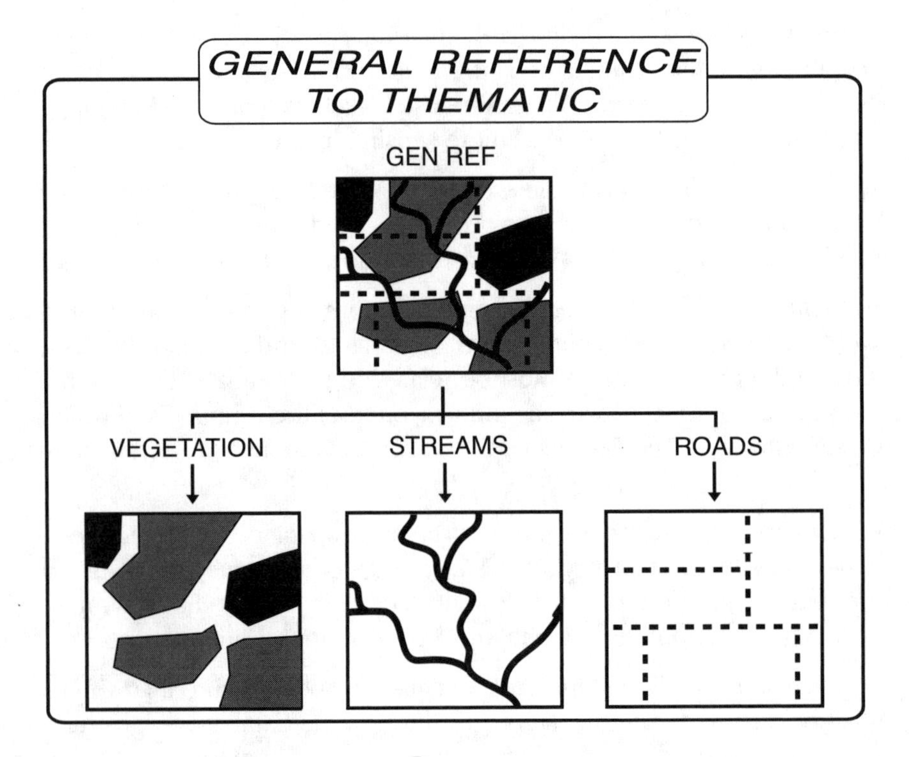
GENERAL REFERENCE TO THEMATIC
GEN REF
VEGETATION
STREAMS
ROADS

General Reference to Thematic Data

The process of making GIS coverages involves changing "real-world" data into specific data files. GIS coverages are normally single-theme data files, so a general reference map would not be appropriate for most GIS work (exceptions always exist). Therefore, each theme is digitized separately and stored as individual files.

Illustrated are three coverages created from the general reference map at top: vegetation, streams, and roads. The goal is to build a database with numerous coverages (layers) of specific themes. Individual coverages stand alone, but can be combined with others. For a simple purpose, the three coverages in the illustration might be the complete data set. It is possible to combine them with others to construct a general reference map. In short, GIS can be a simple mapping technique or a sophisticated data analysis technology.

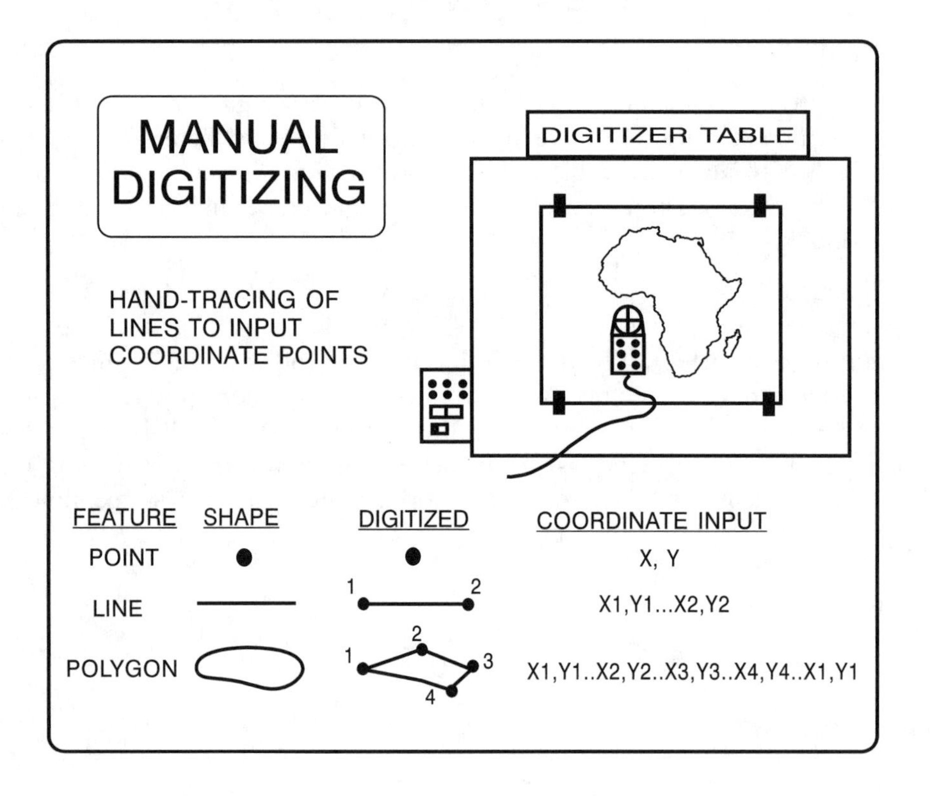
MANUAL DIGITIZING
DIGITIZER TABLE
HAND-TRACING OF LINES TO INPUT COORDINATE POINTS
FEATURE
SHAPE
DIGITIZED
COORDINATE INPUT
POINT
X, Y
LINE
1
2
X1,Y1...X2,Y2
POLYGON
1
2
3
4
X1,Y1..X2,Y2..X3,Y3..X4,Y4..X1,Y1

Manual Digitizing

The most common method of GIS data entry is digitizing maps by hand. Although this is slowly changing, manual digitizing will be with us for some time. It involves placing a map on an electronically sensitive table called a digitizer table (or just digitizer), and then tracing the map features with a mouse device, sometimes called a puck or cursor (after the monitor pointing dot). Number and letter buttons on the device control operations, such as designating inputs as nodes, vertices, and labels to define features. A cross-hair aiming window helps with pointing.

Normally, digitizing is vector and the digitized data are coordinate points that are later connected into chains. Features are recorded as shown at the bottom of the illustration. Spatial points are single, digitized elements having one X-Y position. Lines have end-point nodes, and midstream vertices when there are bends or changes in line direction. Polygons have a single start-end node and midstream vertices.

Each feature receives a label or identity of some type, either during digitizing or after. One method of labeling is to specify the classification code number (a crop-type class, for example) before tracing a given feature. Most GISs automatically assign a sequence identification number (ID) to each feature, so there is at least one way of distinguishing each point, line, or polygon. Every feature in the database will have an ID and a classification label.

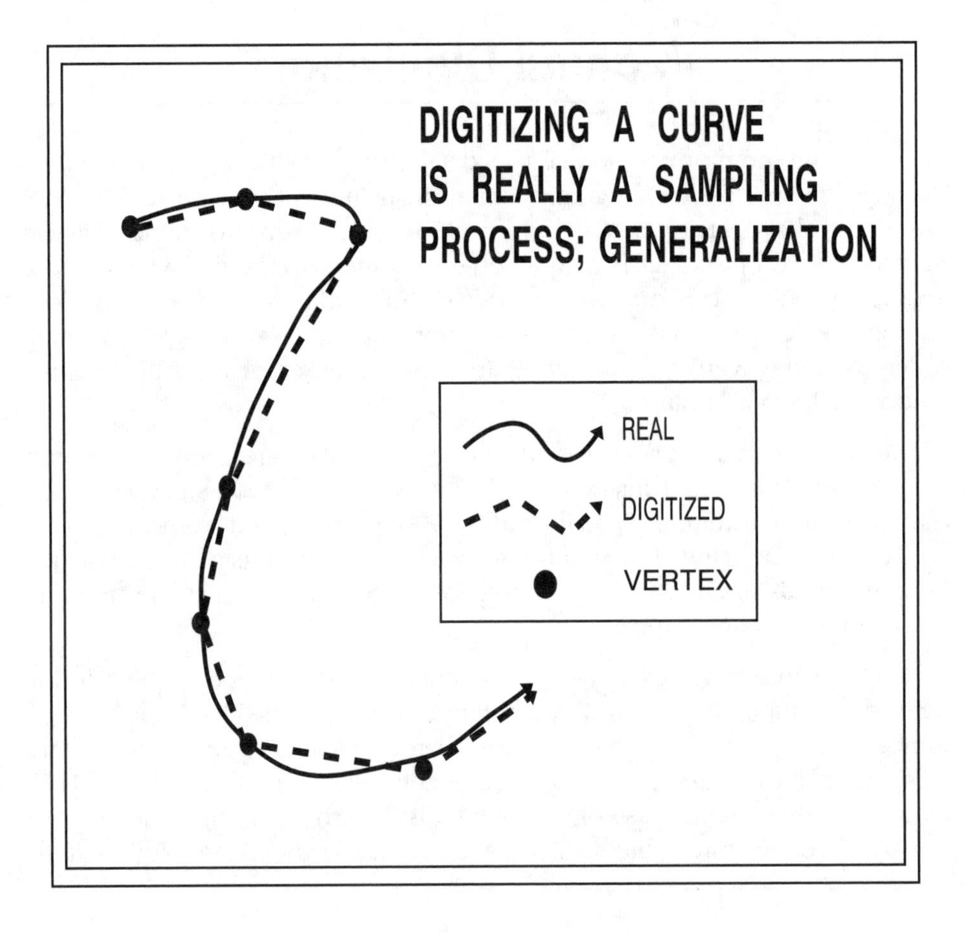
DIGITIZING A CURVE
IS REALLY A SAMPLING
PROCESS; GENERALIZATION
REAL
DIGITIZED
VERTEX

Digitizing a Curve

Digitizing is tracing of map features, which really means entering nodes and vertices, not actually drawing lines. A square feature is easy to outline: touch the corners and press the data entry button. The only problem is being careful to locate nodes and vertices accurately.

Geometric features are fairly simple to enter, but curves have potential inaccuracies. Perfect tracing by hand is very difficult, and vertices input along the arc of the curve's path make straight-line connections to each other, thereby creating a generalized feature. In effect, the curved line is being sampled for representative points, not actually traced along every point on the line. Note the difference between the real and digitized lines on the illustration.

One possible solution to this problem of inaccuracy is the use of a curve generation routine (program) offered by some digitizing packages that can draw a controlled curve between two vertices. This takes careful and slow supervision by the operator, but it may be necessary for high accuracy. This also represents entry of many vertices, but in a more reliable manner than hand control.

Manual digitizing has natural and expected inaccuracy. Normally, there is a need for editing to correct imperfections—a process that can take as long as or longer than the original digitizing. Not fun, but necessary.

An alternate method of entering individual vertices by hand (finger tapping the entry key each time) is to use "stream mode" digitizing, available in most GISs. This method automatically enters a vertex at selected distances, usually in fractions of an inch or centimeter, as the cursor is traced along the feature. But stream mode requires very steady hands to ensure accuracy—a rather difficult requirement for large features. Most users do not employ this option.

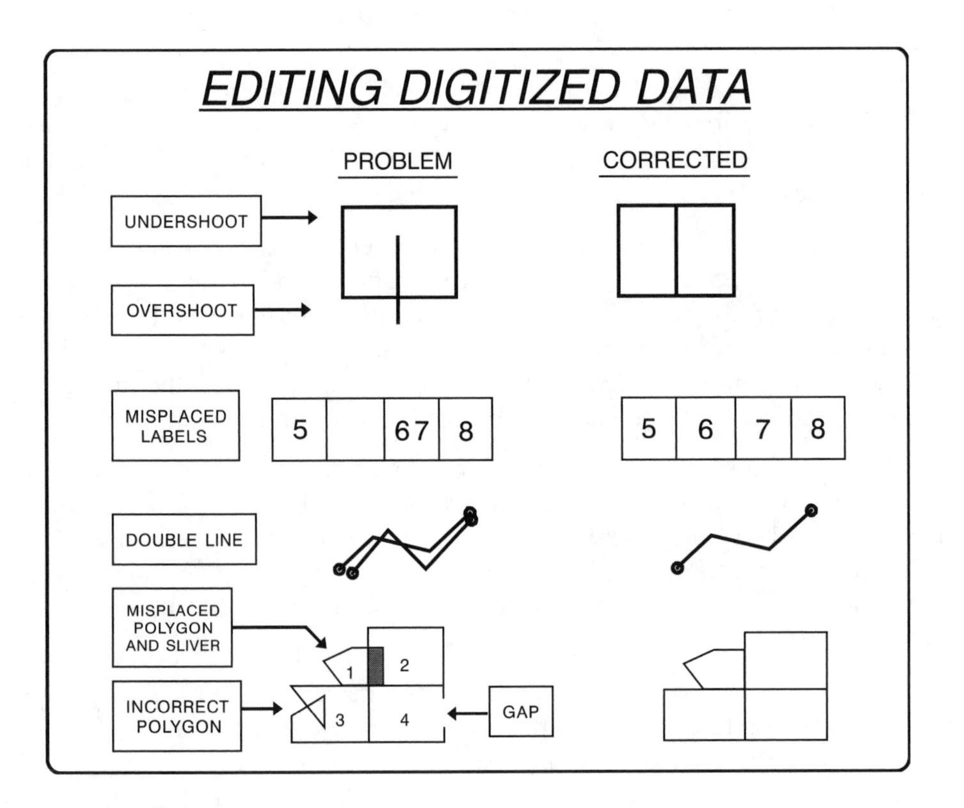
EDITING DIGITIZED DATA
PROBLEM
CORRECTED
UNDERSHOOT
OVERSHOOT
MISPLACED LABELS
5
67
8
5
6
7
8
DOUBLE LINE
MISPLACED POLYGON AND SLIVER
INCORRECT POLYGON
1
2
3
4
GAP

Editing Digitized Data: Problems

Editing is an essential task, although it takes time, and a critical eye for detail and perfection. Good digitizing programs offer error-spotting routines and easy operations for correction of problems. Some of the more common errors are illustrated (problem on the left and correction on the right).

- *Undershoot and Overshoot:* Chains that do not reach other chains as intended, or that extend beyond the targeted location. This requires either extending or reducing the chain length, usually just by pointing to the incorrect chain, then to the target chain spot and pressing a key to instruct the editing program to put them together properly.

- *Misplaced Labels:* Classification labels are entered during or just after digitizing to define features. Some features may receive no label, multiple labels, or incorrect labels. In the illustration, the polygon's label of the second square was accidentally placed into the third square. It is easily moved by pointing to the incorrect label and then to the place where it belongs.

- *Double Line:* It is easy to enter an original data feature twice, particularly from complex maps. In the illustration, one line has been repeated. The least accurate one is typically removed by pointing to it and pressing a delete key.

- *Mistakes in Tracing:* On complex maps it is easy to make mistakes in drawing. In the illustration, the bullet-shaped polygon 1 has been misplaced too far to the right, resulting in a "sliver" inside polygon 2. The sliver can be removed by relocating polygon 1 to the left, in its proper position. Polygon 3 contains errors (perhaps due to hand-control problems), and the line must be corrected, typically by removing the two vertices that define the incorrect part of the chain. Note that if the two inside vertices, near the number 3, are deleted, the line will become straight. A gap, or incomplete chain, is evident on polygon 4, and the closure must be made by pointing to the open end of each chain and instructing a connect, similar to the undershoot repair.

The foregoing types of corrections are usually easy to make. The solution if often as simple as pointing to the area to be corrected and making a single correction command. However, such changes take time.

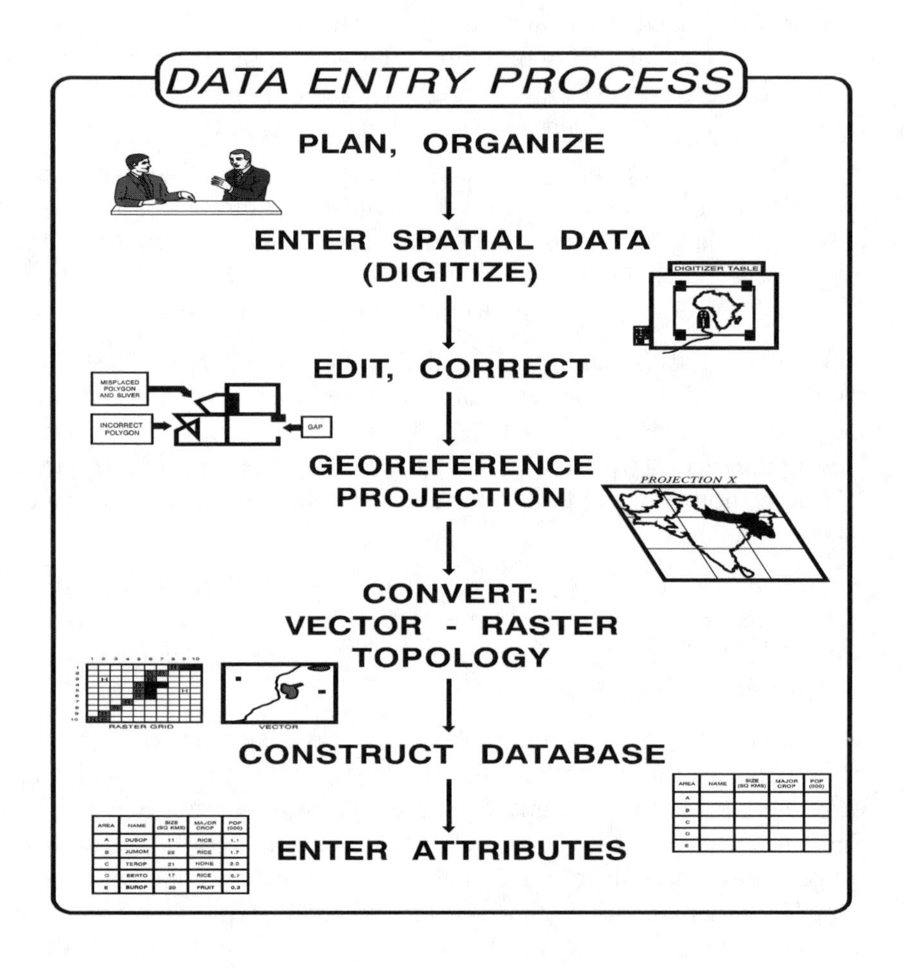
DATA ENTRY PROCESS
PLAN, ORGANIZE
ENTER SPATIAL DATA
(DIGITIZE)
DIGITIZER TABLE
EDIT, CORRECT
MISPLACED POLYGON AND SLIVER
INCORRECT POLYGON
GAP
GEOREFERENCE
PROJECTION
PROJECTION X
CONVERT:
VECTOR - RASTER
TOPOLOGY
RASTER GRID
VECTOR
CONSTRUCT DATABASE
AREA NAME SIZE (SQ KMS) MAJOR CROP POP (000)
A B C D E
ENTER ATTRIBUTES
AREA NAME SIZE (SQ KMS) MAJOR CROP POP (000)
A DUBOP 11 RICE 1.1
B JUMOM 22 RICE 1.7
C TEROP 21 NONE 2.0
D EERTO 17 RICE 0.7
E BUROP 20 FRUIT 0.3

Manual Digitizing Overview

In summary, manual digitizing, the most common process of converting maps to coverages, is

- *Labor intensive*, requiring careful attention and much work.
- *Slow*, usually taking considerable time. In fact, it has been estimated that data entry can take 60 or 70% of the entire project time.
- Filled with *potential inaccuracies*, given the hand tracing process and manual input of associated data. It is nearly impossible to digitize a map without error.

Thus, there is a *need for rigorous editing*, requiring time, patience, and work. Consequently, the process is *not fun!* This is not a minor point. Manual digitizing is an evil necessity—one that quickly becomes a boring and laborious chore. Dissatisfied personnel make mistakes, which cost time, money, and effort, affecting overall quality. A better way to enter data is welcome, and that method is automatic digitizing, a developing technology.

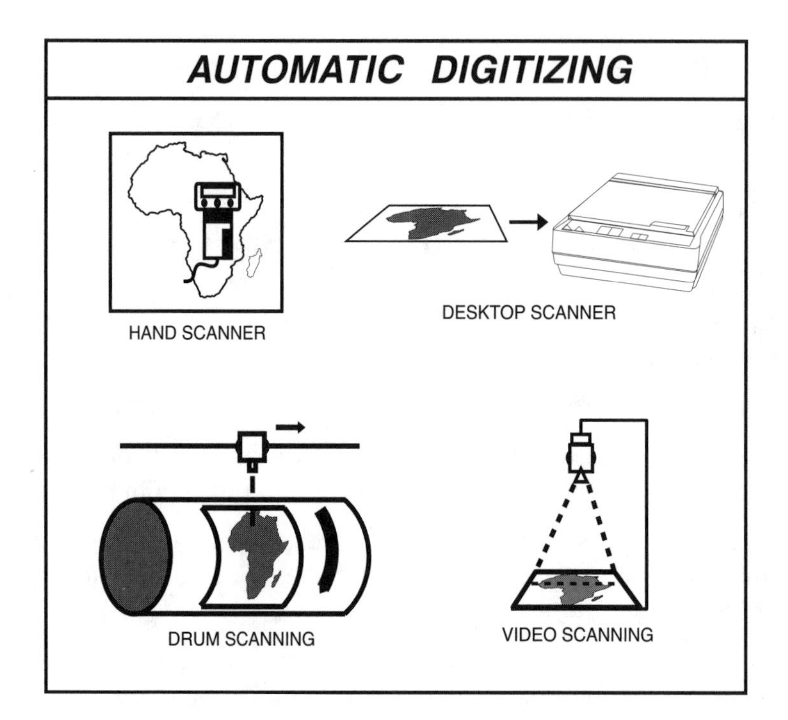

Automatic Digitizing

What Is Automatic Digitizing?

An alternative to manual digitizing is automatic digitizing, which would seem to be a logical and common technique. However, the technology has not yet matured, with factors such as cost and limited capabilities keeping it out of the range of most GIS operations for now. Several types of automated digitizers exist, among which are the following.

◆ *Hand Scanner:* This is a hand-held copying device moved along the map (or graph, tabular data list, or text). These devices are usually too small, however, for maps larger than text page size and are limited in spatial and tonal resolution. However, hand scanners are very inexpensive and may be suitable for simple, small maps, text, and tabular data.

◆ *Desktop Scanner:* A larger and more stable version of the hand scanner, these devices are fast and offer reasonably good quality. However, they are limited to page size (or slightly larger) and are not really meant for map data. They are ideal for converting text, tabular data, and even photography into digital form. The relative low cost is attractive, even for color versions. However, they are not widely used for map digitizing at present. Perhaps this will change in the near future as the technology improves.

◆ *Drum Scanner:* The best map scanners are large and expensive devices that contain a rotating drum and moving scanner. Large maps can be digitized in a fairly short time, with a high degree of accuracy and quality. These, so far, are mostly monochrome (black and white); color presents problems. However, the technology is evolving rapidly, and color may not be a significant problem for long.

◆ *Video Scanner:* A reasonable alternative to drum scanners and other expensive machinery is the inexpensive video scanner. It works much like a normal video camera, making scanning sweeps of the map, gathering tones that are rasterized and converted into numbers. These machines are fast, but tend to generalize data, sometimes creating problems of map feature recognition. Also, their raster format creates spatial generalization (see the raster-vector discussion in Chapter 4). Color can be used through a series of color filters and multiple scans, though the problems of feature recognition may increase. Video scanners are useful, particularly for simple maps, but they do not offer a good replacement to manual digitizing of complex or general reference maps.

◆ *Laser Line Follower:* One type of scanner (not shown in this illustration) uses a laser to trace map feature outlines. These devices can be very accurate and fast, but they are expensive and require human intervention to monitor progress in order to make decisions (see the next illustration). This technology is also advancing rapidly, and laser digitizers may be much more common in the near future.

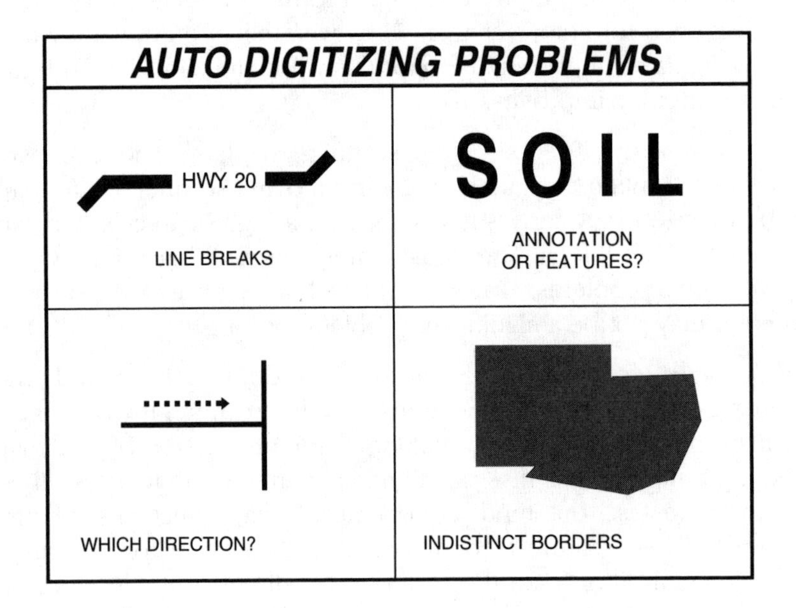
AUTO DIGITIZING PROBLEMS
HWY. 20
LINE BREAKS
SOIL
ANNOTATION
OR FEATURES?
WHICH DIRECTION?
INDISTINCT BORDERS

Problems

Automatic digitizing has problems that keep it an unfulfilled promise for the moment. Human monitoring (in effect, "baby-sitting") is necessary—sometimes as the digitizing occurs, thereby slowing the process and making it less than fully automatic. At best, the process is "semiautomatic." The following are among the common problems associated with automatic digitizing.

◆ *Line Breaks:* Although the human eye dismisses small and logical breaks in a line (e.g., a highway) a line-following machine stops when an interruption occurs. The "baby-sitter" can manually continue the line over the annotation (text) and have the scanner pick it up on the other side. If the system does not permit interruptions, the human editor must ultimately erase the text and manually draw the line over the gap—a tedious process.

◆ *Annotation or Features?:* Scanners can't read, so they may recognize text as features. The "o" in *soil* might be seen as a polygon, and the other letters may be interpreted as lines. Again, the operator must make corrections.

◆ *Which Direction?:* If a laser following a horizontal line from left to right comes to a T intersection, it has no way of knowing which direction is correct. It may stop, awaiting a human decision before continuing. For finished products, the human must decide which parts belong to which features: Is the vertical line a separate feature, one side of a polygon, or should it be broken into two parts as sides of two polygons? This is particularly important when topology is to be attached to the final dig file.

◆ *Indistinct Borders:* A map might have two similar tones or colors that are difficult for the scanner to distinguish. The two polygons shown in the illustration might be made into a single polygon, requiring either another scan operation with a readjusted setting (which might not work with other parts of the map) or a human to manually draw the border. Either way, the process is slowed considerably.

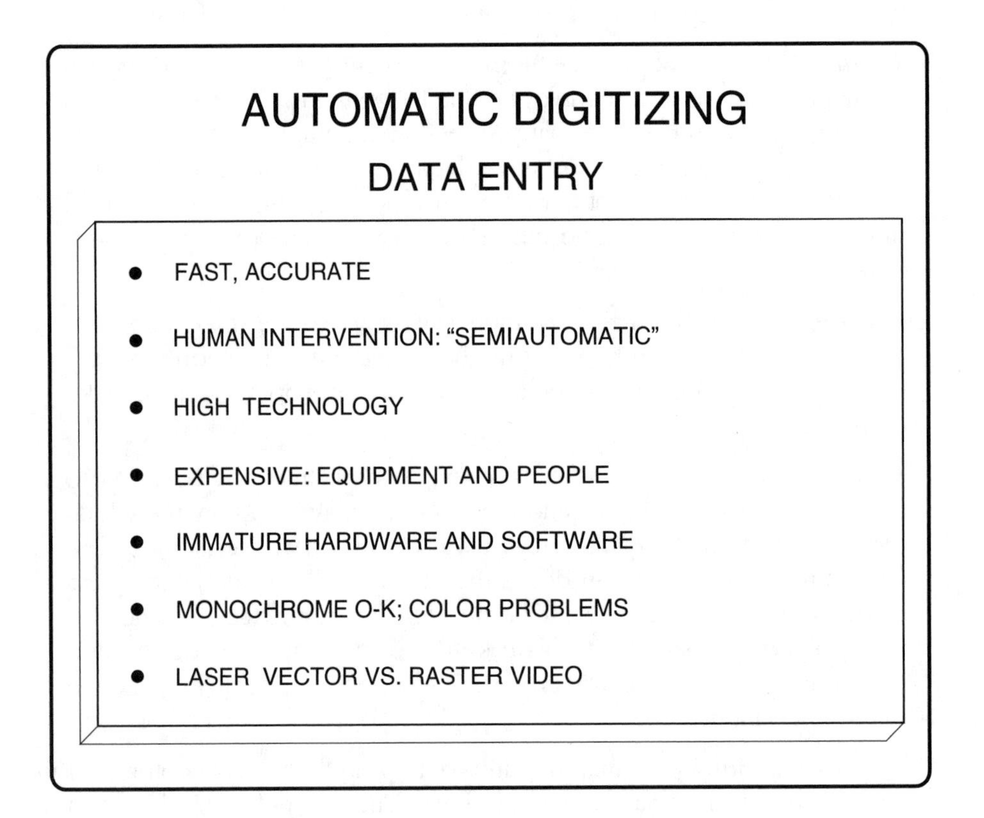
AUTOMATIC DIGITIZING
DATA ENTRY
FAST, ACCURATE
HUMAN INTERVENTION: "SEMIAUTOMATIC"
HIGH TECHNOLOGY
EXPENSIVE: EQUIPMENT AND PEOPLE
IMMATURE HARDWARE AND SOFTWARE
MONOCHROME O-K; COLOR PROBLEMS
LASER VECTOR VS. RASTER VIDEO

Considerations

Truly operational automatic digitizing may be the major GIS innovation of the near future, but today it is not readily available, at least for most GIS operations. In summary, we note the following considerations:

- Automatic digitizing can be very *fast* and very *accurate.* A laser, for example, can follow lines very well, with insignificant deviations. Even tabletop scanners offer the advantages of speed and spatial precision.

- *Human intervention* is needed to guide the process and make decisions, such as choosing which map element to trace. This means that the process is actually *semiautomatic.*

- A good automatic digitizing system implies *high technology.* Low-tech devices are not suitable for map digitizing.

- Such systems are *expensive* in terms of e*quipment and people.* High-tech equipment is costly, but the people to run it and to make decisions cost even more.

- Currently, the *hardware and software* are technologically *immature.* They cannot overcome many of the natural problems that exist with maps and spatial data. Although advances are made continually, affordable and simple automatic digitizers may not be available for some time.

- One aspect of the current immaturity of automatic digitizing is that existing systems are *monochrome*-based; that is, they work only with a single color, typically black-and-white formats. More than one color, or different colors, present problems. On maps where color identifies features, automatic digitizers may have difficulty determining the difference and making the appropriate classifications.

- *Laser* devices are excellent because they work with lines (*vector* format), and are fast and accurate. However, they are not ready for common use. An alternative may be *video scanning*: scanning maps using a television device that converts the image into a *raster* format. Generalization and accuracy problems exist, but these systems are available and relatively affordable. Generating a satisfactory number or density of grid cells is a major concern, but once that has been solved, video digitizing may be an acceptable alternative to manual approaches for some types of maps.

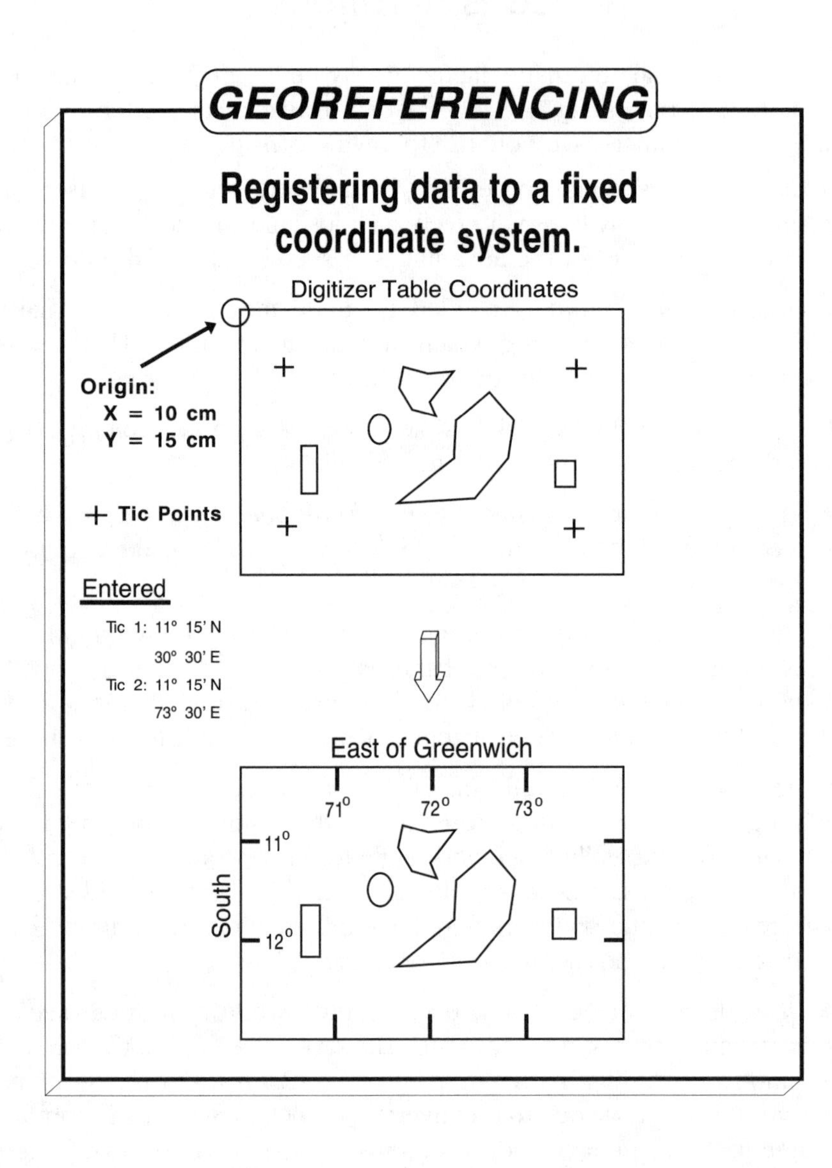
GEOREFERENCING
Registering data to a fixed coordinate system.
Digitizer Table Coordinates
Origin:
X = 10 cm
Y = 15 cm
+ Tic Points
Entered
Tic 1: 11° 15' N
30° 30' E
Tic 2: 11° 15' N
73° 30' E
East of Greenwich
71°
72°
73°
11°
12°
South

Georeferencing

What Is Georeferencing?

GIS data files usually must have a real-world coordinate system if they are to be valid coverages (some exceptions exist, but the general rule is to provide coordinates). The process is termed *georeferencing,* defined as registering, or fixing, data to a standard coordinate system. This is a data transformation process.

The best method of establishing proper georeference is to define at least four "tic points" around the area being digitized (close to the corners if possible), each with a precisely known real-world coordinate position that is typed into the program. When the program "knows" these points, digitized points can be properly located relative to the precise tics.

In the illustration, the original digitized file (top) is in digitizer table centimeters. Note the upper left-hand corner "origin," from which the initial points are measured. Two tic points have been entered, with their actual latitude-longitude coordinates typed into the program (four is usually the minimum number to use). Some digitizing programs require georeferencing before data input, whereas others can take data after the digitizing process.

Georeferencing changes (transforms) the digitized file into the coordinate system of the map (in this case, Latitude-Longitude). This is where computers are necessary; translating one reference system to another is a number-crunching task. The program notes the tic point coordinates and gives the rest of the file a proper georeferenced structure. Good GISs render the chore rather easy by offering selected coordinate systems and then making the computational process invisible to the operator. (The numbers inside the map at bottom are for illustration and are not normally part of the presentation unless asked for.)

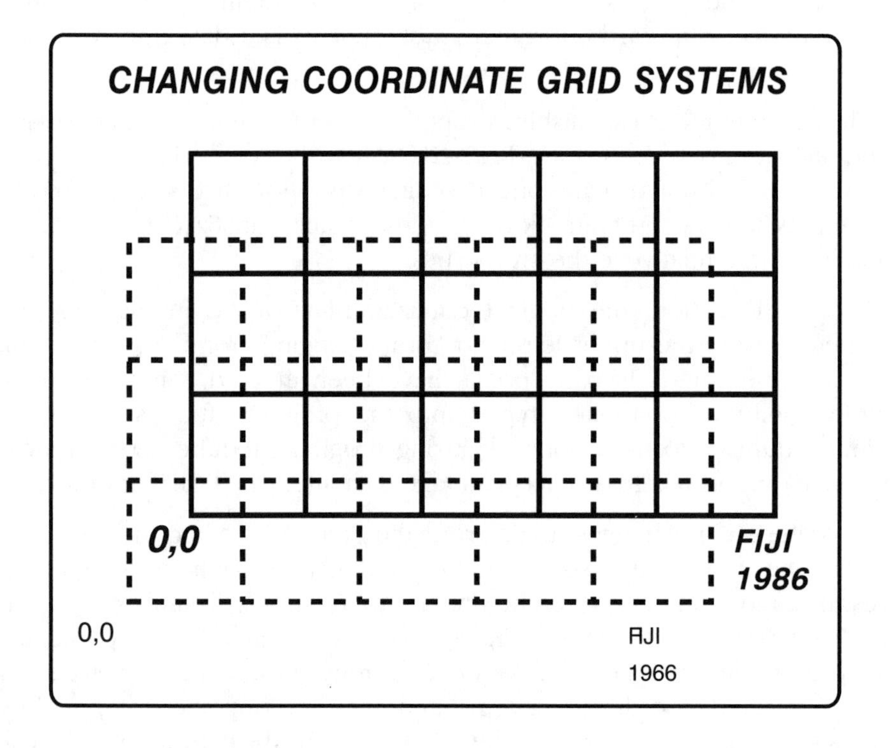
CHANGING COORDINATE GRID SYSTEMS
0,0
FIJI
1986
0,0
FIJI
1966

Changing Coordinate Grid Systems

Latitude-Longitude is a world coordinate system, but smaller ones exist for regional purposes and more accurate positions. Georeference systems are not necessarily stable; they may change over time, for a variety of reasons. Whereas the Latitude-Longitude system remains the same, some of the smaller grid systems can shift position. For example, Fiji's regional grid was moved in 1986 as an update from its original 1966 system. Note its 0,0 origin at the lower left of the illustration.

It is necessary that GIS data consider the date if such a coordinate system shift has occurred. Otherwise, positions can be misplaced and uncoordinated. Special programs exist to change a coverage from one system to another, but they require precise knowledge of old and new coordinate systems. Coordinate system management can be a difficult chore if modifications are needed.

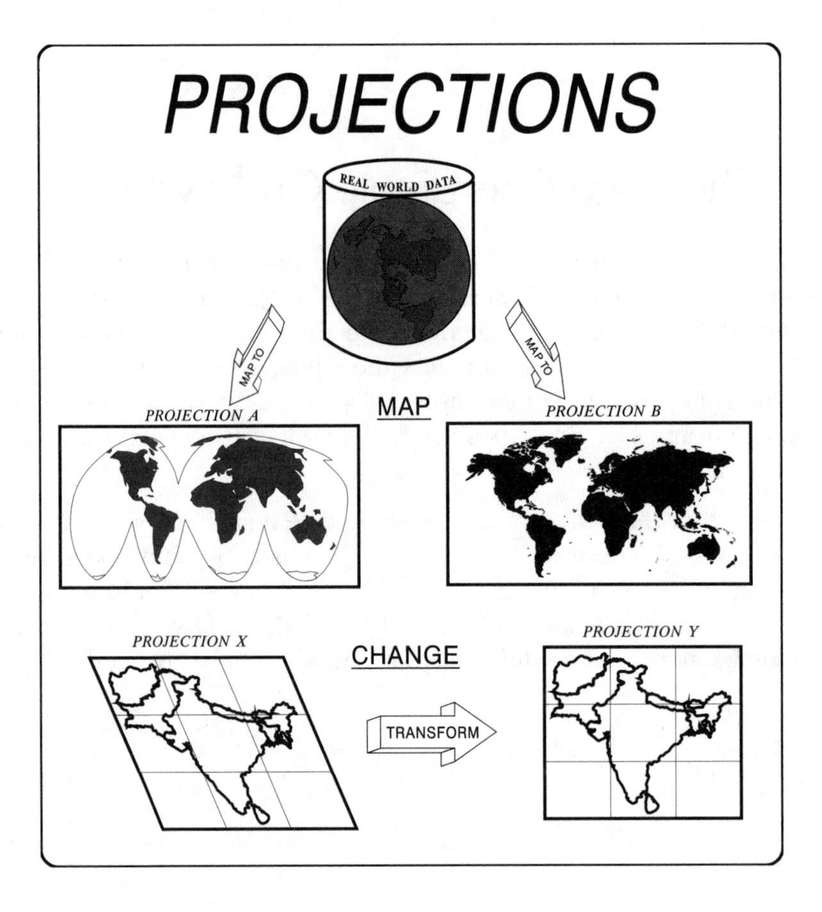
PROJECTIONS
REAL WORLD DATA
MAP TO
MAP TO
MAP
PROJECTION A
PROJECTION B
PROJECTION X
CHANGE
PROJECTION Y
TRANSFORM

Projections

Even when a file has been georeferenced to a specific coordinate system, the GIS still needs to know which map projection to use in order to place and give proper shape to the features. Projections are special configurations used to fit a portion of the globe onto a flat view; that is, spherical data is converted into a 2D presentation. There are many projections, each with certain advantages. Some applications require specific or particular types of projections. A good GIS can convert the digitized file into various projections, or even change from one projection to another as desired.

Shown are two projections selected for the Real World Data: projections A and B (A appears to be a homolosine projection; B may be a Mercator projection). At the bottom of the illustration, an existing projection (X) is transformed to a more appropriate one (Y). These are standard GIS data preparation operations that are usually very easy to perform; for instance, from a menu selection.

Georeferencing and projection selections are the two standard operations applied to digitized data in the data entry process. With preparation, the procedures are not difficult.

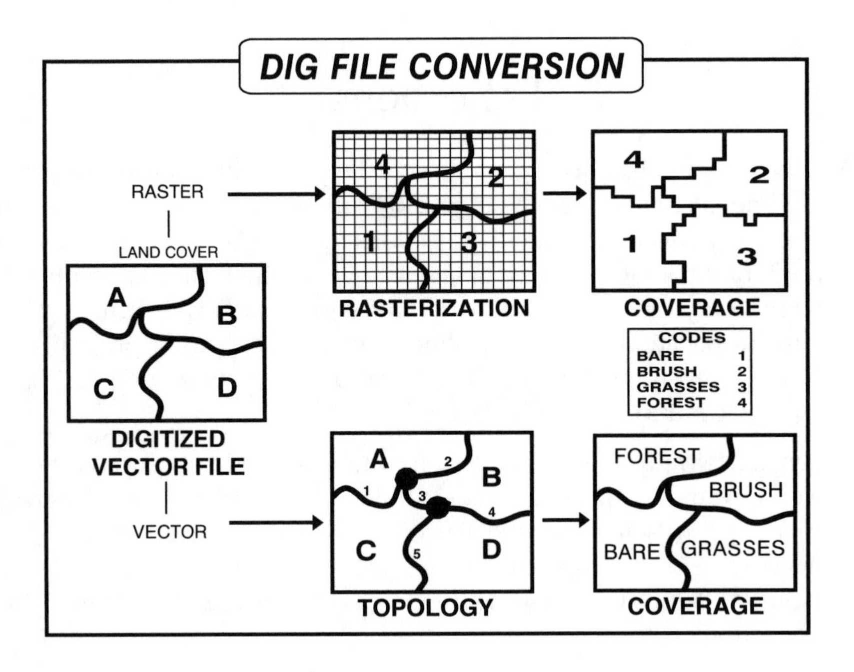
DIG FILE CONVERSION
RASTER
LAND COVER
A
B
C
D
DIGITIZED
VECTOR FILE
VECTOR
4
2
1
3
RASTERIZATION
4
2
1
3
COVERAGE
CODES
BARE 1
BRUSH 2
GRASSES 3
FOREST 4
A
2
B
1
3
4
C
5
D
TOPOLOGY
FOREST
BRUSH
BARE
GRASSES
COVERAGE

Digitized File Conversion

When a dig file has been edited, georeferenced, and given a projection, it is ready to be converted into a GIS coverage. Sometimes this conversion is called attribute assignment because it assigns digitized attributes to a particular data structure. Illustrated is a Land Cover dig file that is to be either a raster or vector coverage.

If the GIS is raster, the file is rasterized by specifying a grid size or resolution and then letting the program change the digitized data into cells and cell values. The resultant coverage is a modified version of the dig file, with noticeably changed polygon borders. Usually, a separate database must be constructed to "describe" cell coding; for example, assigning names to code numbers.

For vector GISs, the dig file shows less apparent change. If the system is topological, a command to establish topology is given. The coverage may look much like the dig file, but is now configured to a coverage with labels, feature numbers, topology, and connected database. These new characteristics are not part of the normal visual display, although they can be viewed if desired. In this illustration, the polygons are given full names rather than numeric codes, but they are not part of the presentation unless labels are displayed.

DATABASE CONSTRUCTION

ID	AREA	NAME	SIZE [SQ KM]	MAJOR CROP	POP [000]

ESTABLISH STRUCTURE

ID	AREA	NAME	SIZE [SQ KM]	MAJOR CROP	POP [000]
1	A	DUROP	11	RICE	1.1
2	B	JUMON	22	RICE	1.7
3	C	TUROP	21	NONE	2.0
4	D	EERTO	17	RICE	0.7
5	E	BUROP	29	FRUIT	0.3

ENTER ATTRIBUTES

Database Construction

Databases are central to GIS applications and their flexibility gives strength to operations. Nonetheless, they should be constructed carefully and logically. Some GISs automatically make an initial database from the coverage, usually with only a few attributes, such as feature ID and perhaps a spatial measure (area and perimeter). From these tables, attributes can be added to make a complete database. New records can also be included. Other GISs require the operator to set up the database, such as establishing anticipated size of fields and other structural items.

The upper illustration shows one database structure as initially established. Perhaps the GIS will provide the feature ID automatically (sequence of feature numbers as they are digitized), but the other fields will be defined and constructed by the operator. Database organization should be developed in the project planning phase; make it correct the first time.

The database at bottom has attributes entered, some possibly during the digitizing stage and others added manually later, depending on software requirements and options. From here, the "real" GIS work can begin.

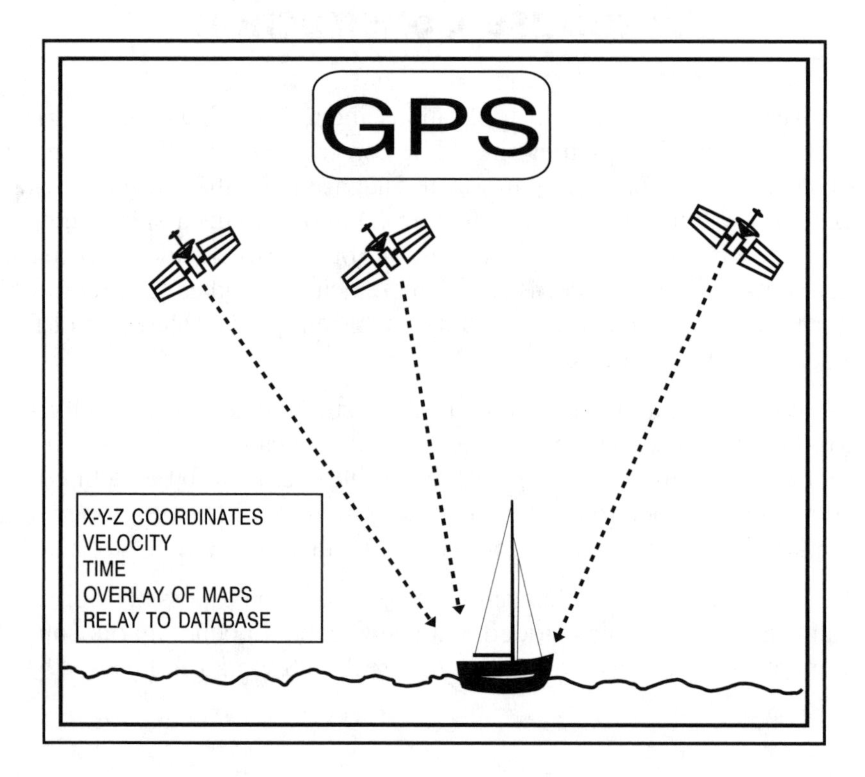
GPS
X-Y-Z COORDINATES
VELOCITY
TIME
OVERLAY OF MAPS
RELAY TO DATABASE

GIS and the GPS

What Is GPS?

One more data entry technique to be considered is the Global Positioning System (GPS). GPS uses a system of satellites that transmit signals to special receivers on the ground for precise determination of X-Y coordinate position. The receivers may be small, hand-held units or larger instruments that can have accuracies within meters (or better). GPS data might also give elevation (Z coordinate), velocity (while moving), time of measurement, and even custom input of selected variables (see next illustration).

The illustration shows a sailboat receiving GPS data. In the middle of an ocean, or a swamp or desert, there is an absence of easily recognizable landmarks for proper location reference. GPS provides highly accurate position almost anywhere on Earth.

Under field conditions, GPS can provide update and modification of coverages; for example, by overlaying the GPS signal on an existing coverage display. GPS data may function as corrections on a coverage, such as road mapping updates. Thus, GPS data can go directly into the GIS, giving immediate and highly accurate data. It is no wonder, given the importance of accurate input data, that GPS is becoming a major technology for many fields, including GIS.

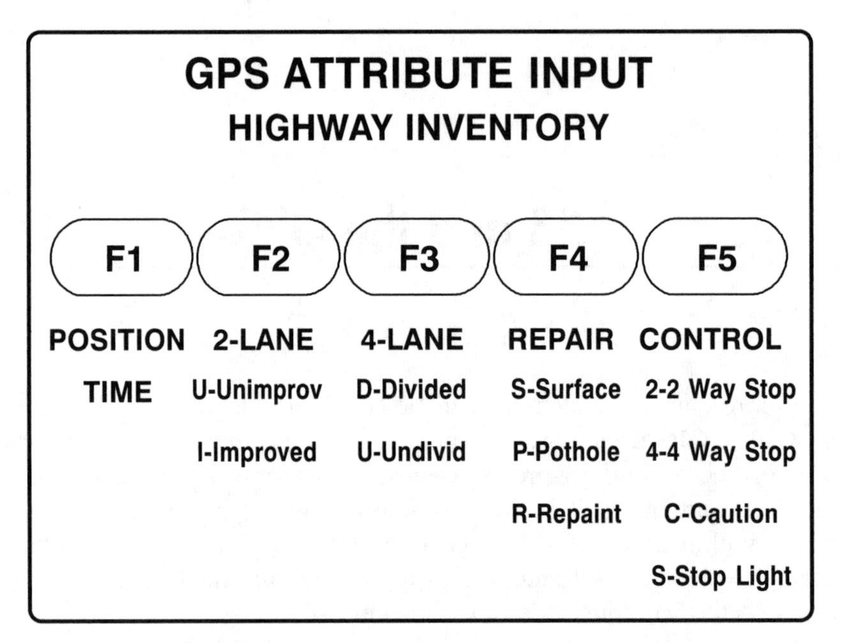
GPS ATTRIBUTE INPUT
HIGHWAY INVENTORY
F1
F2
F3
F4
F5
POSITION
TIME
2-LANE
U-Unimprov
I-Improved
4-LANE
D-Divided
U-Undivid
REPAIR
S-Surface
P-Pothole
R-Repaint
CONTROL
2-2 Way Stop
4-4 Way Stop
C-Caution
S-Stop Light

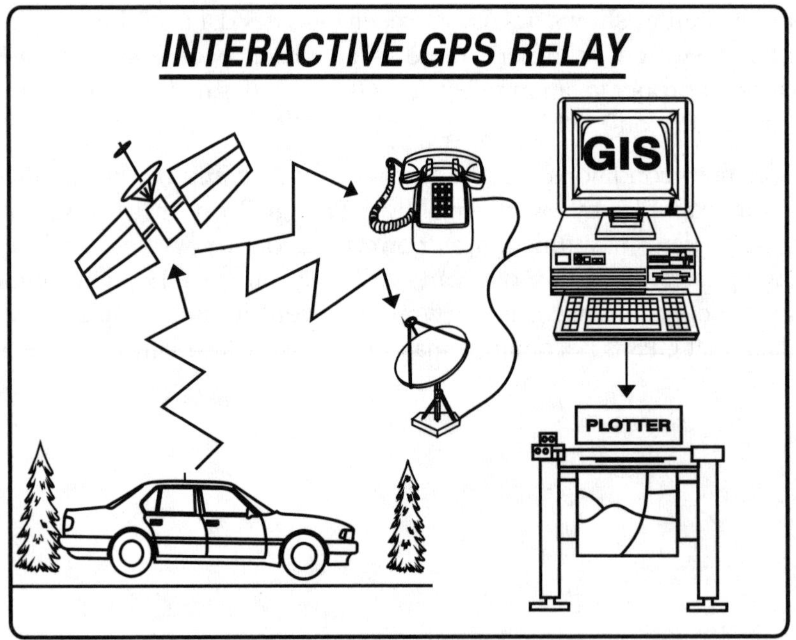
INTERACTIVE GPS RELAY
GIS
PLOTTER

GPS Attribute Input

To show how GPS can be an interactive methodology, the illustration at top represents a highway inventory. The GPS is linked to the GIS while an "interface" program is operating, taking time and locational data from the GPS and feeding them to the GIS. The computer's function keys have been set to record operator input.

As the car moves along the highway (speed is no concern) and a data notation is needed, the operator presses the F1 key to note the position and time. This "freezes" that data until the next time F1 is pressed, giving the operator time to complete the input data while the car is still moving. If the highway inventory data at this point is a repair notice, the operator presses F4 (Repair) and a submenu is presented, asking for the type of repair necessary: S stands for surface problem, P for pothole, and R for repaint. Other options are included in the illustration. Almost any type of data input scheme can be devised. This is a "custom" input system linked to GPS.

GPS offers grand opportunities for GIS, from exact positions to interactive coverage modification. Let's take a look at some other aspects of GPS related to GIS.

Interactive GPS Relay

Not only can GPS operate with GIS interactively, data can be relayed to distant stations (lower illustration). While traveling in the car, data are taken and processed as described in the upper illustration. Data can then be relayed, via a small antenna on the car or by cellular phone, directly to satellites. In turn, receivers on the ground, or a phone system, can take the data and relay it directly to a central GIS, where either further processing occurs or the field GIS data are stored. When the mission is finished (at the last data relay), the central GIS produces a map plot—possibly even before the field technicians have parked the car!

GPS processing and relay can provide very fast and highly accurate data (including maps) from a distant location (even half a world away). This type of technology is changing the way we view, collect, and manage data. As a result, our expectations of what we can do with data and information are also changing.

Further, the finished data can also be relayed by satellite to a customer at a distant third location. We can envision a final product from field to central GIS to customer, all before the field crew finishes packing the equipment at the end of field data collection. Marvelous possibilities exist.

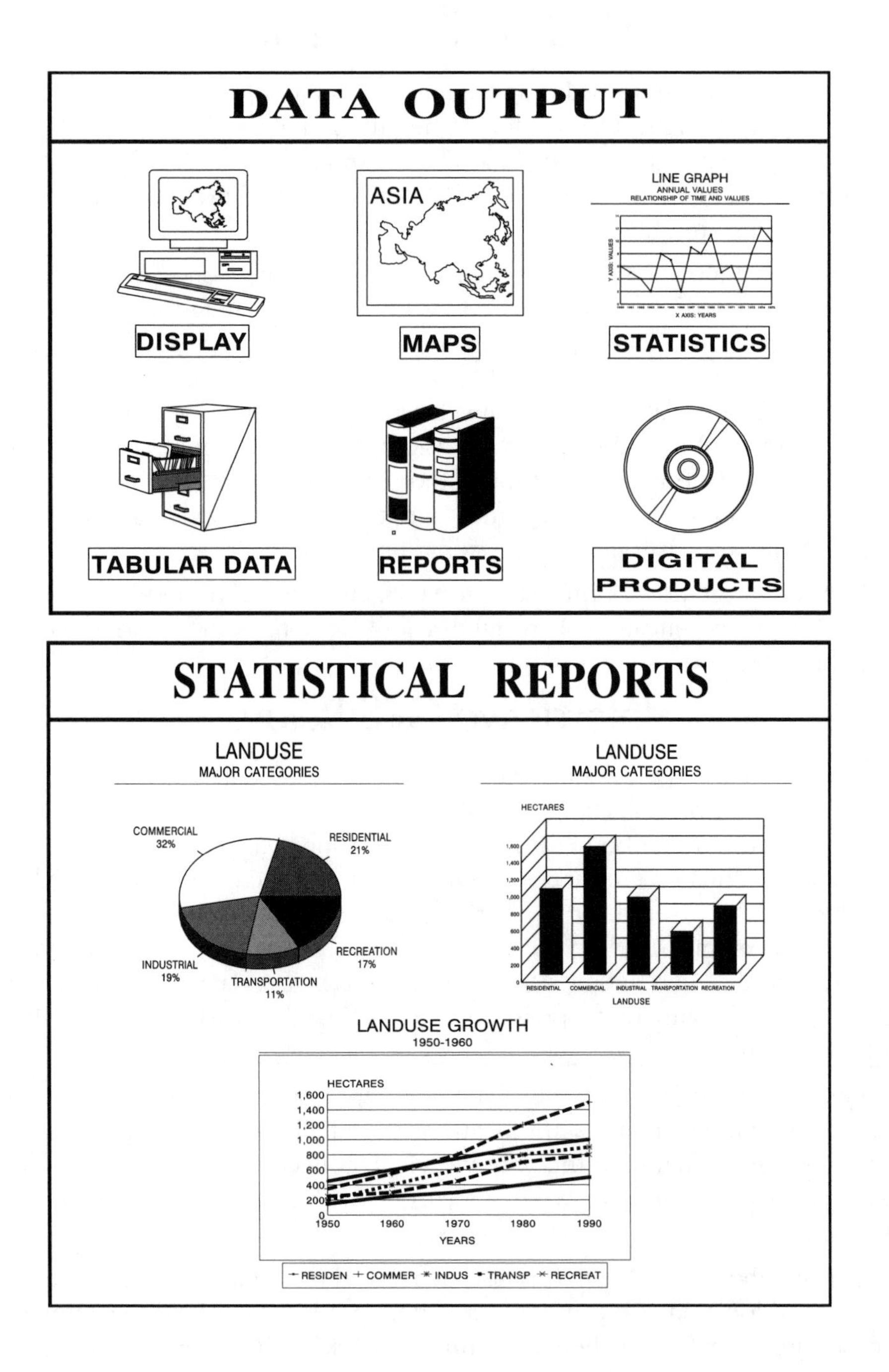
DATA OUTPUT
ASIA
LINE GRAPH
ANNUAL VALUES
RELATIONSHIP OF TIME AND VALUES
X AXIS: YEARS
DISPLAY
MAPS
STATISTICS
TABULAR DATA
REPORTS
DIGITAL PRODUCTS
STATISTICAL REPORTS
LANDUSE
MAJOR CATEGORIES
COMMERCIAL 32%
RESIDENTIAL 21%
RECREATION 17%
INDUSTRIAL 19%
TRANSPORTATION 11%
LANDUSE
MAJOR CATEGORIES
HECTARES
LANDUSE
LANDUSE GROWTH
1950-1960
HECTARES
1,600
1,400
1,200
1,000
800
600
400
200
0
1950
1960
1970
1980
1990
YEARS
RESIDEN
COMMER
INDUS
TRANSP
RECREAT

Data Output

Types of Output

As mentioned, the monitor display is the most common mode of GIS data presentation. Because GIS is very flexible, capable of showing various versions of a coverage or set of results, it is practical to view these options on a screen rather than printing them out every time they need to be viewed.

Maps are the most common form of permanent output product. Most applications eventually need hard-copy maps because they are inexpensive and portable. In addition, almost everyone relates to maps and can use them fairly easily.

Statistics or quantitative reports usually accompany GIS project results. Maps and coverages are excellent data visualization devices, but detailed data are also needed. Graphs that visually represent important statistical summaries or analysis are also highly efficient and useful.

Tabular data might be a compilation of project data. Census tables, for example, can provide an important frame of reference for an area. Most GIS projects use and generate a great deal of data that aren't needed in a final presentation but are necessary in completing the "total picture." Similarly, reports might be thick books of text, but useful supporting material for many projects. Therefore, GIS makes it possible to include scanned or hand-input text in databases.

Many GIS projects are now presenting data and results in digital format, including actual coverages and statistics. The media for this technology might be computer disk, CD-ROM, computer tape, or even network electronic transmission. We can anticipate increased use of digital products, and although hard-copy maps will undoubtedly be used less in the future, they will probably never become obsolete.

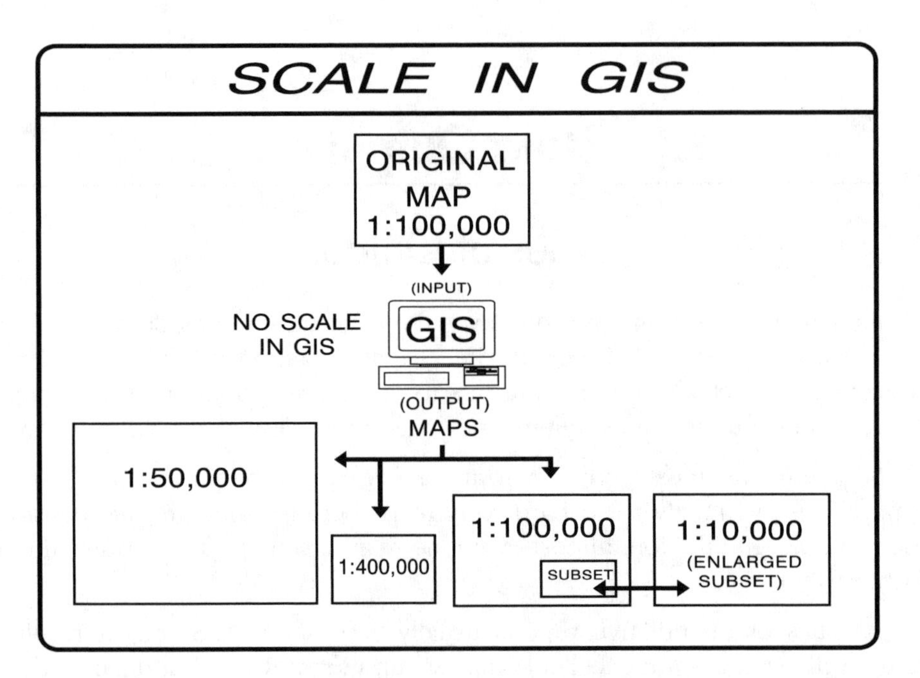
SCALE IN GIS
ORIGINAL
MAP
1:100,000
(INPUT)
NO SCALE
IN GIS
GIS
(OUTPUT)
MAPS
1:50,000
1:400,000
1:100,000
SUBSET
1:10,000
(ENLARGED
SUBSET)

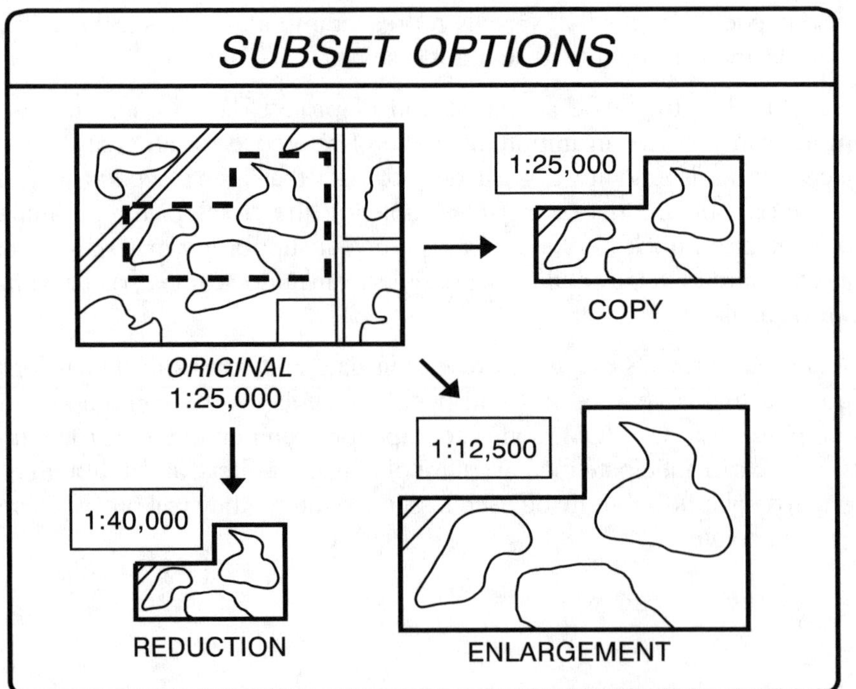
SUBSET OPTIONS
ORIGINAL
1:25,000
1:25,000
COPY
1:12,500
ENLARGEMENT
1:40,000
REDUCTION

Scale

Map scale is a common concern in traditional geographic output. GIS does not actually use scale directly in its operations, but it is important in determining data quality. GIS products can produce almost any desired map scale, but care must be taken to maintain acceptable quality.

GIS can receive a variety of scales and types of data, integrate them, and produce a variety of products. The upper illustration shows a 1:100,000-scale map digitized into the GIS. Output can be in a range of scales, such as 1:10,000 (rather large for the original data scale in this example) to 1:400,000 (or smaller, if needed). Even production at small scales requires caution because too many details can make a small-scale map cluttered and thus unreadable.

Subset Options

Because GIS is not scale dependent, entire coverages can be made into multiple-scale maps. Even portions of a coverage can be selected, made into a subset coverage, and then produced as maps at various scales. A good GIS permits interactive drawing of an area within a coverage, separation of the selected area into another coverage, and printing or plotting at various scales. GIS includes powerful features that produce a wide variety of output for many types of applications.

INVENTORY
OPERATIONS
• BASIC DATABASE WORK
• MEASUREMENT
• COVERAGE MANIPULATIONS
• GRAPHIC EDITS
• RECODING

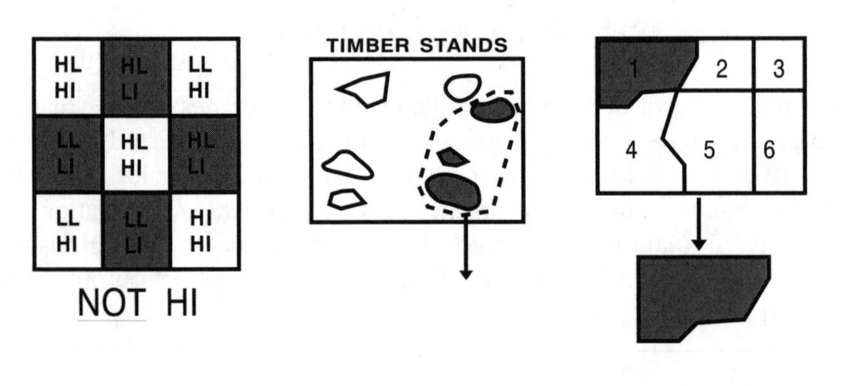

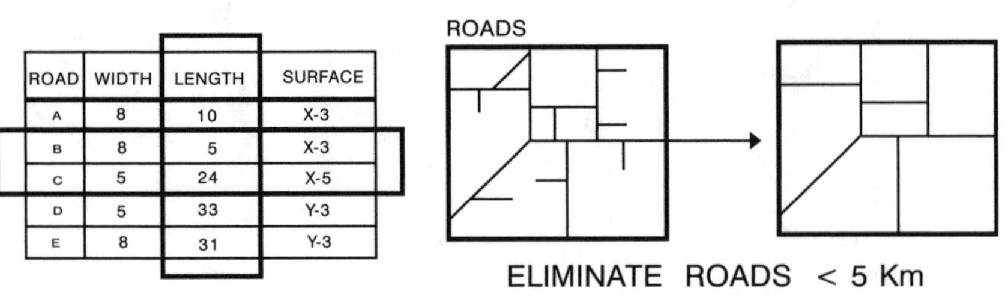

ROAD	WIDTH	LENGTH	SURFACE
A	8	10	X-3
B	8	5	X-3
C	5	24	X-5
D	5	33	Y-3
E	8	31	Y-3

7

Inventory Operations

Introduction

One of the most simple yet most useful sets of GIS operations is termed *Inventory,* which includes steps to extract basic data and information, simple work on coverages, and a few other operations. Discussed in this chapter are some of the more common and important inventory options among many in various GISs.

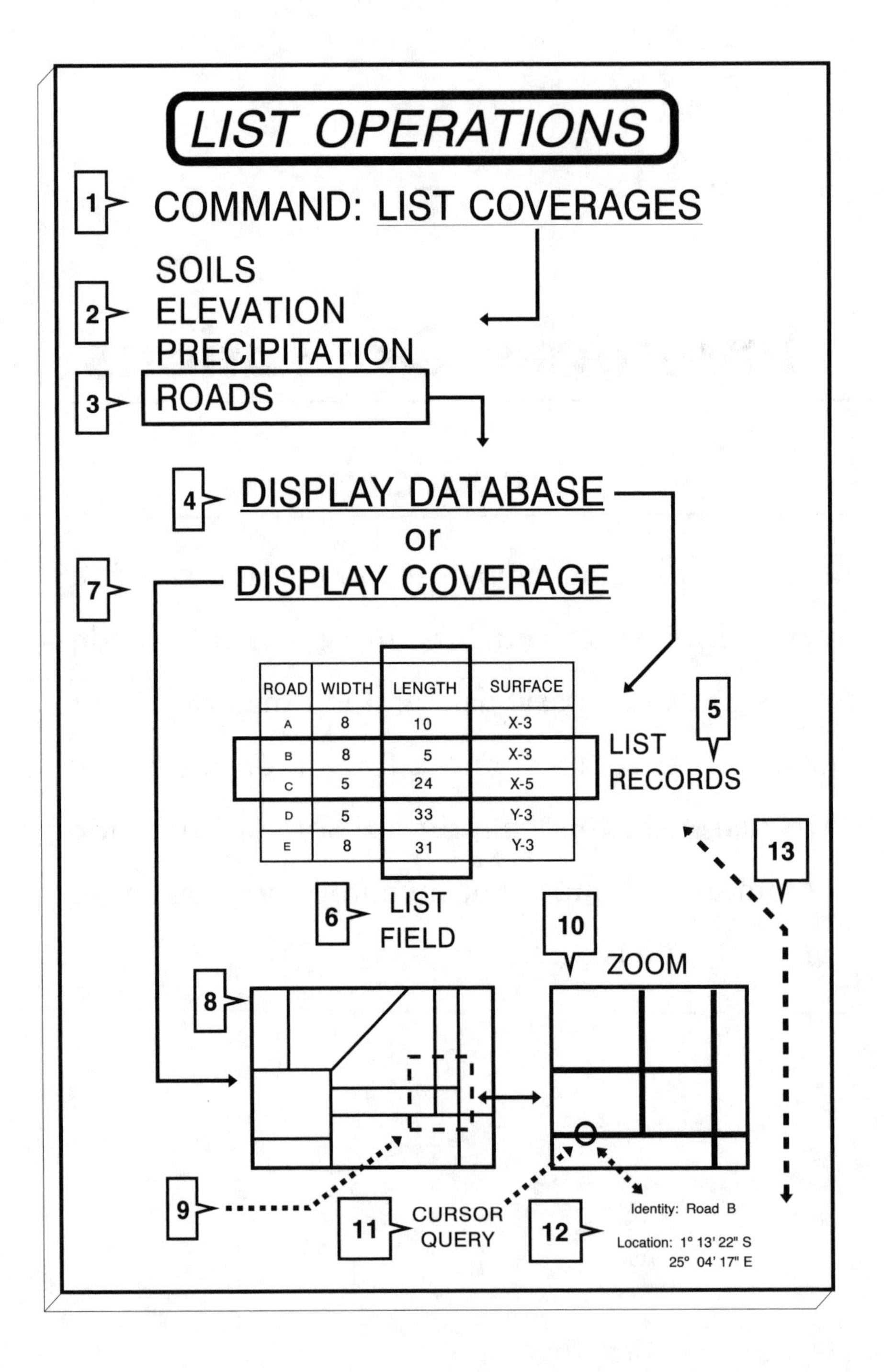

LIST OPERATIONS
1 COMMAND: LIST COVERAGES
SOILS
2 ELEVATION
PRECIPITATION
3 ROADS
4 DISPLAY DATABASE
or
7 DISPLAY COVERAGE
ROAD WIDTH LENGTH SURFACE
A 8 10 X-3
B 8 5 X-3
C 5 24 X-5
D 5 33 Y-3
E 8 31 Y-3
5 LIST RECORDS
6 LIST FIELD
13
10 ZOOM
8
9
11 CURSOR QUERY
12 Identity: Road B
Location: 1° 13' 22" S
25° 04' 17" E

List Operations

Typically, initial GIS operations deal with viewing files and coverages, normally requiring only a simple command or two. Each GIS uses various commands, operations, and names. Some are run by command line entry (typing the operation and needed details), some by pull-down menus, and others by mouse point-and-click techniques. A "generic" system is used here for convenience of illustration. It does not necessarily mirror a system you will use. Follow the numbers on the illustration.

The List Coverages command is given (1) and the list of GIS data files (Soils, Elevation, Precipitation, and Roads) is presented (2). The Roads file is chosen (3), and the next option is to view either the database or the graphic coverage. If the database is selected (4), it is presented on screen (or printed if desired). Usually there are options to view selected records (rows, 5) or attribute fields (columns, 6).

Even at this early stage, valuable work can be accomplished. Useful data can be extracted, isolated, and translated into information. Many GIS tasks are performed at the database, without having to view coverages (7). Displaying the coverage is another major option (8). Most GISs have a zooming, or enlargement, capability. A box is drawn or moved to the selected location (9), and the magnified view is displayed (10).

Other options may include feature identification and location by pointing the cursor (11) at a selected road and reading the data (12). This is a very nice link between the graphics and the database (13), which reads the coordinates from the graphical position and then reaches into the database to assign attribute information, such as the name.

GIS DATABASE
ORIGINAL & CLASSIFIED DATA

LANDSCAPE FEATURE	LANDUSE LEVEL 1	LANDUSE LEVEL 2	VALUE (OOO)
FOOD STORE	URBAN	COMMERCIAL	250
BREAD SHOP	URBAN	COMMERCIAL	75
CLOTHING SHOP	URBAN	COMMERCIAL	100
PRIMARY SCHOOL	URBAN	INSTITUTION	---
SECONDARY SCHOOL	URBAN	INSTITUTION	---
TARO	AGRIC.	FOOD CROP	50
LIME	AGRIC.	FOOD CROP	75

GIS DATABASE
PROPERTY DATA

PROPERTY NUMBER	AREA (Ha)	OWNER	TAX CODE	SOIL QUALITY
1	100,000	TUCATU	B	HIGH
2	50,100	BRAUDO	A	MEDIUM
3	90,900	BRAUDO	B	MEDIUM
4	40,800	ANUNKU	A	LOW
5	30,200	ANUNKU	A	LOW
6	120,200	SILIMA	B	HIGH

1. WHAT ARE THE ATTRIBUTES OF PROPERTY 2?
2. HOW MUCH PROPERTY DOES BRAUDO OWN?
   ANSWER: 141,000 Ha
3. WHO OWNS HIGH SOIL QUALITY PROPERTIES?
   ANSWER: TALATU, SILIMA
4. WHO OWNS THE LARGEST SINGLE PROPERTY?
   ANSWER: SILIMA

Database Capabilities

Data Categories

As noted, databases are central to GIS operations and applications. Typically, a GIS database holds original data, data that have been classified, and a mix of qualitative and quantitative formats.

The Landscape Feature column in the upper database in the illustration holds the original field observation data—specific identification of features. This column should never be changed because it represents the best detail and accuracy.

The two Landuse Level columns are landuse classifications for features. They are generalizations of the original data that will be used in analysis and mapping. The Value column is quantitative and is an associated attribute; in this case, probably collected at City Hall from another database.

The lower database in the illustration shows that extraction of data is easy, whether requesting attribute information or specific records. It shows relatively simple questions (queries) and their answers. The first question involves a simple record reading. Question two needs the Owner attribute listing to collect all of Braudo's records and then read those records' Areas. For the third question, all Soil Quality "Highs" are read and then compared with the Owner for a final answer. Question four uses the same approach: find the highest Area and then read the Owner.

GISs with attached databases can address these types of questions easily, giving them powerful inventory capabilities. Let's now take a look at the role of relational database queries.

DATABASE QUERIES

PROPERTY NUMBER	AREA (SQ M)	OWNER	TAX CODE	SOIL QUALITY
1	100,000	TULATU	B	HIGH
2	50,100	BRAUDO	A	MEDIUM
3	90,900	BRAUDO	B	NO MEDIUM
4	40,800	ANUNKU	A	LOW
5	30,200	ANUNKU	A	LOW
6	120,200	NO SILIMA	B	HIGH

CONDITIONAL QUERY

Ex: The best property has:
- Area: >40,000 sq m
- Owner: not Silima
- Tax Code: B
- Soil Quality: High

Answer: Property number 1

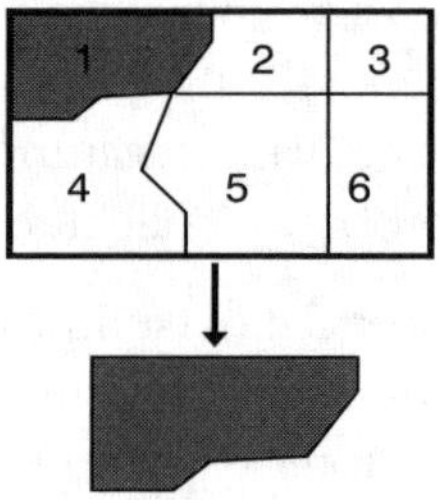

Relational Database Queries

The Relational Database offers significant advantages over "flat-file" databases, which are little more than simple tables. The relational model allows flexible queries (ways of asking for information) and multiple conditions (characteristics). The illustrated query shows how the program finds the set of conditions in each attribute column, selects the records satisfying these conditions (or sifts out the unneeded records), and eventually arrives at those meeting all of the expressed conditions. Follow the steps in the illustration.

A good GIS also has a direct link between its database and graphics so that when the final selection is made results can be seen on the coverage display. In this illustration, property 1 is the only one that meets all criteria of the conditional query, and thus is highlighted on the coverage display.

EASY UPDATING

DISTRICT POPULATION

CURRENT CENSUS

NAME	1970	1980	1990	2000
DISTRICT A	1254	1545	~~1650~~ =	**1640**
DISTRICT B	2555	3006	3200	
DISTRICT C	3412	3644		
DISTRICT D	1400	1350		
DISTRICT E	2350	2300		

Database Update

A good database is accessible for easy editing, permitting changes to be made. The illustration shows a population database containing 1970 and 1980 data. Census data from 1990 is being entered. Data can be typed in directly or supplied through direct input of digital sources (e.g., tape, disk, or CD). Errors should be easy to correct. For example, District A has a mistake that is corrected by typing. The remainder of the 1990 column is to be filled in.

A new column for the year 2000 has been created, perhaps to receive trend data from the previous three censuses. Making new columns and entering data from existing column calculations are very useful database capabilities.

Once these tasks have been completed, a full database serves as the core of GIS data and operations. Therefore, it is very important to construct a flexible and complete database.

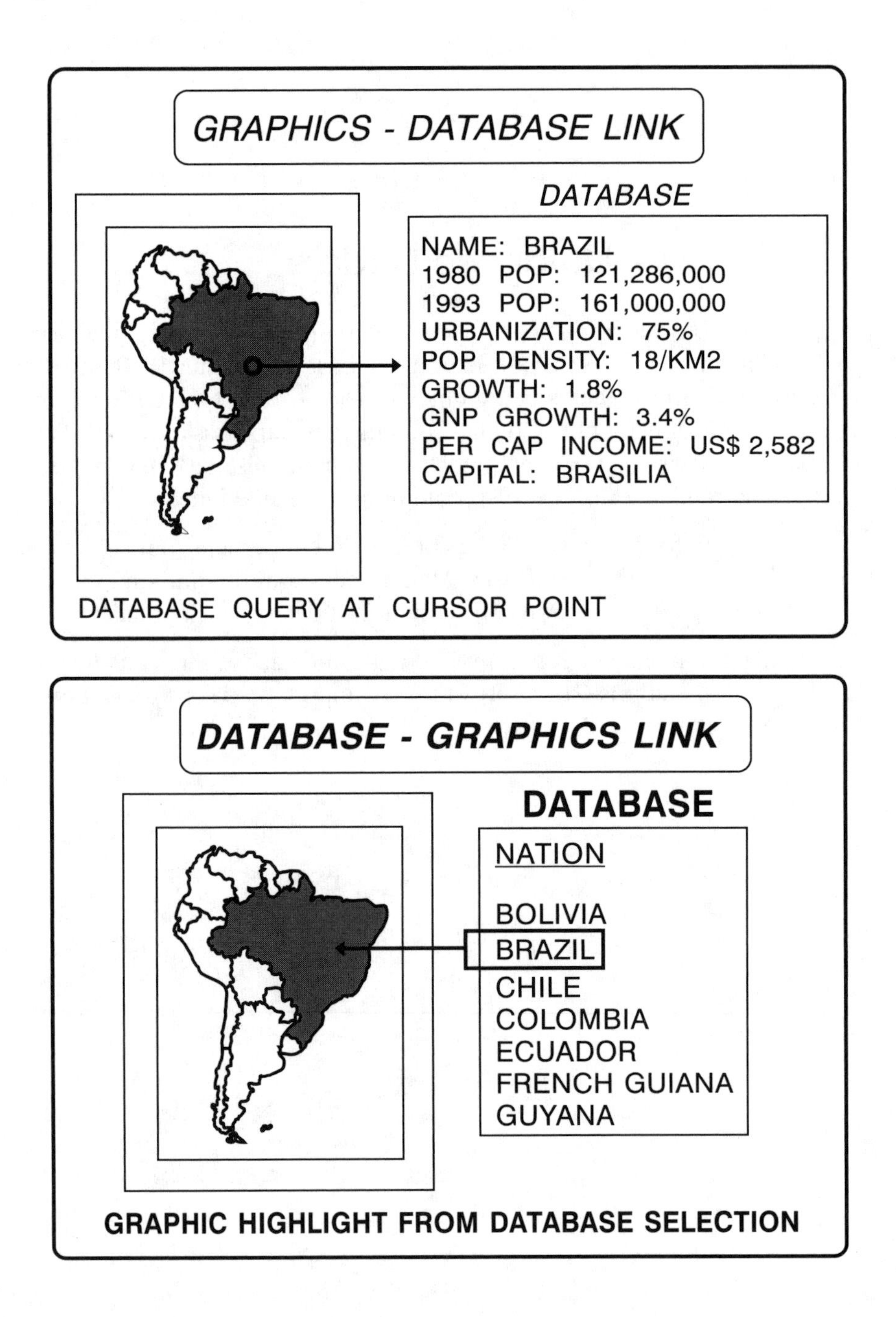
GRAPHICS - DATABASE LINK
DATABASE
NAME: BRAZIL
1980 POP: 121,286,000
1993 POP: 161,000,000
URBANIZATION: 75%
POP DENSITY: 18/KM2
GROWTH: 1.8%
GNP GROWTH: 3.4%
PER CAP INCOME: US$ 2,582
CAPITAL: BRASILIA
DATABASE QUERY AT CURSOR POINT
DATABASE - GRAPHICS LINK
DATABASE
NATION
BOLIVIA
BRAZIL
CHILE
COLOMBIA
ECUADOR
FRENCH GUIANA
GUYANA
GRAPHIC HIGHLIGHT FROM DATABASE SELECTION

Linking

Graphics-Database Link

GIS is meant to link (connect) graphics and databases, giving it an advantage unavailable in most data management systems. The nature of the connection ranges from a simple list on paper (very low-end GISs) to transparent, seamless (invisible to the operator), fully functional integration of data and graphics. A good GIS permits pointing to a selected part of a coverage and viewing its data in a database presentation. In the upper illustration, the cursor points to Brazil, and the database box displays all of the fields and record entries for that nation. The database may be presented on a split-screen overlay of the coverage display, on a second screen (flip-"pages"), or on a second monitor. It may also be printed. Regardless of display method, printing the database is a useful option.

Database-Graphics Link

A good GIS allows the user to select records or attributes in the database and to view the results on a coverage display. Working at the database, the operator selects Brazil (in the lower illustration) or goes through a conditional query on the relational database. The country is then highlighted on the graphics.

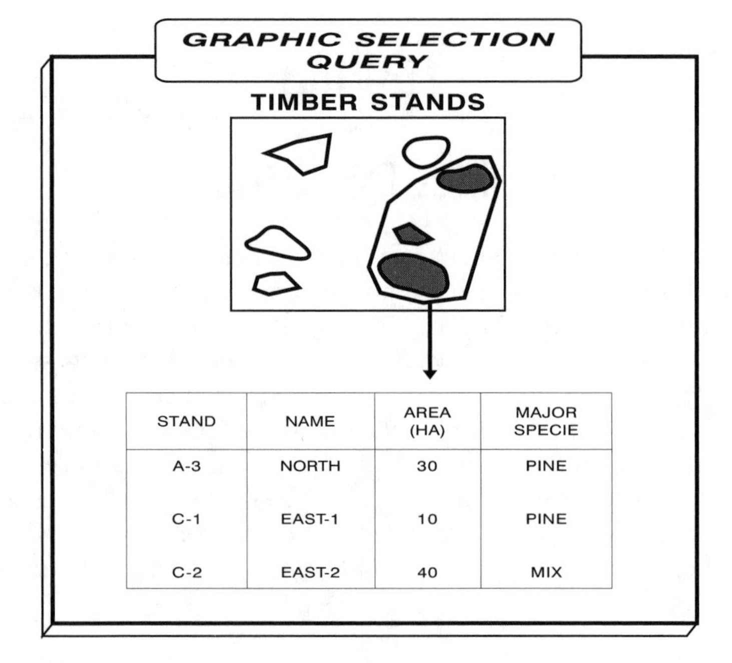

STAND	NAME	AREA (HA)	MAJOR SPECIE
A-3	NORTH	30	PINE
C-1	EAST-1	10	PINE
C-2	EAST-2	40	MIX

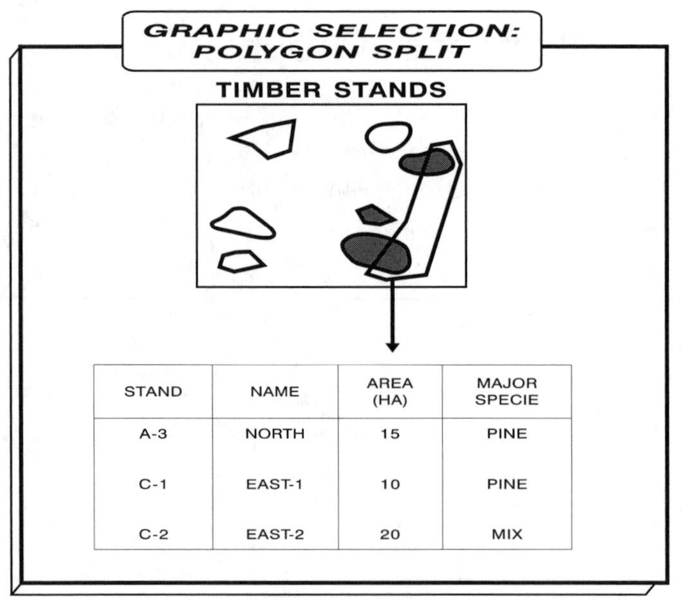

STAND	NAME	AREA (HA)	MAJOR SPECIE
A-3	NORTH	15	PINE
C-1	EAST-1	10	PINE
C-2	EAST-2	20	MIX

Graphic Selection Query

Selection of features is not limited to single records. Drawing an area on the graphics display may serve as the selection pointer. For example, in the upper illustration, an area is drawn around the three timber stands, which are listed in the corresponding database. In some systems, the stands would be simply highlighted on the existing database; in a different system, they might be grouped into a smaller version of the database (as illustrated). Some GISs allow pointing and clicking on a group of separated features for selection.

This example shows selection of three features and their records, but some powerful GISs have options to select only parts of records. For example, the lower illustration shows a selection area that bisects (splits) two of the previously chosen timber stands (A-3 and C-2). The resultant database reports the same attributes, but the Area numbers are about half. This type of operation is difficult and requires a sophisticated GIS.

As seen in the previous illustration, the South America–Brazil graphics-database link, many GISs permit point-and-click database selections, but only whole polygons will be chosen. There is no easy way to select only part of a polygon from a database and have that part highlighted on graphics.

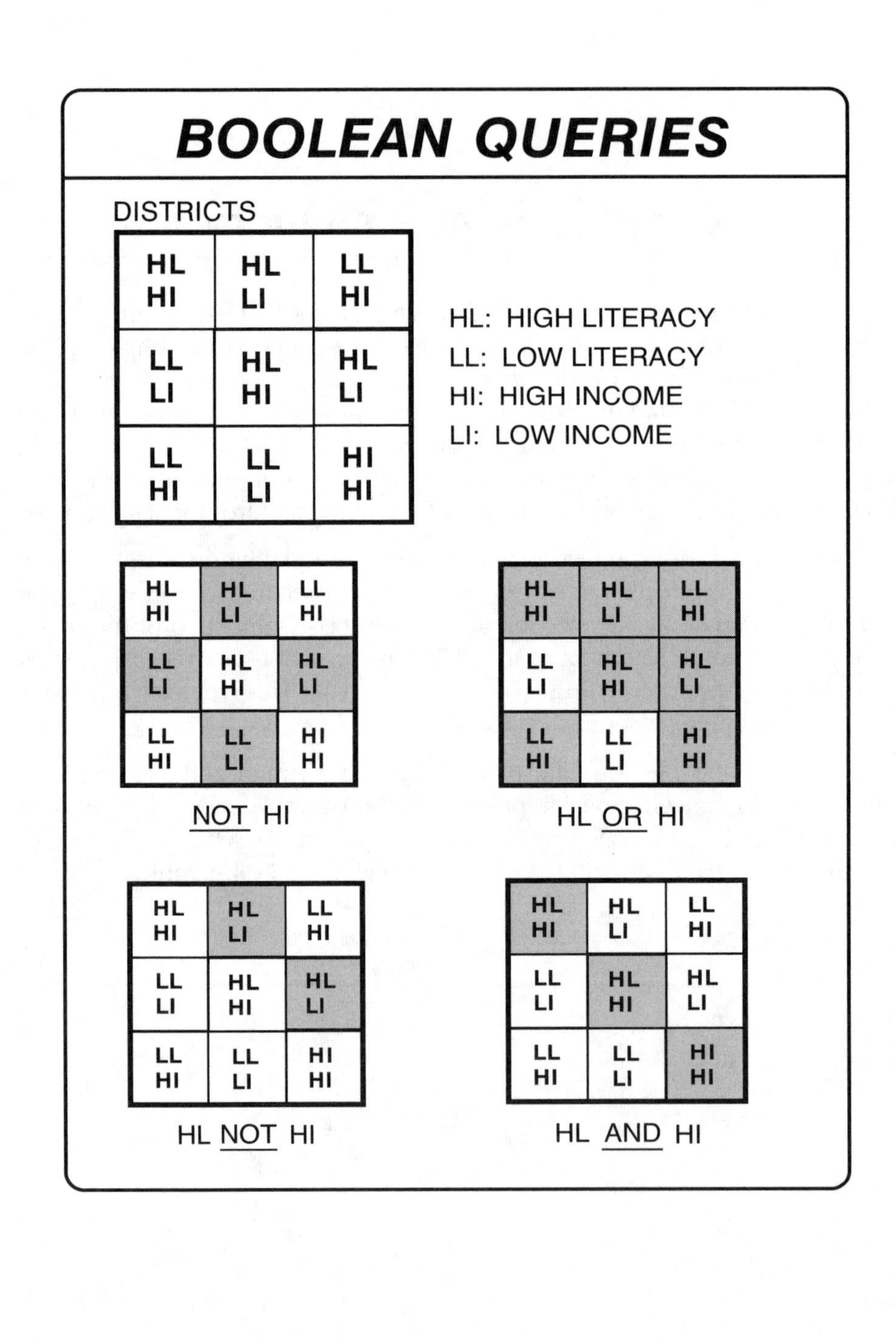
BOOLEAN QUERIES
DISTRICTS
HL HI
HL LI
LL HI
LL LI
HL HI
HL LI
LL HI
LL LI
HI HI
HL: HIGH LITERACY
LL: LOW LITERACY
HI: HIGH INCOME
LI: LOW INCOME
NOT HI
HL OR HI
HL NOT HI
HL AND HI

Boolean Queries

George Boole was a mid-1800s mathematician who attempted to combine properties of logic with mathematics. Known as Boolean logic, this method basically determines the true-false, yes-no, or presence-absence properties of objects. That is, the operation determines if one or if two choices exist, which is a fairly simple calculation for a computer. This is called a binary decision (one of two choices).

Boole's methods include comparison between two or more sets of Boolean properties in a given feature, using *and, or, and/or, not,* and other combining language. For example: "Does feature X have trees [Yes-No] *and* a river [Yes-No]?" GIS is very good at Boolean logic. Illustrated are examples of testing the relationship between literacy and income, shown as nine square districts (map data, not tabular).

Rather than using numbers and magnitudes of features, as in the relational database approach, Boolean logic reduces measures to some binary condition. In this example, a district is said to have either high or low literacy (HL or LL), and either high or low income (HI or LI).

Note the larger map in the upper left that describes each district's level of literacy and income (this is the original data recording for each district). The four shaded maps are illustrations of Boolean query results; they show particular conditions or sets of literacy-income properties, as stated in the text below each one. The first coverage looks at each district and asks the GIS to show all districts that do *not* have high income (NOT HI). Literacy is not considered in this query. The four darkened districts meet the Not High Income condition.

The second coverage queries districts having high literacy *or* high income; that is, one or the other (HL OR HI). All but two districts have at least one high of either literacy or income.

The lower-left coverage shows districts having high literacy but not high income (HL NOT HI), with only two meeting these conditions. The last coverage asks for districts having both high literacy *and* high income (HL AND HI).

Boolean logic queries are very useful in GIS analysis. This concept could be extended to more complicated sets of Boolean properties, such as High-Medium-Low. Most GISs have some form of Boolean query. Some systems use Boolean logic to simplify the results of overlay operations, a topic presented later in this chapter.

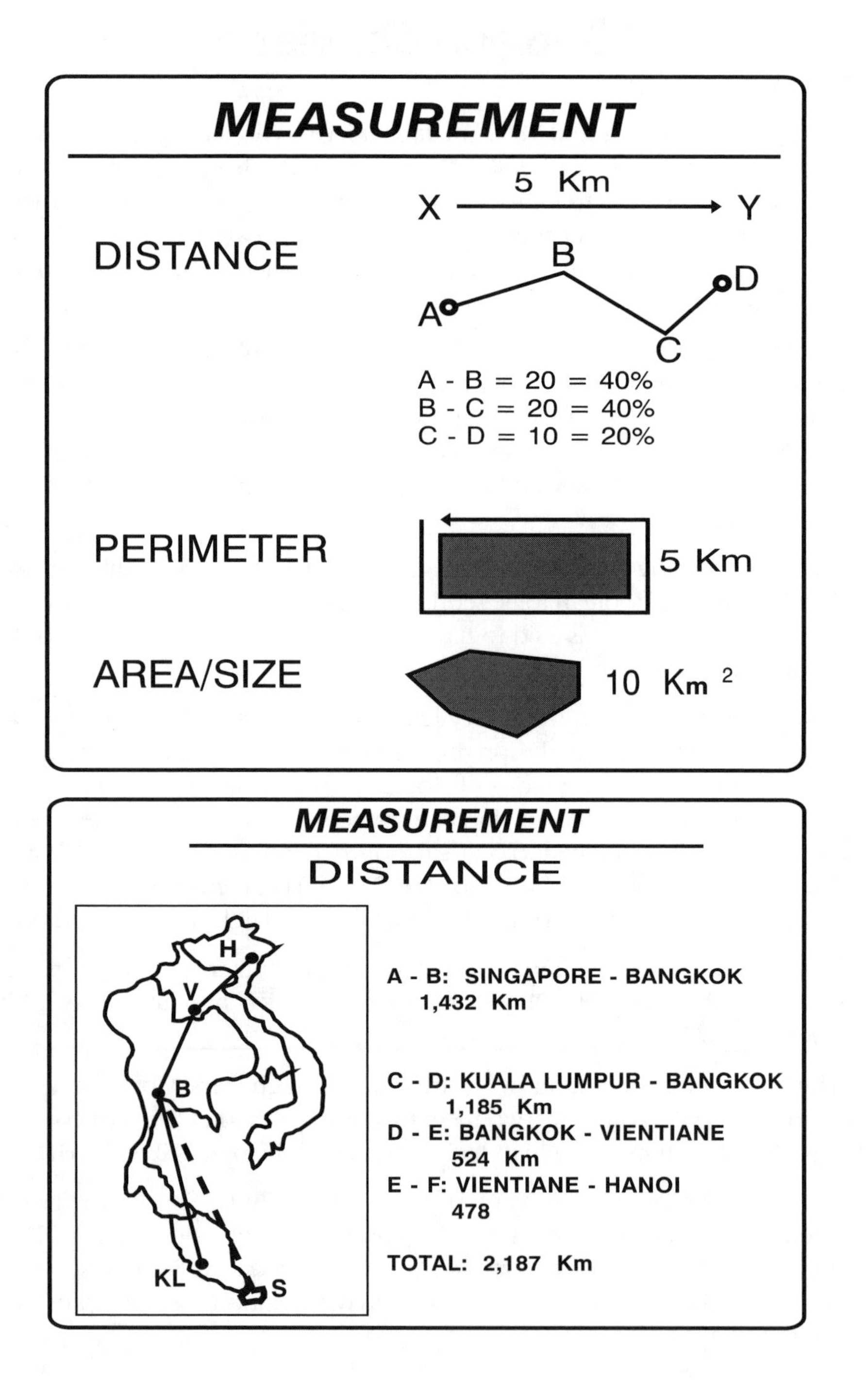
MEASUREMENT
DISTANCE
X 5 Km Y
A B C D
A - B = 20 = 40%
B - C = 20 = 40%
C - D = 10 = 20%
PERIMETER
5 Km
AREA/SIZE
10 Km 2
MEASUREMENT
DISTANCE
H
V
B
KL
S
A - B: SINGAPORE - BANGKOK
1,432 Km
C - D: KUALA LUMPUR - BANGKOK
1,185 Km
D - E: BANGKOK - VIENTIANE
524 Km
E - F: VIENTIANE - HANOI
478
TOTAL: 2,187 Km

Measurement

Types of Measurement

One simple but highly useful operation to apply to GIS data is measurement. This might be done on the coverage display by using a mouse to draw lines (beginning, end, and mid-points) and polygons. Most GISs will give real-world distances. However, raster systems may present the number of raster cells from one point to another, which then have to be converted into distance.

In vector systems, the database normally includes area for polygons; in raster systems, they are expressed as the number of cells per feature. Topological systems may also have length of each chain and perimeter of polygons. A good topological GIS can determine the distance from one node or vertex to another. Illustrated at top is a complex line that has chain lengths and proportions. Remember that perimeter and area are automatic feature attributes in some GISs.

Distance Application

The example in the bottom illustration shows a simple application of city distances in Southeast Asia (map letters correspond to cities on the side text). The region is a standard coverage (nations and cities), and either database query or point-and-click selections chose the route. City-to-city distances were requested and a total trip length was displayed.

If the database contains appropriate attributes, numerous Cost, What-If, and Cost-Benefit analysis queries can be performed easily. For example, airlines (and each type of aircraft) have fixed cost per distance and varying landing and ground charges for each city; therefore, the partial and total trip costs can be calculated quickly. Then, various passenger (income) scenarios can be tried in order to estimate cost-benefit trips, such as projected number of passengers boarding and exiting at each city. Perhaps a Singapore–Hanoi route may be more cost efficient with only a stop at Bangkok, excluding Kuala Lumpur and Vientiane. GIS can be very useful in these types of applications.

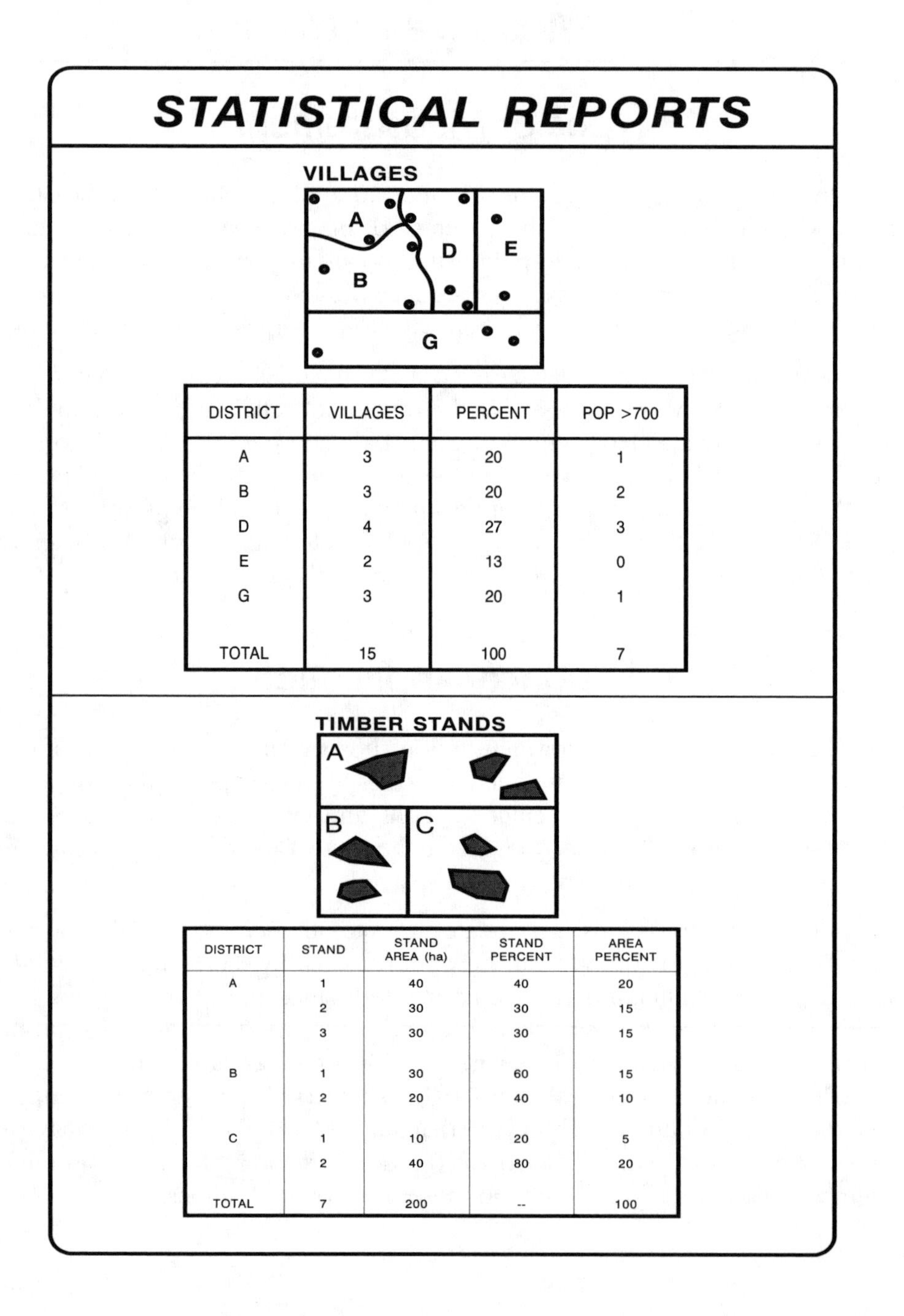

DISTRICT	VILLAGES	PERCENT	POP >700
A	3	20	1
B	3	20	2
D	4	27	3
E	2	13	0
G	3	20	1
TOTAL	15	100	7

DISTRICT	STAND	STAND AREA (ha)	STAND PERCENT	AREA PERCENT
A	1	40	40	20
	2	30	30	15
	3	30	30	15
B	1	30	60	15
	2	20	40	10
C	1	10	20	5
	2	40	80	20
TOTAL	7	200	--	100

Statistical Reports

Useful descriptive statistics can be produced easily, often as a standard report option. These provide measurements, including total number of all or selected features on a coverage. In the upper illustration, the total number of villages (point features) in each mapped district is given, along with the percentage of villages in each district and the number of villages with populations over 7,000 (a measure needed by the user). This is an example of the beginning of spatial analysis of villages—recognizing the relationship between villages and districts.

The statistics in the lower illustration give the area of each stand, its percentage in the district, and its percentage in the area. Each data field is useful for various timber management activities. These types of reports are excellent inventory data, which could later be used in more complex GIS analysis.

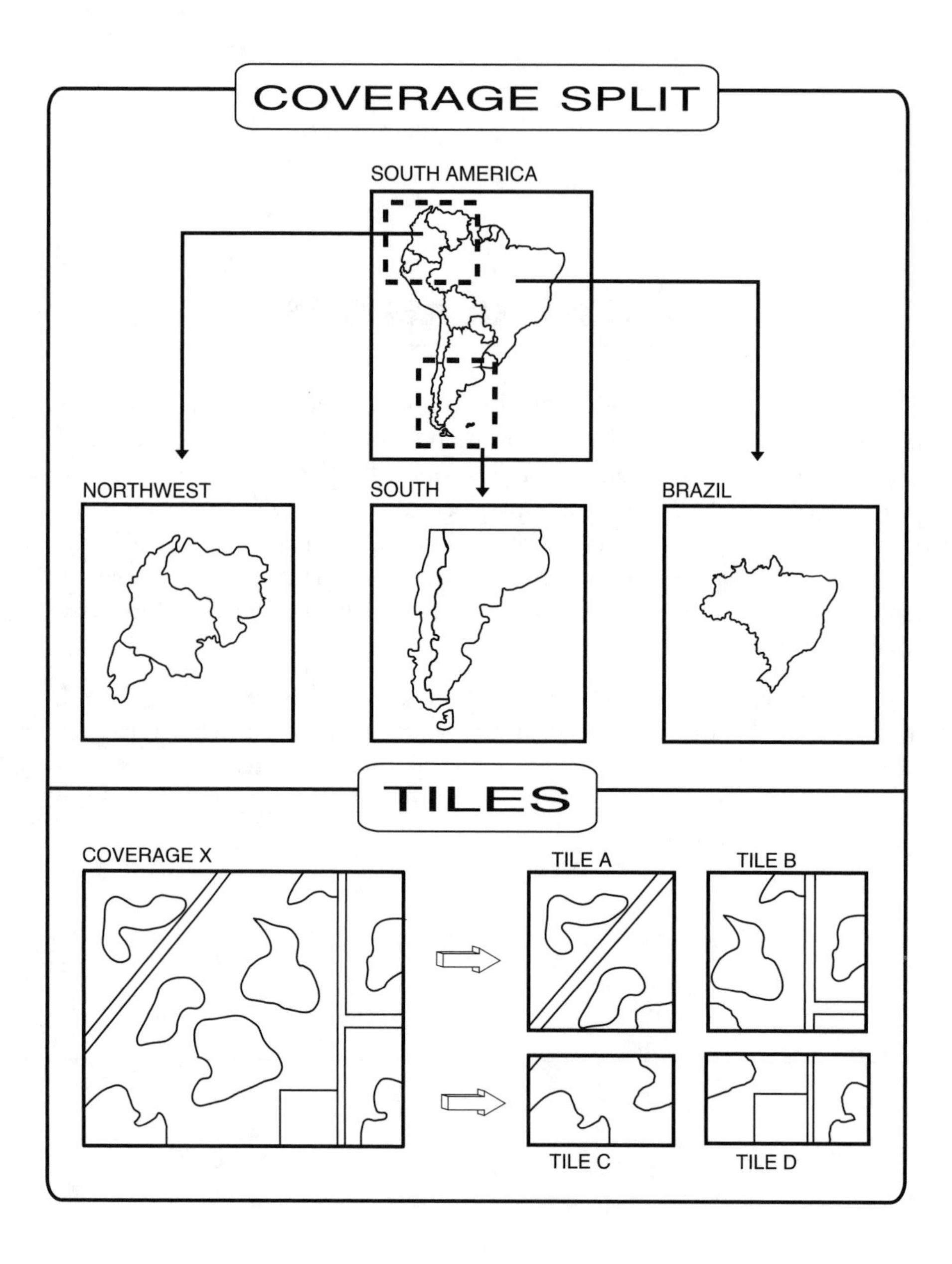
COVERAGE SPLIT
SOUTH AMERICA
NORTHWEST
SOUTH
BRAZIL
TILES
COVERAGE X
TILE A
TILE B
TILE C
TILE D

Coverage Modification

Coverage Split

It is normal to have a large area as the base map or base coverage, but often it may be unnecessary to apply GIS operations to the entire area when only a part is under investigation. All of South America, for example, need not be used if the current project deals only with Brazil. Reducing the coverage supports time, effort, computer processing, project storage demands, and cost.

Splitting the coverage for a particular user task is a common operation. Drawing the appropriate lines and instructing a split may be all that is required (though many GISs use more steps). Illustrated at top are three versions of splitting. On the right, Brazil has been selected, typically just by pointing to it as a single feature and making the GIS operation command. The database is automatically reconstructed for Brazil as a separate coverage.

On the left, the three northeast countries are selected. They were the only complete features within the box (or window) drawn around them. The middle shows Argentina and Chile split by the interactive box, a process that can be difficult for the program—splitting features and recalculating their records to reflect only the selected area. This is similar to the part-polygon timber-stand query previously discussed.

Tiles

The lower illustration shows the concept of "tiles." When a large base area needs to be split into separate areas of responsibility or work, it can be divided into "sub-base" coverages or tiles. Perhaps the entire Coverage X is too much work for one operator, so reduction to four sections becomes a more appropriate project resource allocation.

Tiles are considered subsets of a larger area, but they can be treated separately, as individual GIS projects, later to be reunited into the original large area if necessary. A good GIS automatically constructs a database for each tile so that work can continue normally.

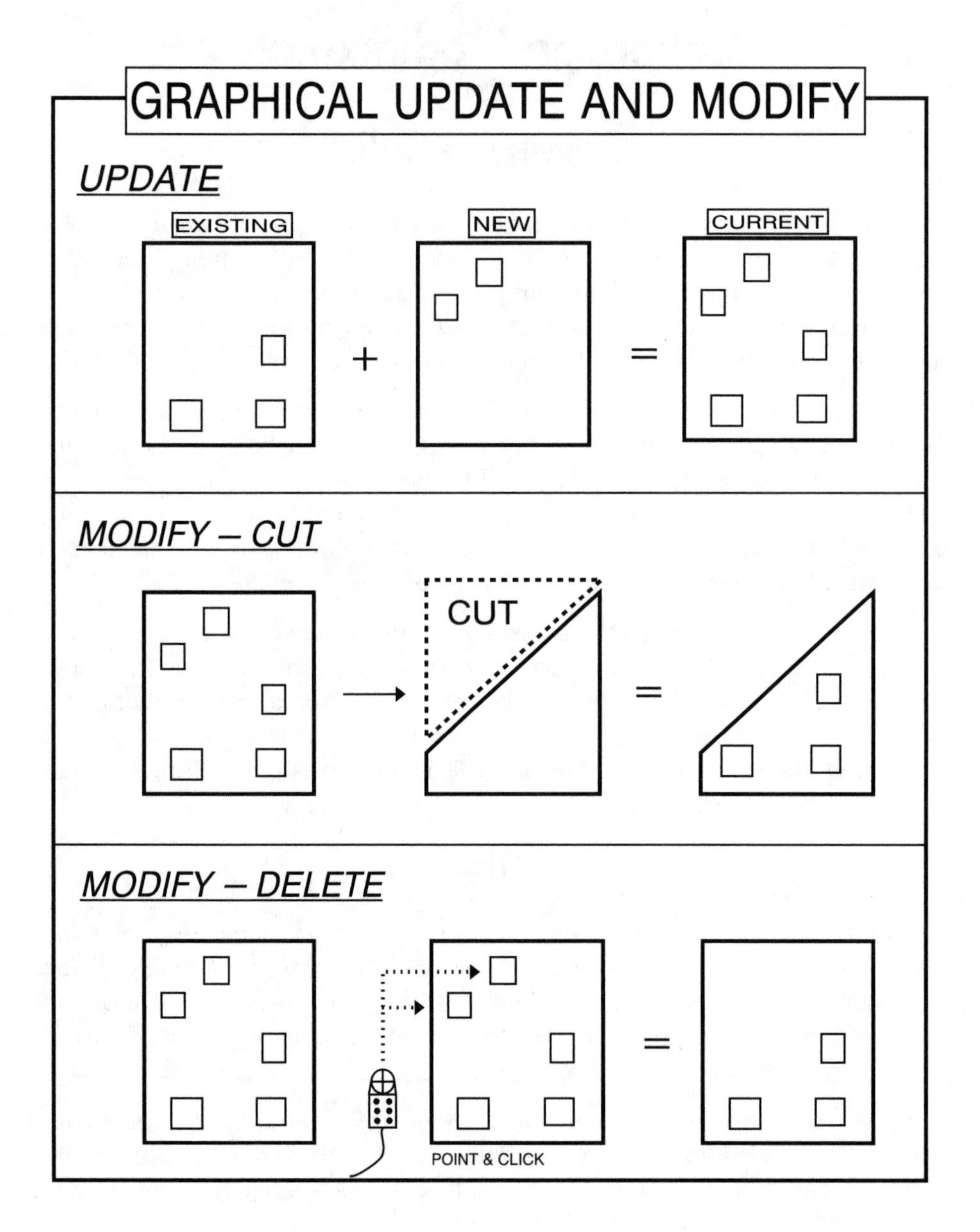
GRAPHICAL UPDATE AND MODIFY
UPDATE
EXISTING
NEW
CURRENT
MODIFY – CUT
CUT
MODIFY – DELETE
POINT & CLICK

Graphics Update and Modification

Making changes directly on graphics is a fairly simple procedure in some GISs. The Update example in the illustration shows how two square features are added. They may be placed by entering the new features into the database or by point-and-click positioning.

Other GISs do this by combining two coverages to make a third. The overlay procedure (see Chapter 8) combines two coverages (Existing and New) into a third (Current).

Modification by removal (cut) of features or an area is also rather easy. The five-square coverage at left middle is modified by outlining an area to be deleted, resulting in an odd, triangular coverage with only three squares. The illustration at bottom shows modification of the coverage accomplished by point-and-click on the features and a command to eliminate, much like spatial delete, discussed in the following section. In a good GIS, the features are deleted in the database at the time of modification. However, some GISs require a separate database delete operation. Really good GISs will make a temporary delete but not really excise the data until told to do so—a safety feature.

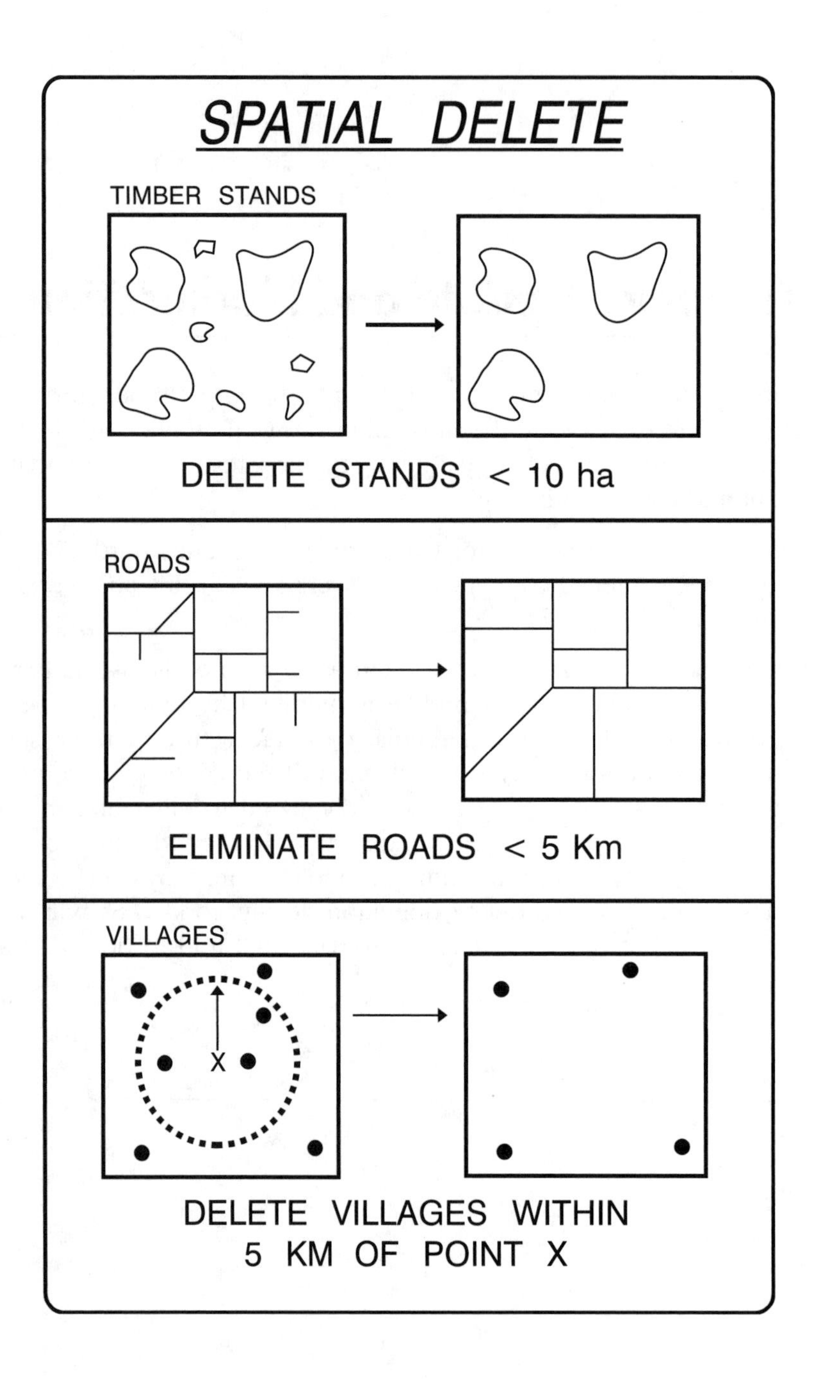
SPATIAL DELETE
TIMBER STANDS
DELETE STANDS < 10 ha
ROADS
ELIMINATE ROADS < 5 Km
VILLAGES
X
DELETE VILLAGES WITHIN
5 KM OF POINT X

Spatial Deletes

Inventory operations are relatively simple procedures used to read or select data for particular purposes. Deleting unimportant or confusing features can be very useful. Three "spatial deletes" are shown. The illustration at top keeps only the large timber stands. Those smaller than 10 ha are not needed and the command to erase them may be as simple as stated: "Delete stands <10 ha." This can occur at the database or the graphics.

The middle example is very similar: deleting roads that are less than 5 km in length. Some GISs allow point-and-click selection for deleting. In a Delete mode, the user points to each offending road and clicks. The selections are then deleted from the graphics and the database.

The example at bottom shows the use of a designated area around a selected point, typically called a "buffer zone." A 5-km circle around the X position has been designated, and all villages within that area are deleted from consideration, leaving only those outside. The process is simple: point to the center spot, ask for a circular 5-km zone, and then give the command to delete the villages within.

NOTE: *The original data file should never be modified by these delete operations. Any changes are saved under a new name.*

Other types of spatial deletes are offered by most GISs. Let's turn our attention to a function similar to deletion, called dissolve.

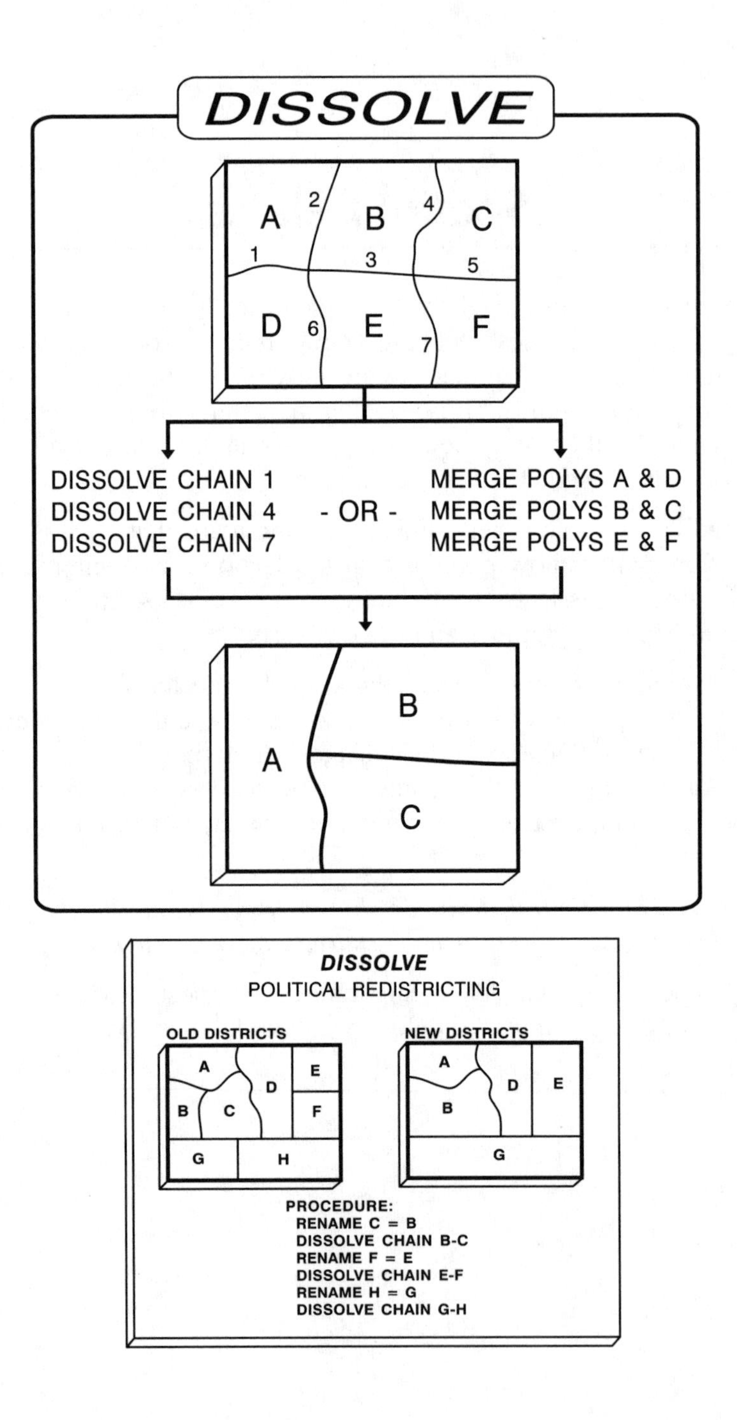
DISSOLVE
A
B
C
D
E
F
1
2
3
4
5
6
7
DISSOLVE CHAIN 1
DISSOLVE CHAIN 4
DISSOLVE CHAIN 7
- OR -
MERGE POLYS A & D
MERGE POLYS B & C
MERGE POLYS E & F
A
B
C
DISSOLVE
POLITICAL REDISTRICTING
OLD DISTRICTS
A
B
C
D
E
F
G
H
NEW DISTRICTS
A
B
D
E
G
PROCEDURE:
RENAME C = B
DISSOLVE CHAIN B-C
RENAME F = E
DISSOLVE CHAIN E-F
RENAME H = G
DISSOLVE CHAIN G-H

Dissolve

Often in GIS there is a need to merge or combine features, such as small districts into larger provinces. The upper illustration shows the process of reducing six districts into three new districts, using two similar methods. Follow the steps. The first is to dissolve specific borders. Districts A and D are combined by dissolving chain 1, B and C are joined by dissolving chain 4, and chain 7 is dissolved to combine E and F.

A more direct method would be to instruct the GIS to integrate districts by merging selected polygons. This may involve no more than pointing to each district and giving the command to merge (which, in effect, dissolves the borders). As above, A and D, B and C, and E and F are merged. The results are shown in the A-B-C map. As expected, a good GIS will reconstruct an appropriate database for each new district.

A political redistricting application is shown at the bottom of the illustration, where existing districts are redesigned into a new set of districts. This is often needed because of shifting populations and other considerations. Follow the steps, which are similar to the procedures just discussed.

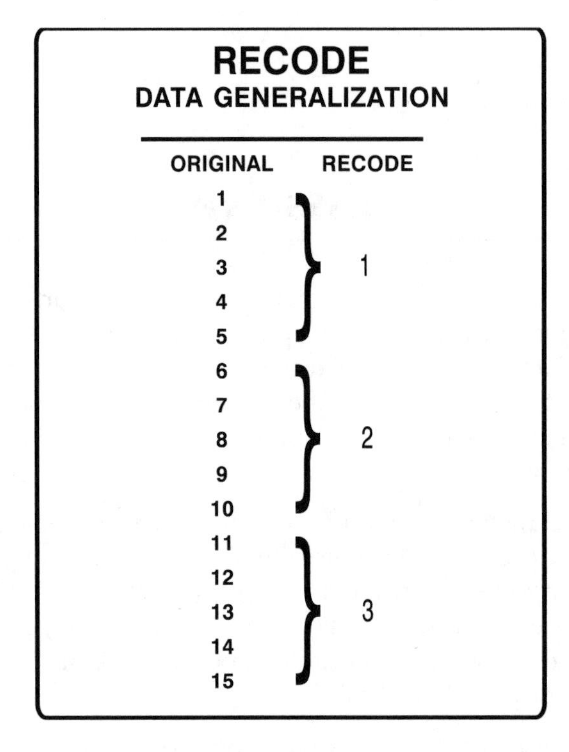
RECODE
DATA GENERALIZATION
ORIGINAL
RECODE
1
2
3
4
5
1
6
7
8
9
10
2
11
12
13
14
15
3

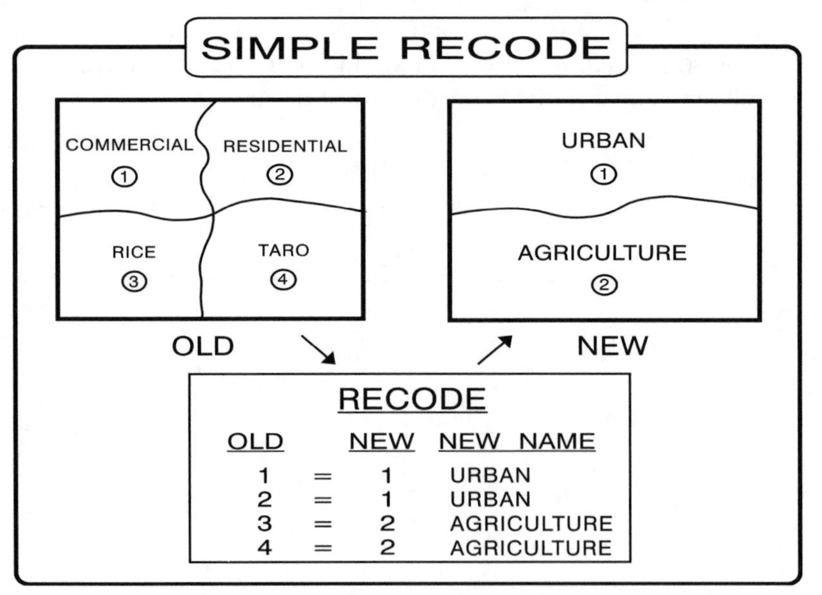
SIMPLE RECODE
COMMERCIAL
1
RESIDENTIAL
2
RICE
3
TARO
4
OLD
URBAN
1
AGRICULTURE
2
NEW
RECODE
OLD
NEW
NEW NAME
1 = 1 URBAN
2 = 1 URBAN
3 = 2 AGRICULTURE
4 = 2 AGRICULTURE

Recode

Modifying data by Recode is one of the most useful and most common GIS operations. It is an easy procedure that reduces "data clutter" and promotes a better understanding of the coverage.

In principle, Recode is a simple step: changing attribute numbers or names. The upper illustration is a small data generalization—perhaps fifteen detailed soil types organized into three general categories. Codes 1 through 5 are recoded to 1, codes 6 and 7 are changed to 2, and 11 through 15 are changed to 3. Now a three-class soil coverage exists that may be less confusing and more useful for some purposes than a fifteen-class coverage. Whether vector or raster system, changing numbers or names is rather easy to accomplish.

The lower illustration is another example, using both numbers and names. The old coverage has four classes of landuse, but they are generalized to only two in the new coverage, changing both codes and names. The Recode operation is shown, which sometimes may be no more complicated at the keyboard than this. Remember that the old data still exist under the original names; the new coverage, created from the same data, simply has a new, unique name.

Taking Stock

Inventory operations gather GIS data and graphics for initial reading and analysis. They range from simple listing and viewing of data to recoding of coverage features. Basic database work is the foundation of GIS, but links to graphics add the data visualization advantages. A relational database, using Boolean queries, is a strong GIS capability. Inventory operations include measurements and statistical reporting. Modifications are easy, such as splitting coverages, and updating databases and graphics with additions and deletions. Many inventory steps are progressions into the operations discussed in Chapter 8, Basic Analysis.

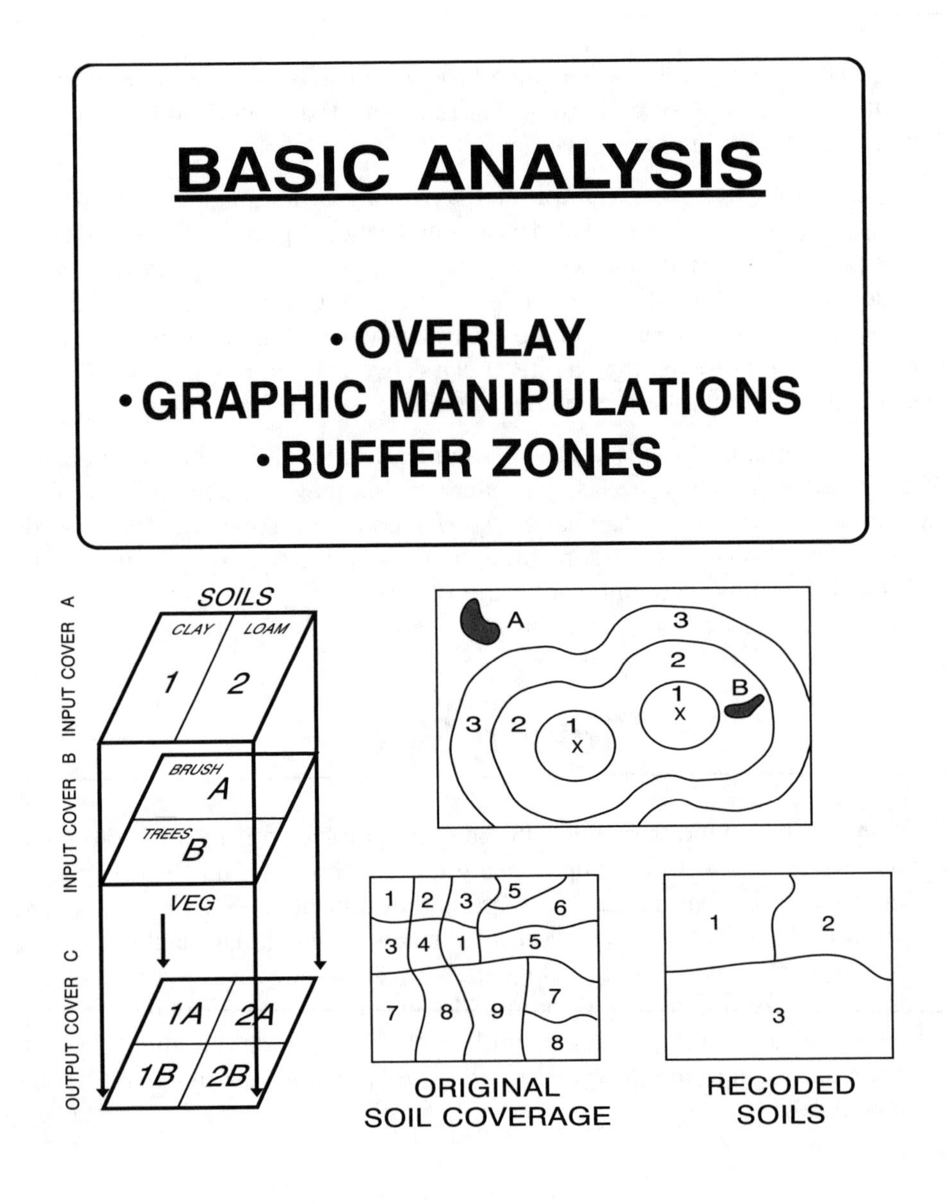
BASIC ANALYSIS
•OVERLAY
•GRAPHIC MANIPULATIONS
•BUFFER ZONES
SOILS
CLAY
LOAM
1
2
BRUSH
A
TREES
B
VEG
1A
2A
1B
2B
INPUT COVER A
INPUT COVER B
OUTPUT COVER C
A
B
3
2
1
X
ORIGINAL
SOIL COVERAGE
RECODED
SOILS

8

Basic Analysis

Introduction

After inventory operations, the next step in using GIS data is to perform basic analytical operations. However, there is no sharp division between inventory and basic analysis because both yield a variety of data and information. Several fundamental GIS procedures that set up many analytical operations, notably overlay and buffers, are presented in this chapter.

DATABASE RECODE
DATA GENERALIZATION
SOIL DATABASE

ATTRIBUTE CODE	SOIL TYPE	ATTRIBUTE RECODE	NEW SOIL
1	A1Z	1	A
2	A2Z	1	A
3	A3Z	1	A
4	A3Y	1	A
5	B1Y	2	B
6	B2Y	2	B
7	C6H	3	C
8	C6J	3	C
9	C7J	3	C

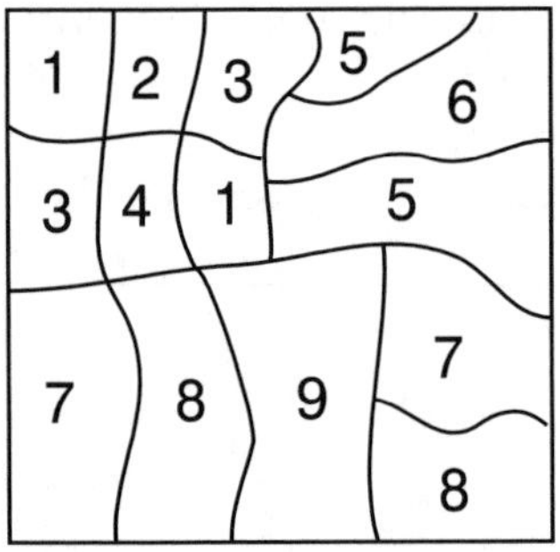

ORIGINAL SOIL COVERAGE

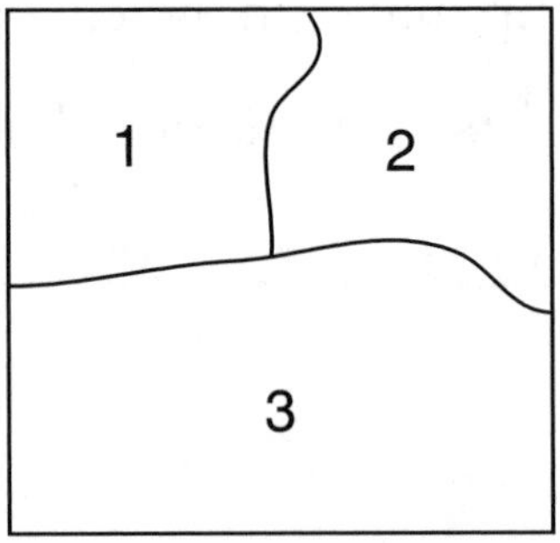

RECODED SOILS

Database Recode

Recoding and Analysis

As discussed in the section on inventory, recoding can be used as an initial analytical device. A coverage with many classes and features may serve as a good database, but its order (or organization) may not be apparent. By reducing the classes into a manageable list (and rearranging into a logical sequence), there are better chances of understanding the data, perceiving new relationships, and gaining new information.

In the illustration, a complicated coverage (Original Soil Coverage) has nine classes of soils scattered over an area. The distribution (or patterns), if any, are almost impossible to detect. Note in the database, however, that three primary soil types exist, designated A, B, and C. By recoding the nine categories to three according to the first letter, perhaps a better view of the major soils can be seen. This effectively simplifies the coverage database and graphics.

The Attribute Recode and New Soil (top box) fields (columns) show that all A soils are recoded to 1, B to 2, and C to 3 in the recode operation.

NOTE: *Some GISs depend on numbers, whereas others can use just letters. This is also known as "data reduction"—reducing the data to a more generalized classification.*

The resultant Recoded Soils coverage (lower right) shows an easily understood spatial relationship. The major soil types occur together as groups and their spatial distribution is evident after generalizing the data.

Recoding is a standard preparatory step for many GIS operations, especially for overlaying and buffering, as will be seen. It is one of the most used and useful GIS procedures.

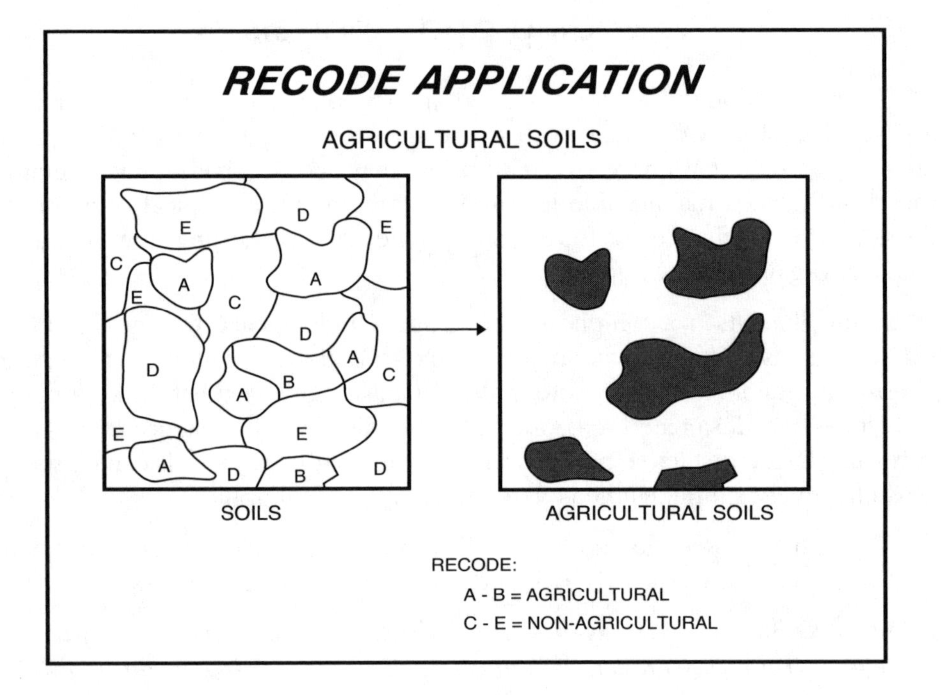
RECODE APPLICATION
AGRICULTURAL SOILS
E
D
E
C
A
A
E
C
D
D
A
C
B
A
E
E
A
D
B
D
SOILS
AGRICULTURAL SOILS
RECODE:
A - B = AGRICULTURAL
C - E = NON-AGRICULTURAL

Recode Application: Agricultural Soils

A simple yet very useful application of recoding is to reduce a spatially complex coverage to a generalized one. In the illustration, the simple task of recoding all A and B soils as agricultural and soils C through E as nonagricultural results in a coverage/map that is much easier to interpret than the original puzzle of soils. The Agricultural Soils coverage presents the soils suitable for cropping much more clearly than does the original Soils coverage. Also, the new database summarizes results so that quantitative management can begin, such as planning for fertilizer application. As noted, the steps involved in these tasks are rather simple, but the conclusions can be highly valuable.

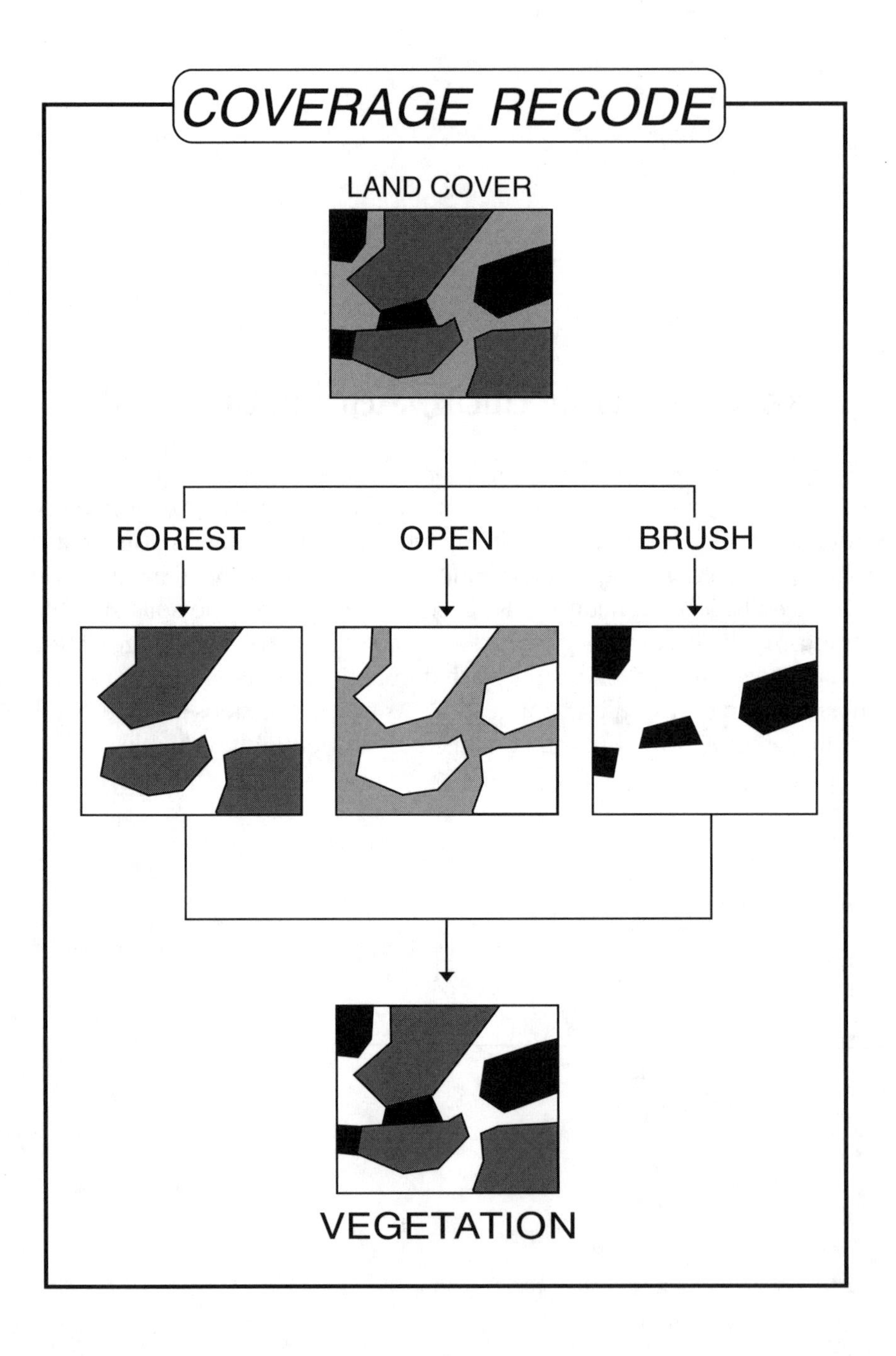
COVERAGE RECODE
LAND COVER
FOREST
OPEN
BRUSH
VEGETATION

Coverage Recode

Reduction of a complex coverage to a more simple coverage (including databases) is a relatively simple recode operation. Illustrated is the generalization of a Land Cover coverage to individual ecosystem themes. This is normally accomplished by recoding all features except the one of interest to 0, or by deleting them altogether. A common alternative is to move the selected feature to a new empty coverage, which is not a direct recode operation, but it produces the same effect.

Each theme can be viewed without data and visual clutter, permitting basic analysis of distribution, sizes, and relationships between each feature within a theme. Also, two selected feature types can be combined to test relationships between them, such as Forest and Brush, merged here to produce a new coverage, called Vegetation. These are basic analytical steps that can be very useful.

The original data file is not changed or lost (obviously an important consideration), but other coverages are made from it. Each coverage stands alone or can be used with other data sets as needed. A synergy of data is produced: all data in one coverage and separate files for each type of feature. Again, this can be a major preparation for more complex analysis.

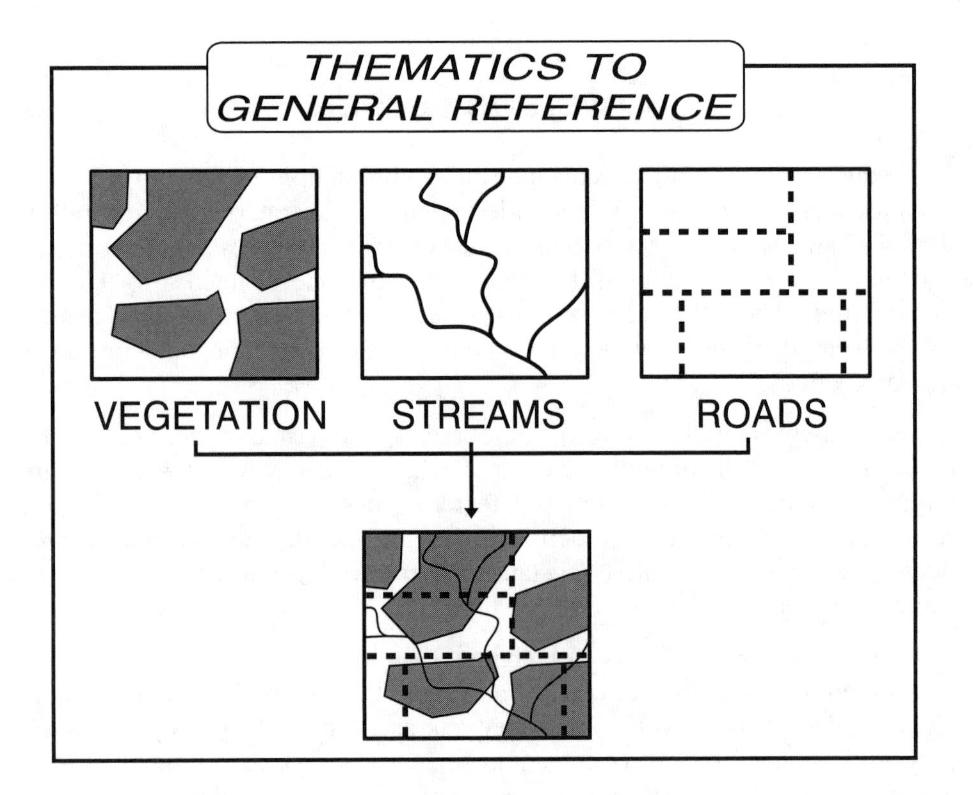
THEMATICS TO
GENERAL REFERENCE
VEGETATION
STREAMS
ROADS

Thematics to General Reference

This illustration shows the merging of various coverages to produce a multi-feature or multiple-theme coverage, even a general reference file. The most common method of merging or combining files in GIS is to overlay them, putting one onto another. Note in the illustration how the linear features, streams and roads, are on top of the polygons, so that all features can be seen. A GIS is very flexible and can show the features in almost any configuration. The following section is a discussion of overlay, probably the most common GIS operation.

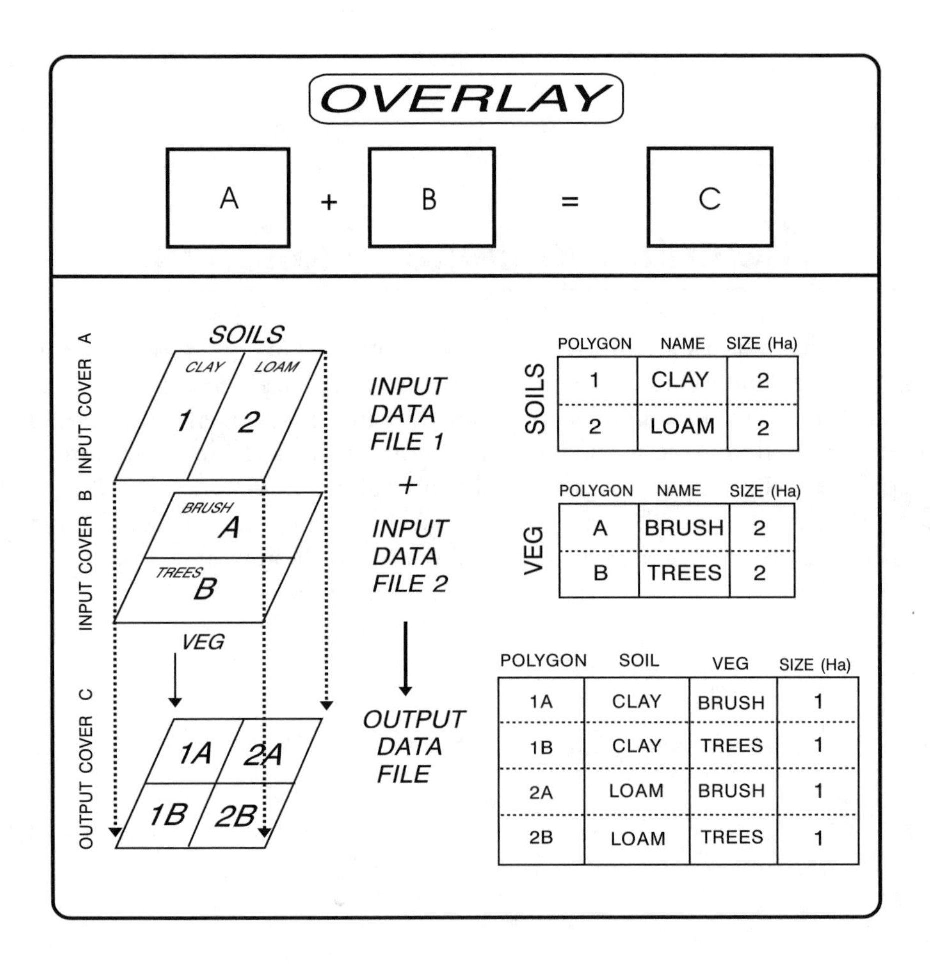
OVERLAY
A
+
B
=
C
INPUT COVER A
INPUT COVER B
OUTPUT COVER C
SOILS
CLAY
LOAM
1
2
BRUSH
A
TREES
B
VEG
1A
2A
1B
2B
INPUT DATA FILE 1
+
INPUT DATA FILE 2
OUTPUT DATA FILE
SOILS
POLYGON NAME SIZE (Ha)
1 CLAY 2
2 LOAM 2
VEG
POLYGON NAME SIZE (Ha)
A BRUSH 2
B TREES 2
POLYGON SOIL VEG SIZE (Ha)
1A CLAY BRUSH 1
1B CLAY TREES 1
2A LOAM BRUSH 1
2B LOAM TREES 1

Overlay

The process of overlay is straightforward. It means, simply, laying one coverage over another, as in coverage A + coverage B = coverage C (see illustration). As will be seen, there are several "tricks" that can be used in the process, but the A + B = C illustration at the top shows the basic operation.

The lower illustration shows the more detailed procedure. Follow the steps and compare coverages and databases. An overlay of Soils (input cover A, input data file 1) and Vegetation Cover (input cover B, input data file 2) should show the relationship between the two themes (output cover C, output data file). When soil types 1 and 2 (Clay and Loam) are combined with vegetation types A and B (Brush and Trees), the result is one of each possible combination, as seen in the output cover C: 1A, 2A, 1B, 2B.

The database results are equally important to the graphics. Note that the Input Data File databases identify each polygon and give their corresponding attribute name and size. The Output Data File database, however, presents the combinations that resulted: Clay-Brush, Clay-Trees, Loam-Brush, Loam-Trees. Note also the new, correctly recalculated sizes. A good GIS offers these products automatically, but some low-end GISs force the operator to do some of the database work.

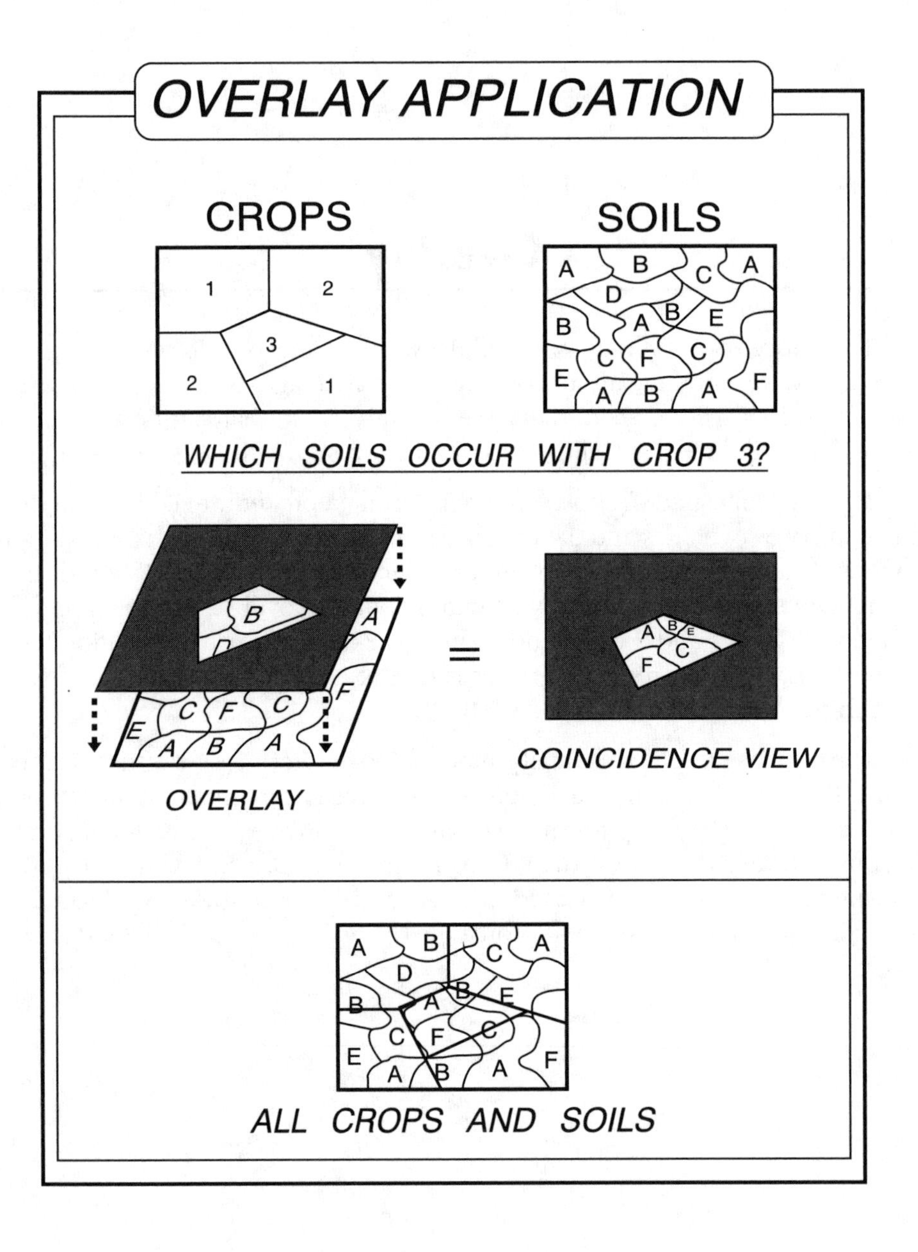
OVERLAY APPLICATION
CROPS
1
2
3
2
1
SOILS
A
B
C
A
D
B
A
B
E
C
F
C
E
A
B
A
F
WHICH SOILS OCCUR WITH CROP 3?
B
A
D
C
F
C
E
A
B
A
F
=
A
B
E
C
F
OVERLAY
COINCIDENCE VIEW
ALL CROPS AND SOILS

Overlay Application

There are many applications for GIS overlay, some of which test or try to find spatial relationships between and among features. The illustration deals with two coverages: Crops and Soils. The question "Which soils occur with Crop 3?" is answered by overlaying Crops on Soils. Several methods could be used, but one common technique is to isolate a feature (Crop 3) and then compare it with another set of features on a second coverage (Soils).

First, the Crop coverage is recoded to make a "window" of the Crop 3 area (usually code 0), leaving all other crops opaque (unseen). Then a simple overlay reveals the soils in the Crop 3 area—the spatial "coincidence" of the two coverages.

Most GISs have several overlay options, and some will be explored in the next few pages. The lower illustration shows an overlay of both coverages, using all of their features. Results can be rather complex, but GIS is good at this type of operation. For instance, a vector topological system may not need to use codes and recodes, but merely matches the two coverages and puts all combinations in a database. The next page shows an example.

OVERLAY OF CROPS AND SOILS

COINCIDENCE TABLE

POLYGON ID	CROP CODE	SOIL CODES	AREA (Ha)
1	1	A	10
		B	10
		C	5
		D	8
2	2	A	6
		B	7
		C	7
		D	3
		E	1
		F	4
3	2	A	6
		B	2
		C	8
		E	11
4	3	A	8
		B	3
		C	9
		E	6
		F	11
5	1	A	14
		B	6
		C	4
		F	11

Database Approach

The previous illustration showed a graphics approach to overlay, which is useful for some applications and often serves as data visualization for additional overlay information. Look at the original Crops coverage in that illustration and note, from top left, polygons with crops 1, 2, 2, 3, and 1. Each polygon will have a unique ID number in the database (probably in order of digitization), from 1 to 5 in this case. Many GISs do not need the graphics or visual part of the coverage to combine files; they can do the "overlay" at the database.

The database approach to overlay can provide more quantitative data. Often, no recoding is necessary to achieve overlay. Sometimes a simple command to combine the two coverages is all that is required (particularly in vector topological systems).

A list of all polygon coincidences (spatial matches) of the overlay is produced in a new database or table. The overlay table in the illustration lists each crop polygon by its ID number, along with the crop code (in the sequence previously mentioned). The overlay matches each polygon with the soils that occur within it. Also listed is the area of each match. Finding the soils for crop 3 involves only browsing the table and reading the details. A coverage could be generated that would include every crop-soil match and list each as a unique polygon; or, if selected, just a crop 3 polygon with its matching soils.

The graphics and database approaches to overlay are not mutually exclusive; they often occur together. Both are essential in a good GIS.

NOTE: *This is one reason polygon IDs are important. There are times when each polygon should be considered separately, regardless of its attribute classification or code.*

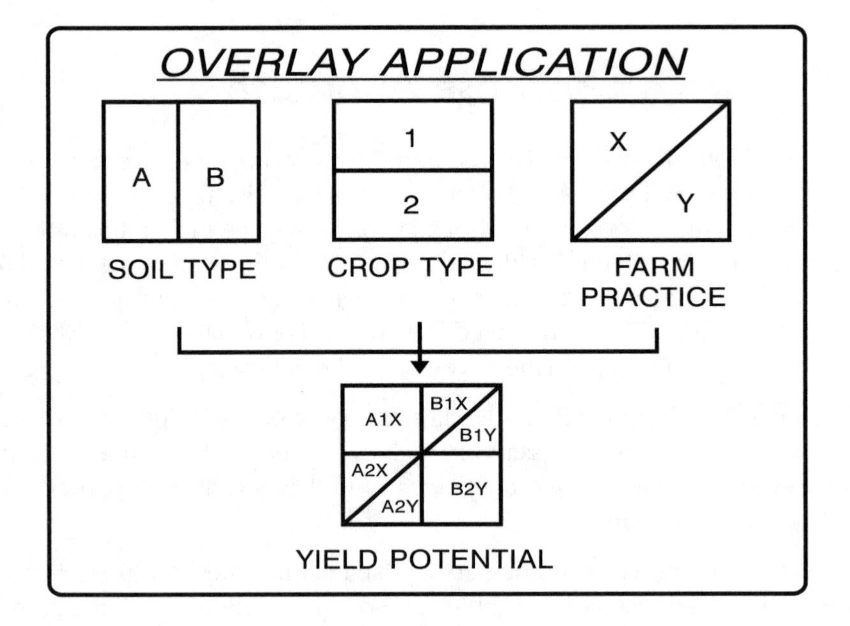
OVERLAY APPLICATION
A
B
SOIL TYPE
1
2
CROP TYPE
X
Y
FARM PRACTICE
A1X
B1X
B1Y
A2X
A2Y
B2Y
YIELD POTENTIAL

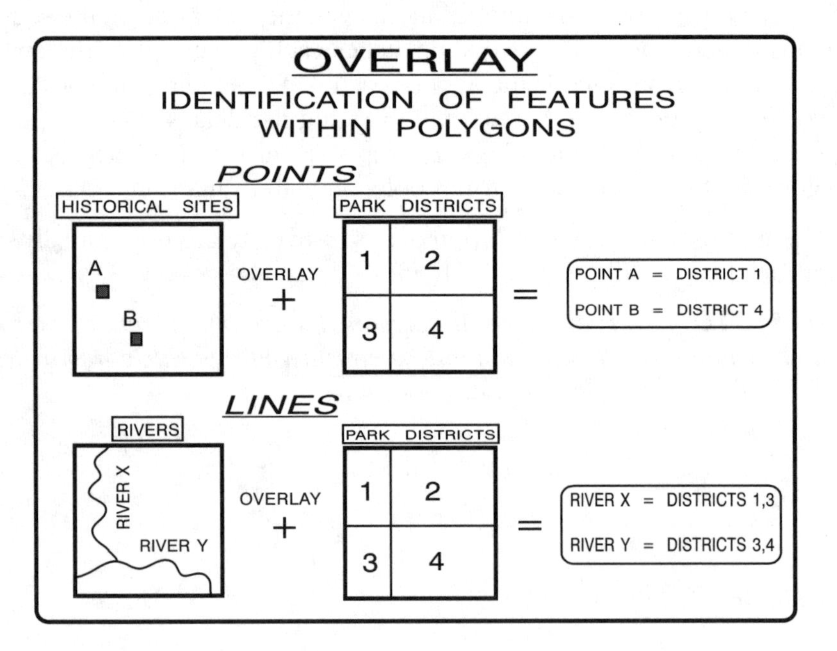
OVERLAY
IDENTIFICATION OF FEATURES WITHIN POLYGONS
POINTS
HISTORICAL SITES
A
B
OVERLAY
+
PARK DISTRICTS
1
2
3
4
=
POINT A = DISTRICT 1
POINT B = DISTRICT 4
LINES
RIVERS
RIVER X
RIVER Y
OVERLAY
+
PARK DISTRICTS
1
2
3
4
=
RIVER X = DISTRICTS 1,3
RIVER Y = DISTRICTS 3,4

Overlay Application: Yield Potential

Overlay is not limited to two coverages, although some GISs can manage only one pair at a time. Here is a three-coverage overlay that combines soils, crop, and farm practices to predict or to help understand yield potential. Presumably, the user has a good idea (or at least is testing) how the three factors interact.

The output Yield Potential coverage shows the spatial coincidence of the three factors, which can then be used in the analysis of their influence on and control of yield. An accompanying database will provide additional details.

NOTE: *It must be remembered that GIS is only a tool and that knowledge of the applications is more important than knowing the technicalities of GIS operations. Without such an understanding, overlay is merely an exercise in fancy graphics. Overlays are simple to do and appear sophisticated, yet the driving reason for doing them is in the applications, not the computer operations. This overlay operation, for example, probably makes little sense to anyone but the researcher, who is trying to find which factors affect yield.*

Identify Features Within Polygons

One effective application of overlay is the identification of features from one coverage within polygons of a second coverage (lower illustration). Mixing points, lines, and polygons usually does not present problems. Illustrated first is the matching of historical sites to park districts. In this simple example it may be easy to pair them visually, but can you easily tell where historic site B goes? However, consider using 50 historical sites and 10 districts, and the need for GIS becomes clear. The real value of this overlay is not just its visual aspect but its help in establishing a digital identification of which historic sites occur in which districts (i.e., spatial relationships for further use).

The second diagram is similar—in this case for use in identifying which districts each river flows through. River lines may cross several polygons and the overlay product will reflect the multiple-feature relationships. As noted, this type of overlay can be accomplished with the database method, resulting in a table of data rather than a new coverage. The coverage is better visual information, whereas the database gives the best quantitative data. Both are useful.

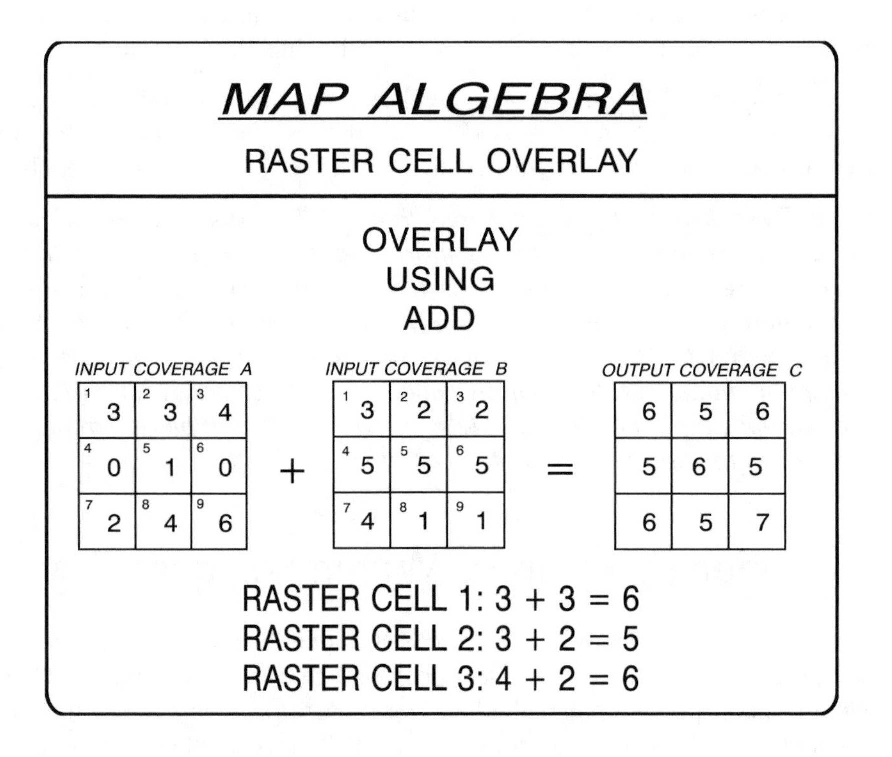
MAP ALGEBRA
RASTER CELL OVERLAY
OVERLAY
USING
ADD
INPUT COVERAGE A
1 3
2 3
3 4
4 0
5 1
6 0
7 2
8 4
9 6
+
INPUT COVERAGE B
1 3
2 2
3 2
4 5
5 5
6 5
7 4
8 1
9 1
=
OUTPUT COVERAGE C
6 5 6
5 6 5
6 5 7
RASTER CELL 1: 3 + 3 = 6
RASTER CELL 2: 3 + 2 = 5
RASTER CELL 3: 4 + 2 = 6

Map Algebra and Overlay

Map Algebra: Raster Cell Overlay

So far, overlay has been discussed as simply putting one coverage on another. In powerful topological vector systems this may be all that is needed because detailed results are automatically produced. This is a major advantage to such systems and many users gladly pay the higher price for more capable operations.

But systems that use coded attribute values, particularly raster structures, may need additional work. If numbers are used (or names that can be converted to numbers), instructions must be given on how to deal with the interaction of each cell pair. That is, when cell 1 of coverage A meets cell 1 of coverage B, a mathematical operation is necessary in order to assign a code for cell 1 of coverage C, the output file. The use of mathematical operations is called map algebra and usually includes Add, Subtract, Multiply, Divide, Exponentiate, and others. We will see that this is useful in GIS modeling.

Illustrated is a simple view of nine raster cells from two coverages and their interaction when overlaying with an Add map algebra operation. The Add operation means that input coverage code values are added to produce an output coverage sum in each cell. Think about that: the value of cell 1 in the first coverage is added to the cell 1 value of the second, overlaid coverage to produce a cell 1 sum in the output third coverage. Input coverage A cell 1 value plus input coverage B cell 1 value equals output coverage C cell 1 value.

The small number in each cell's upper left is the cell number (sometimes referred to by Row-Column sequence), and the large number is the attribute value. Coverage A and coverage B are paired (overlaid) and the attribute value in each is added, with the sum showing in the corresponding cells of output coverage C. For example, in cell 1, coverage A's 3 is added to coverage B's 3 to produce coverage C's sum of 6. Study the other cells. Raster overlays are very efficient and fast because the program has only the simple task of adding two numbers.

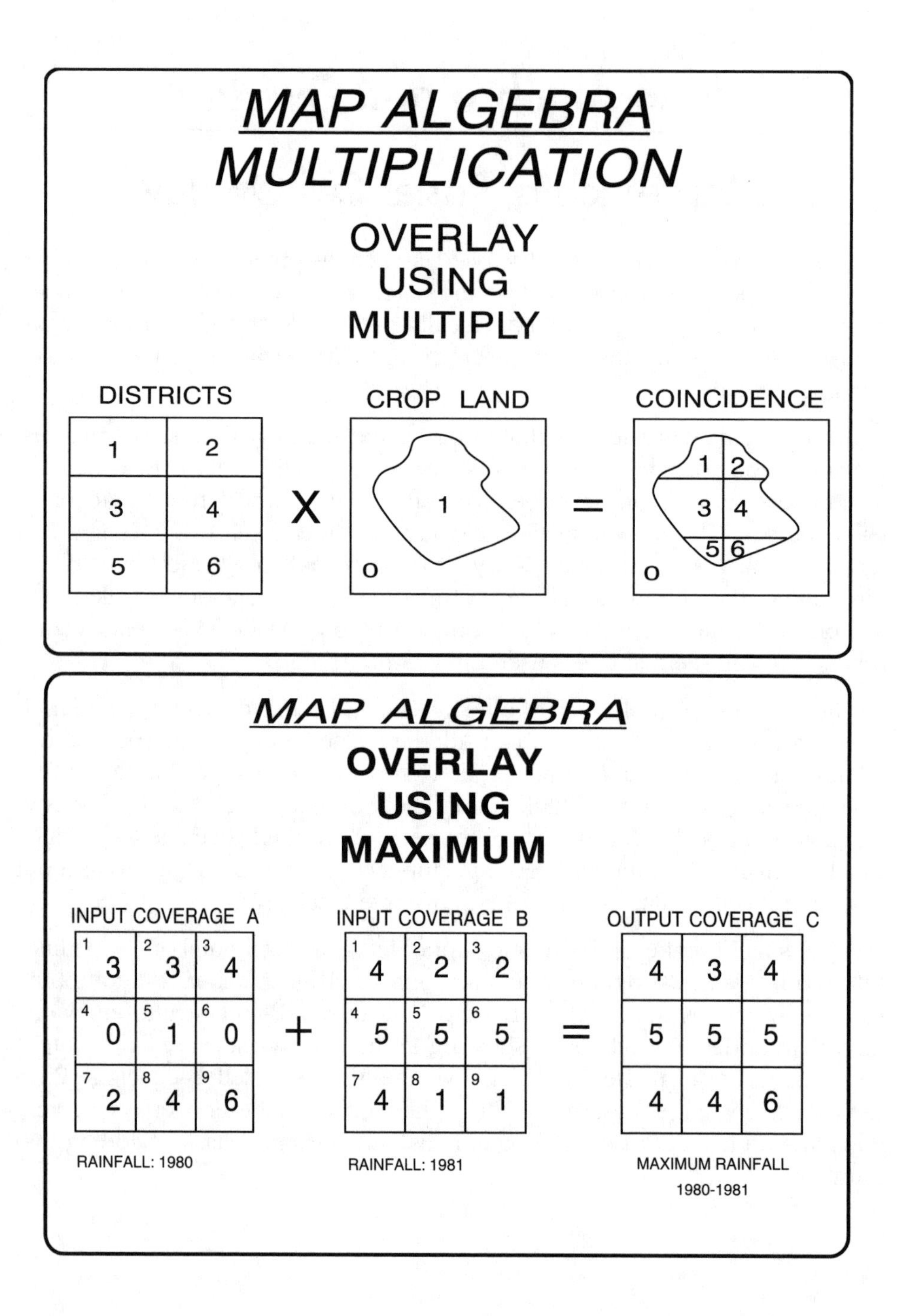
MAP ALGEBRA
MULTIPLICATION
OVERLAY
USING
MULTIPLY
DISTRICTS
1
2
3
4
5
6
X
CROP LAND
1
O
=
COINCIDENCE
1 2
3 4
5 6
O
MAP ALGEBRA
OVERLAY
USING
MAXIMUM
INPUT COVERAGE A
1 3 2 3 3 4
4 0 5 1 6 0
7 2 8 4 9 6
RAINFALL: 1980
+
INPUT COVERAGE B
1 4 2 2 3 2
4 5 5 5 6 5
7 4 8 1 9 1
RAINFALL: 1981
=
OUTPUT COVERAGE C
4 3 4
5 5 5
4 4 6
MAXIMUM RAINFALL
1980-1981

Map Algebra: Multiplication

Thinking in terms of map algebra can make the overlay process efficient and convenient. Consider the illustrated application (top) of showing crop land in political districts. The districts are numbered 1 through 6 (perhaps codes for a set of names) and crop land is coded 1 (non crop land = 0). There are several possible approaches, but the simplest in this case is to overlay using Multiply. Anywhere a district meets non crop land (0), the obvious product will be 0, but because crop land is value 1, each district retains its original number (N x 1 = N). The results are easily interpreted: where there was no crop land (code 0), the districts are not noted; where there is crop land (code 1), the corresponding number of the district results. Pretty clever stuff.

Other math options could be used, such as Add or Subtract, but an additional recode of results may be needed. (Work out an Add to test this statement.)

➡ **NOTE:** *Map Algebra permits overlay of more than two coverages, but Multiply and Divide can get difficult when using three or more. Add and Subtract usually work best in such cases.*

Map Algebra: Maximum

Determining the maximum number between two or more cells is very easy for a computer. Overlay using Maximum is useful for numerous applications. It compares the coverages cell by cell, finding the highest number and placing it in the output coverage value. Study the lower illustration to see how easy this is. It shows only nine raster cells of rainfall coverages (usually there are thousands of cells, making overlay by computer necessary), resulting in the maximum recorded rainfall for the period used. Of course, there could be more than two coverages in this operation, but the computer merely selects the maximum value for a given cell. Easy.

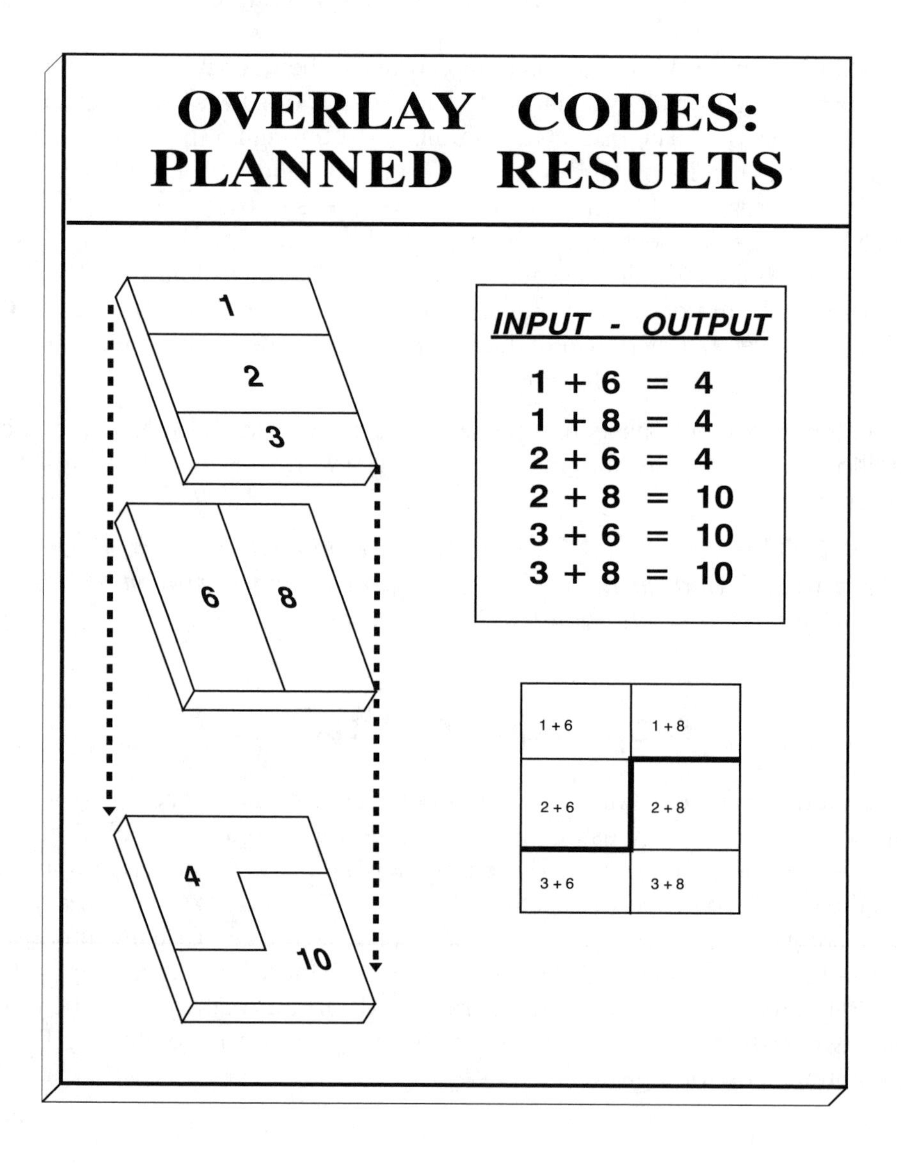
OVERLAY CODES:
PLANNED RESULTS
1
2
3
6
8
4
10
INPUT - OUTPUT
1 + 6 = 4
1 + 8 = 4
2 + 6 = 4
2 + 8 = 10
3 + 6 = 10
3 + 8 = 10
1 + 6
1 + 8
2 + 6
2 + 8
3 + 6
3 + 8

Overlay Codes: Planned Results

Some better GISs permit assignment of overlay results rather than relying on map algebra to perform calculations and determine results. That is, the GIS presents a set of choices such as the following:

Coverage A code 1 + coverage B code 1 = ?

Coverage A code 2 + coverage B code 1 = ?

Coverage A code 3 + coverage B code 1 = ?

Coverage A code 1 + coverage B code 2 = ?

Coverage A code 2 + coverage B code 2 = ?

Coverage A code 3 + coverage B code 2 = ?

At the "?" the user enters the *desired* output codes, and the program performs the work (isn't that what computers are for?). The user may use one output code as many times as needed.

Illustrated is a generic application, with the top coverage coded 1 through 3 (perhaps they are increasing population densities), and its overlay "partner" coded 6 and 8 (magnitudes of poverty). For some reason, the user needs codes 4 and 10, probably for additional analysis later in the project. The Input-Output box shows the operation, with the user providing the 4 and 10 codes when prompted by each possible combination of overlay matches.

Results will be as desired rather than having to conform to algebraic limitations. This avoids a final recode from map algebra codes to the desired codes; it is a simple process.

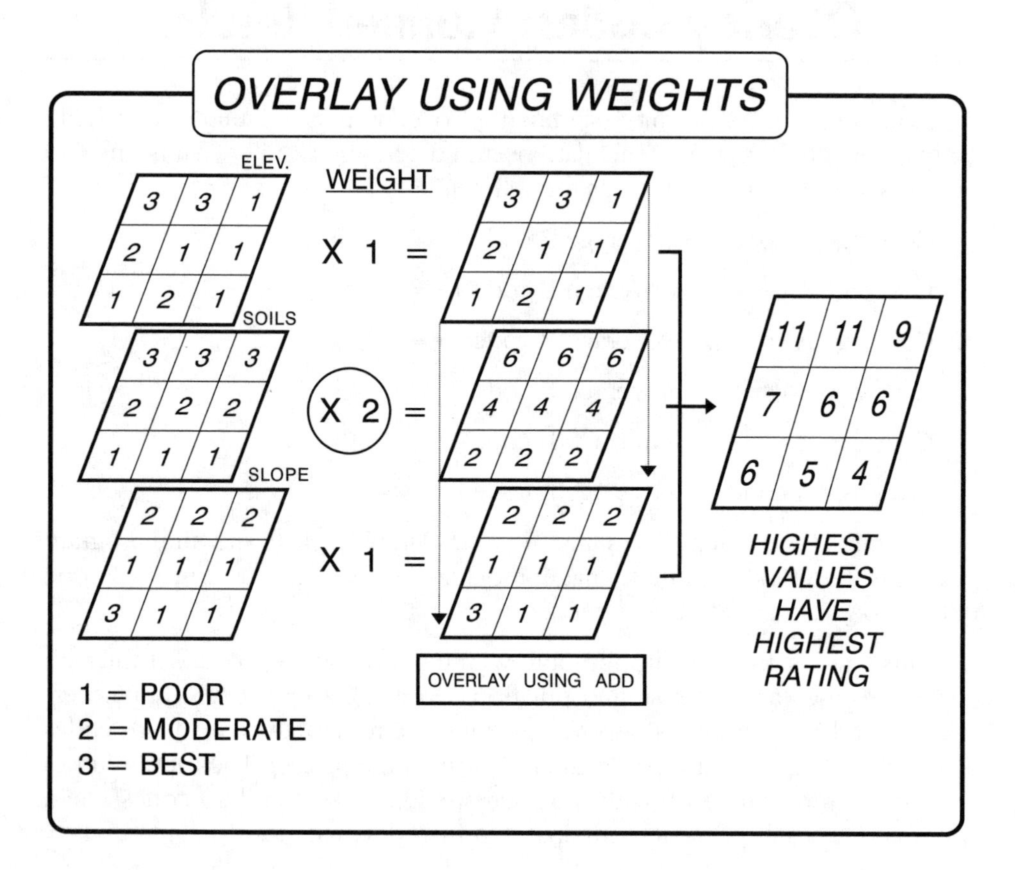
OVERLAY USING WEIGHTS
ELEV.
3 3 1
2 1 1
1 2 1
SOILS
3 3 3
2 2 2
1 1 1
SLOPE
2 2 2
1 1 1
3 1 1
WEIGHT
X 1 =
X 2 =
X 1 =
3 3 1
2 1 1
1 2 1
6 6 6
4 4 4
2 2 2
2 2 2
1 1 1
3 1 1
OVERLAY USING ADD
11 11 9
7 6 6
6 5 4
HIGHEST VALUES HAVE HIGHEST RATING
1 = POOR
2 = MODERATE
3 = BEST

Overlay Using Weights

Giving extra importance to some coverages or features is often necessary, such as when one theme is more important than others in a particular overlay application. The user could recode the important coverage to an appropriate value to get the proper attribute value "weight," but that involves additional time-consuming, error-prone work. Why not let the computer provide the adjustment quickly and easily? Many GISs allow a weighting factor in connection with overlay.

In the illustration, an area is under investigation for possible agricultural development, and the best sites need to be found. All input coverages have been recoded to Poor, Moderate, and Best categories (designated 1 through 3, respectively, on each coverage). However, Soils are more important than Elevation or Slope, and therefore its values are given twice (2x) the weight in the overlay (circled). The coverages are added, and results show that the highest values have the highest ratings. The 2x weighting of Soils gave that theme extra influence in the overlay process.

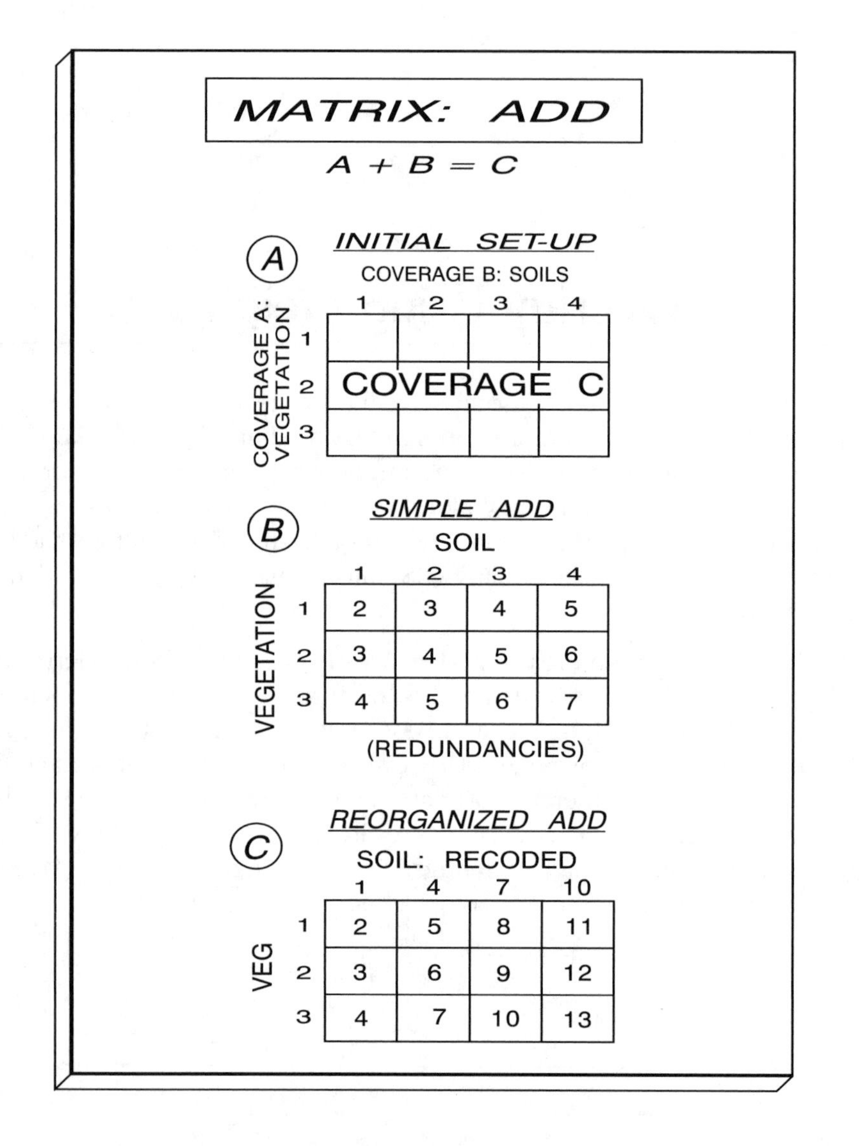

Matrix: Add

Overlaying raster coverages that use number codes is rather easy, yet it requires careful planning to understand results. We've used overlay with Multiply and Maximum, and you can see how the process works. To control or predict results, an organized approach is needed.

In this illustration, we are preparing to overlay, using Add, two coverages—A (Vegetation) and B (Soil)—to produce new coverage C. We need to make sure that C's numbers make sense and that we understand how they are made. A matrix table that pairs the two coverages is written on paper to see the various possible combinations. Figure A shows the basic setup: coverage A, Vegetation, is presented as rows, and B, Soil, as columns. Coverage C resultant numbers will go in the matrix. Follow the numbers.

Vegetation is coded 1 through 3, and soils 1 through 4. If we do a simple overlay with Add, the results of each combination are presented in matrix B. For example, all soil code 2s overlaying vegetation code 1 will sum to code 3 in matrix C. Examine the matrix for the other sums.

The problem here is that there will be redundancies—identical numbers for two or more input combinations. Coverage C code 3 can come from either Vegetation 1 + Soil 2 or Vegetation 2 + Soil 1. If code 4 is seen in C, we cannot be sure which of three overlay combinations it came from: Soil 1 + Vegetation 3; Soil 2 + Vegetation 2; or Soil 3 + Vegetation 1. Better analysis would result if each C code were unique, a sum of only one possible soil-vegetation pair.

Therefore, we must recode one of the input coverages so that unique codes are produced for output coverage C. In this case, we use the matrix to help decide the most efficient new set of numbers for soils (vegetation will not be recoded). Follow the logic and numbers carefully: we put the desired numbers in C and set (recode) the input coverage numbers. Let's try a simple sequence of numbers in C, starting with 2 and filling in the matrix column by column, up to 13.

The first column remains untouched—the numbers progress from 2 to 4. If we want a 5 in C, the new soil code number should be 4 (so we recode Soil 2 to 4). Note how 4 plus each Vegetation number sums to the desired C numbers. To get 8 in C, we must recode Soil 3 to 7. Using the same procedure, old Soil 4 will be recoded to 10 in order to finish the proper C-matrix numbers.

Thus, the Soil recode will be: 1 = 1 (no change), 2 = 4, 3 = 7, and 4 = 10. The matrix shows that coverage C will have individual numbers based on unique combinations, and we will know the origin for each output code. Examine the matrix at bottom; the process and results are not as confusing as they sound.

The use of Matrix is fairly simple but requires care. It can be used with most overlay options.

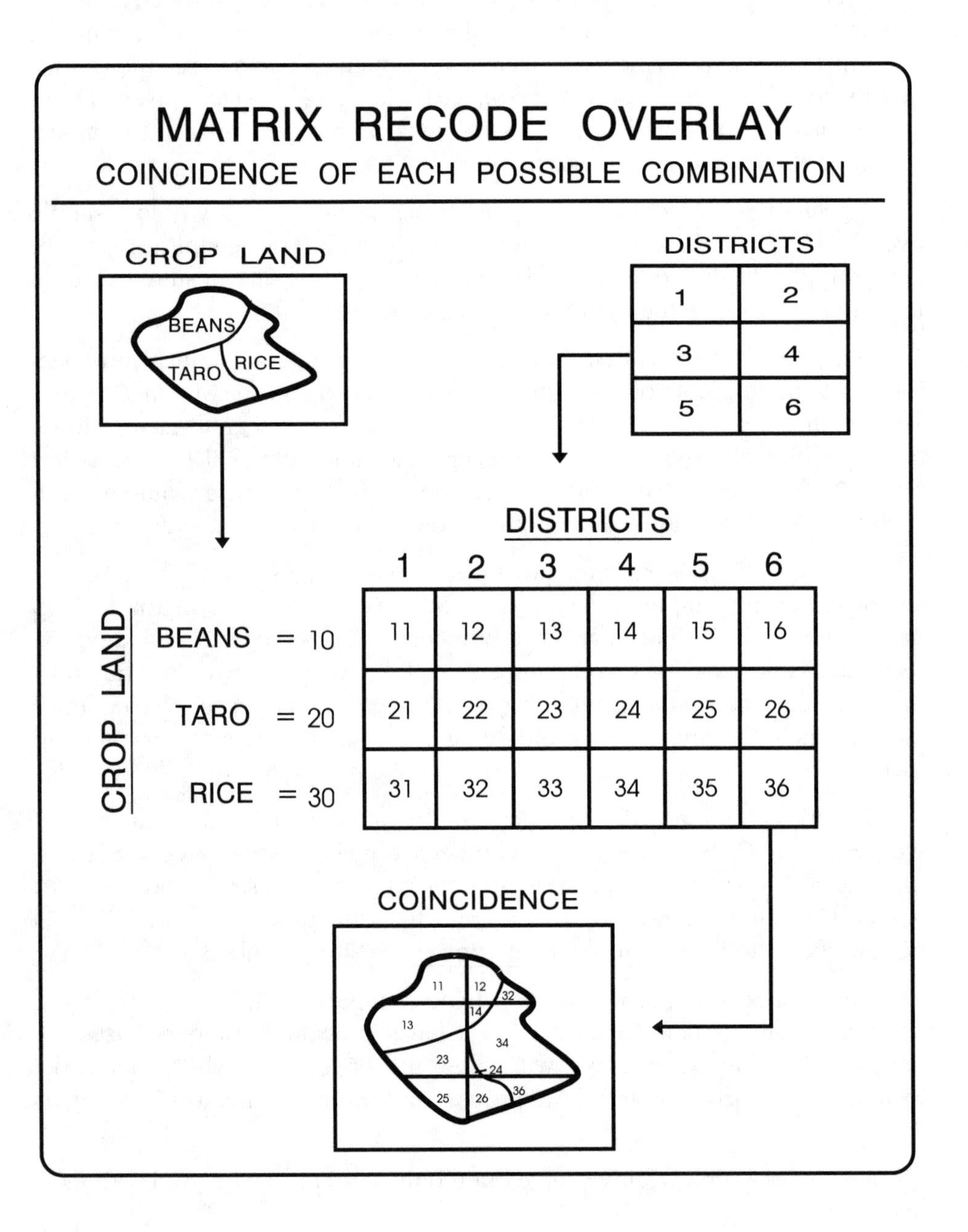
MATRIX RECODE OVERLAY
COINCIDENCE OF EACH POSSIBLE COMBINATION
CROP LAND
BEANS
TARO
RICE
DISTRICTS
1
2
3
4
5
6
DISTRICTS
1 2 3 4 5 6
CROP LAND
BEANS = 10
TARO = 20
RICE = 30
11 12 13 14 15 16
21 22 23 24 25 26
31 32 33 34 35 36
COINCIDENCE
11
12
32
14
13
34
23
24
25
26
36

Matrix Recode Overlay

The illustration is an example of an applied Matrix—overlay of Crop Land with Districts. The districts are six rectangles (from a previous illustration using overlay with Multiplication), and we wish to know which crops occur in each. The matrix is set up with crops as rows and districts as columns. Overlay with Add is used, and the results are shown at the bottom of the illustration.

A regular sequence of numbers could be tried as before, but a more clever system is used. To ensure fast and easy interpretation of crop-district combination in the final product, we will recode crops to 10, 20, and 30. This shows that each tens number (11 through 16) came from crop 1, each twenties number (21 through 26) from crop 2, and each thirties number (31 through 36) from crop 3. Also, the last digit in those numbers reflects the district number. It is a clever coding: first number = crop, last number = district. Thus, 25 is crop 2 and district 5.

This is a good system and the computer easily uses 18 numbers, but the human eye has difficulty tracking that many colors or shades on a monitor or map. One tactic to employ in this case might be to combine members of a logical set, such as coding all Beans to a shade, perhaps blue, with or without a corresponding symbol (e.g., cross-hatch lines), Taro to red, and Rice to yellow. This allows the eye to see all major colors as crops, and perhaps particular shades or symbols as individual districts.

Conversely, if the emphasis is on the district, they could be specific colors and the crops could each be a special pattern (easy when only three crops are used). Output presentation considerations are very important, particularly if data visualization is a part of project production. But again, the computer has no more difficulty with 180 distinct codes than it does with 18; a change may not be needed.

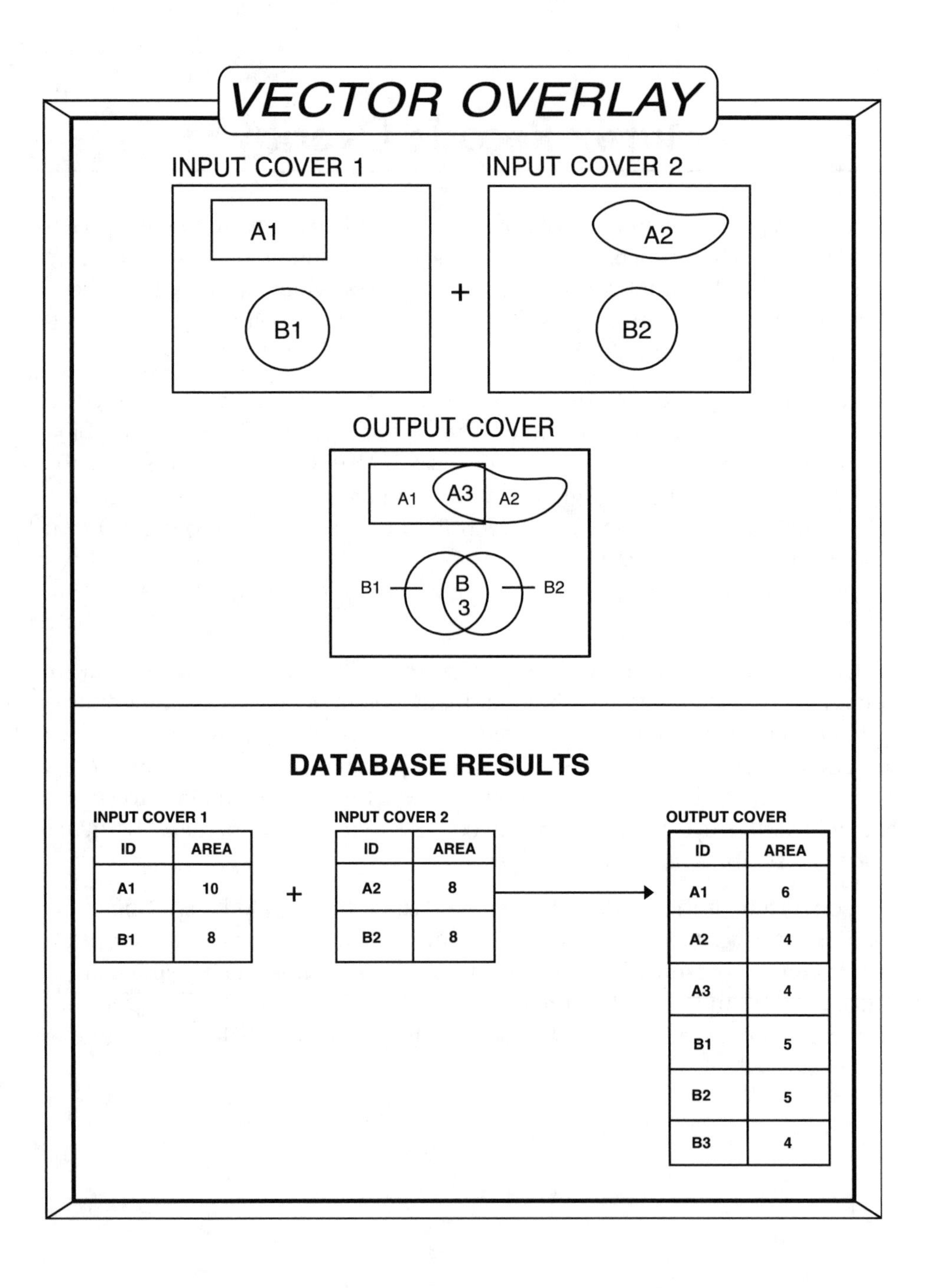

INPUT COVER 1

ID	AREA
A1	10
B1	8

INPUT COVER 2

ID	AREA
A2	8
B2	8

OUTPUT COVER

ID	AREA
A1	6
A2	4
A3	4
B1	5
B2	5
B3	4

Vector Overlay

We've been involved with raster coding, which requires some thought and preparation. Many GISs use that system, but most topological vector systems do not require number conversions and matrix preparations. Usually vector overlays are simply executed and the results automatically recognized, including new polygons.

The illustration shows input cover 1 overlaid with input cover 2 to produce the output cover. Where features from the input coverages overlap, new polygons are created in the output coverage. A combination of A1 and A2, for example, is produced as A3. An accompanying database adds the new polygons and adjusts attributes for the original polygons if needed, such as area and perimeter.

You can appreciate the ease and convenience of this type of overlay, whereby troublesome numerical calculation and recoding are avoided. This is one reason many GIS users prefer vector topological systems despite their higher cost and other demands. But also remember the advantages of raster systems, such as easy use of remote sensing imagery. Some applications use enough imagery to make a raster GIS the best choice. Also, because raster systems use numbers, they are better for numeric modeling.

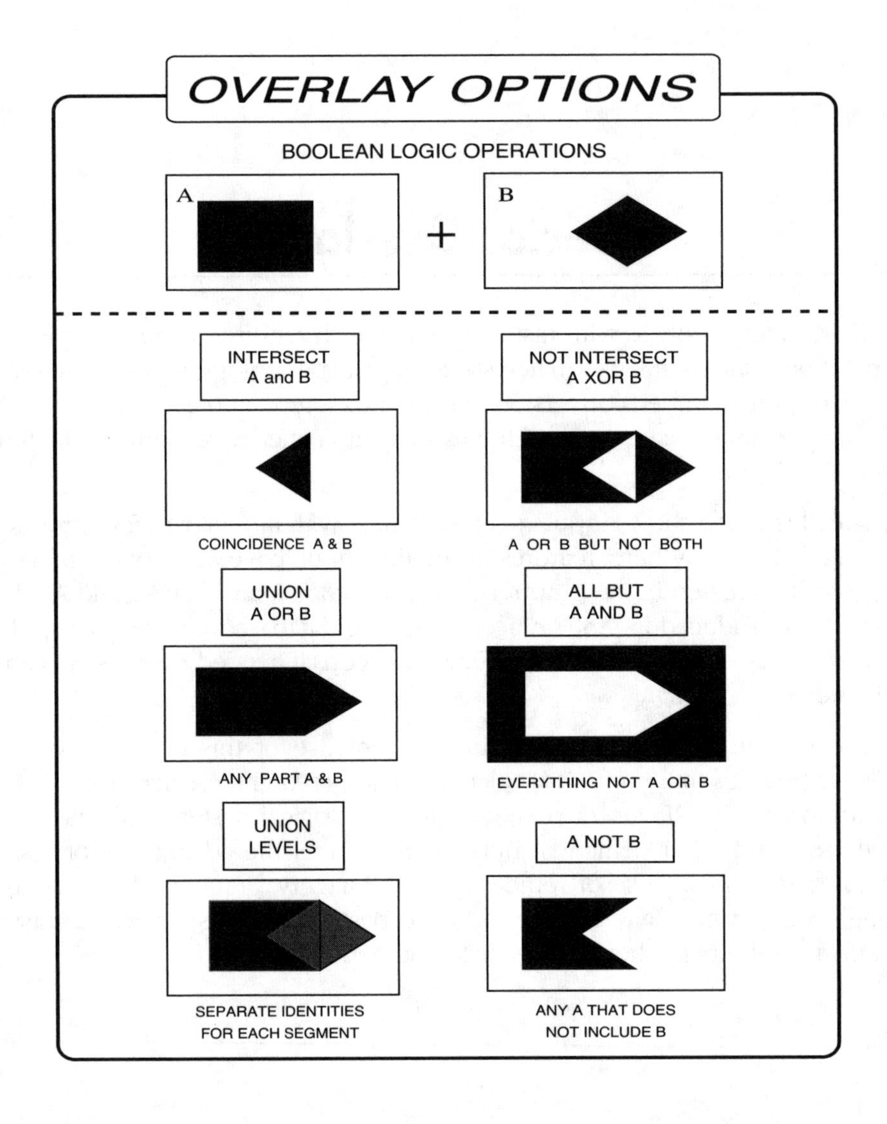

Overlay Options

There are several ways of viewing and working with overlays, and a major concern is how features on each coverage match or relate to one another. Recall the Boolean concept, wherein features are designated as Yes-

No; that is, having a particular attribute or not having it (a binary decision). This can be used in overlay planning. Numerous Boolean logic overlay options exist, and six of the more useful are illustrated. These are usually available in vector systems rather easily, but raster systems may take more preparation, such as recoding.

First, we presume that each coverage has been recoded or reduced to only those features needed for analysis—a common preparation for overlaying. In the upper illustration, feature A is an agricultural field and B is the application area of a special fertilizer. Here the coverages have been simplified to a single feature in each, but in reality there may be many features of a given theme—many fields and several fertilizer sprayings. Note that the fertilizer has been applied to only part of the agricultural area; feature A only partially matches (coincides with) the area of feature B. The Boolean overlay options include the following:

- *Intersect; A and B:* Where the fertilizer hit the crops. In some cases it is important to know where features in coverage A coincide, or intersect, with those in coverage B. To understand the effects of the fertilizer on plant growth, we need to know which areas received it and which did not. Therefore, the Intersect option shows the small triangle area of agriculture that received the fertilizer.

- *Not Intersect; A or B:* A or B, but not both. This includes where two features exist, but does not include where they intersect—the opposite of Intersect. That is, Agriculture and Fertilizer, but not where both occur together.

- *Union; A or B:* Any part of A and B. All of any features that touch or intersect, even partially. This could include all properties that have received some fertilizer, even if the total area of the properties did not get covered.

- *All But A or B:* Everywhere except the union properties.

- *Union Levels:* A, B, and their intersect. These are separate identities for each segment of the union. The application and effect of fertilizer on each part may be different; therefore, each should be identified.

- *A Not B:* Any part of A that does not include B. All crops (A) that have not received fertilizer (B).

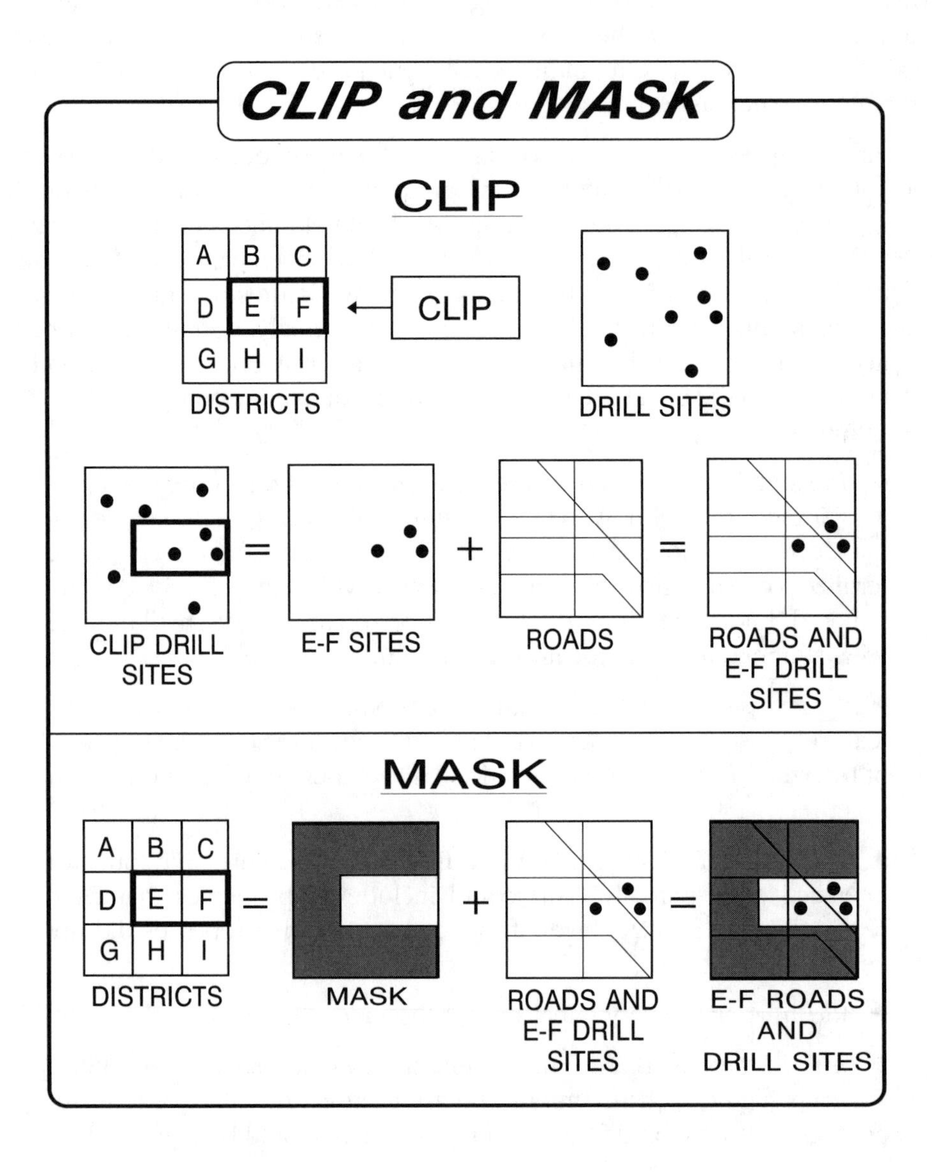
CLIP and MASK
CLIP
A
B
C
D
E
F
G
H
I
CLIP
DISTRICTS
DRILL SITES
=
+
=
CLIP DRILL SITES
E-F SITES
ROADS
ROADS AND E-F DRILL SITES
MASK
A
B
C
D
E
F
G
H
I
=
+
=
DISTRICTS
MASK
ROADS AND E-F DRILL SITES
E-F ROADS AND DRILL SITES

Clip and Mask

Clip

Clip is a standard GIS option that removes part of a coverage for analytical operations without making a separate coverage. Think of it as a temporary coverage. This is useful when only a small part of a coverage is needed temporarily. Clip provides a "subset" of a coverage to use as a "mold" or "cookie cutter." Think: what if you need a portion of a coverage (maybe to overlay onto a second coverage) that had features of the same code or name as features in other parts of the coverage? You can't do a recode operation to separate the needed features unless you have a powerful database and the patience to find each listing in order to type in changes. Simply drawing a box around your needed area will be much easier. This is known as clipping.

The top row of the upper illustration shows a Districts coverage, with districts E and F to be clipped. Petroleum Drill Sites are in the accompanying coverage. We want to separate the drill sites of districts E and F and use them later for environmental management purposes.

The second row shows an overlay of the Districts E-F clip on the Drill Sites (Clip Drill Sites), making a separate E-F Sites coverage. Then an overlay of it with Roads makes a Roads and E-F Drill Sites coverage.

In some GISs, Clip offers Include or Exclude options. These allow the user to include an area and features of an outlined area, as in the illustration, or to exclude them, leaving the remainder of the coverage.

Mask

Mask is a type of Clip, where the designated section or piece is converted to a transparent window and the remaining area is opaque. Overlay with a mask results in only the window showing features. This would normally involve coding the window area to 0, and the remainder of the area to a high value blocking out surrounding features.

The lower illustration shows the E-F Clip converted to a Mask. Then it's overlay with the Roads and E-F Drill Sites coverage makes an E-F Roads and Drill Sites product to isolate the selected window area. Follow these steps carefully and you should see a relatively simple process: Clip and then proceed with overlays. The Crop 3/Soils illustration under the previous Overlay Application section is another example of Mask.

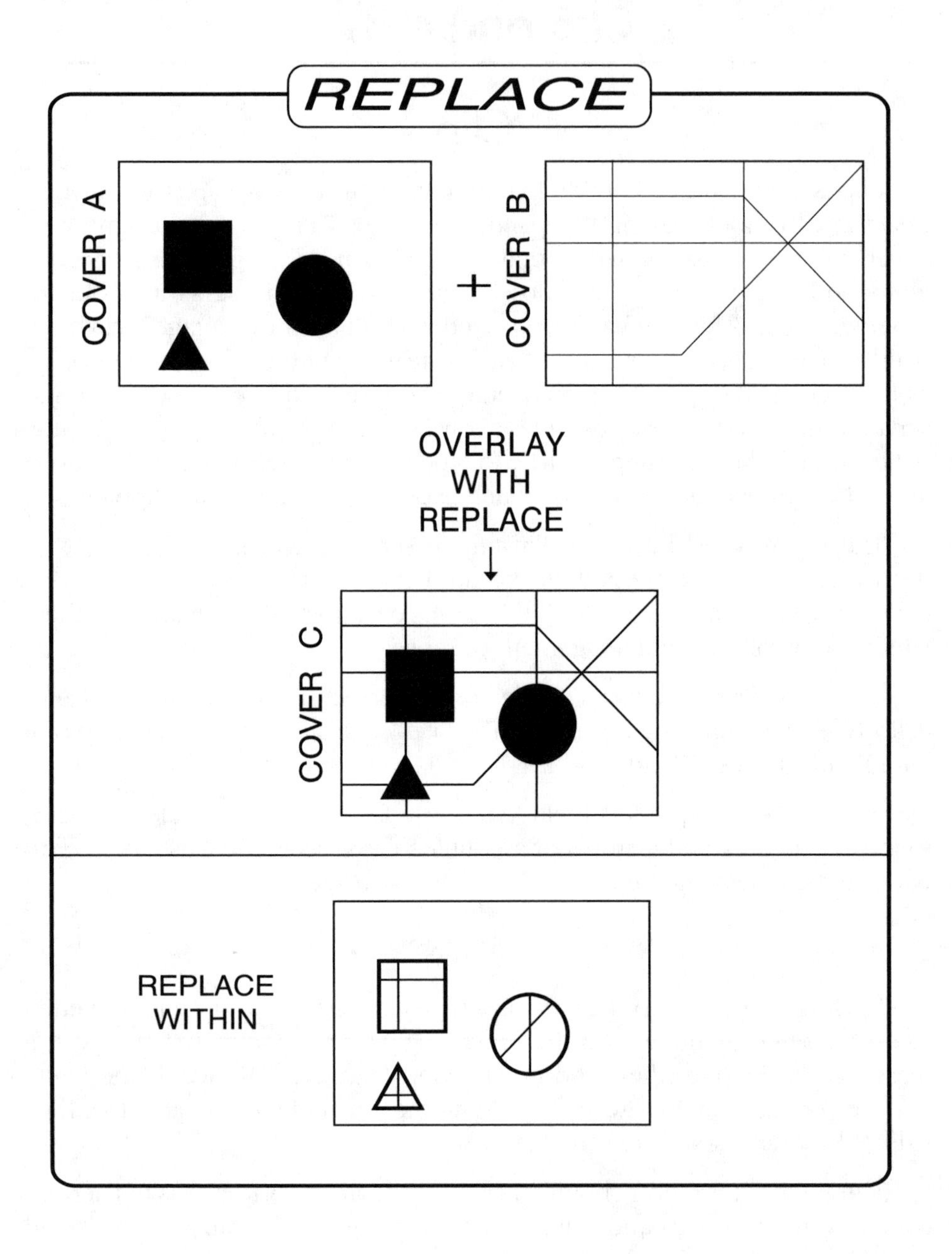
REPLACE
COVER A
+
COVER B
OVERLAY
WITH
REPLACE
COVER C
REPLACE
WITHIN

Replace

When all features of one coverage are needed on another, they can be transferred with Replace (also called Cover in some GISs). In the upper illustration, the features of coverage A go onto coverage B, and recoding is not necessary. The selected features of A are placed onto B, replacing any features on B they overlay. Replace saves a number of steps for some overlay tasks. Study the steps. Note that coverage A cannot be completely covered with features; otherwise, coverage C would simply become a copy of A.

The lower illustration shows a reverse of replacing features by using the selected features as a mask to show what is within. This is an example of Replace Within.

NOTE: *A recode is not necessary, particularly in vector systems, but a bit of confusion can result from identical feature code numbers from both input coverages. If the square in cover A (agriculture) is coded 1 and some roads in cover B are also code 1, then cover C code 1 will be both agriculture and roads, normally not an acceptable data situation. There will be some resultant confusion in identification and statistics. In this case, a recode of one of the input coverages is suggested in order to keep all features unique.*

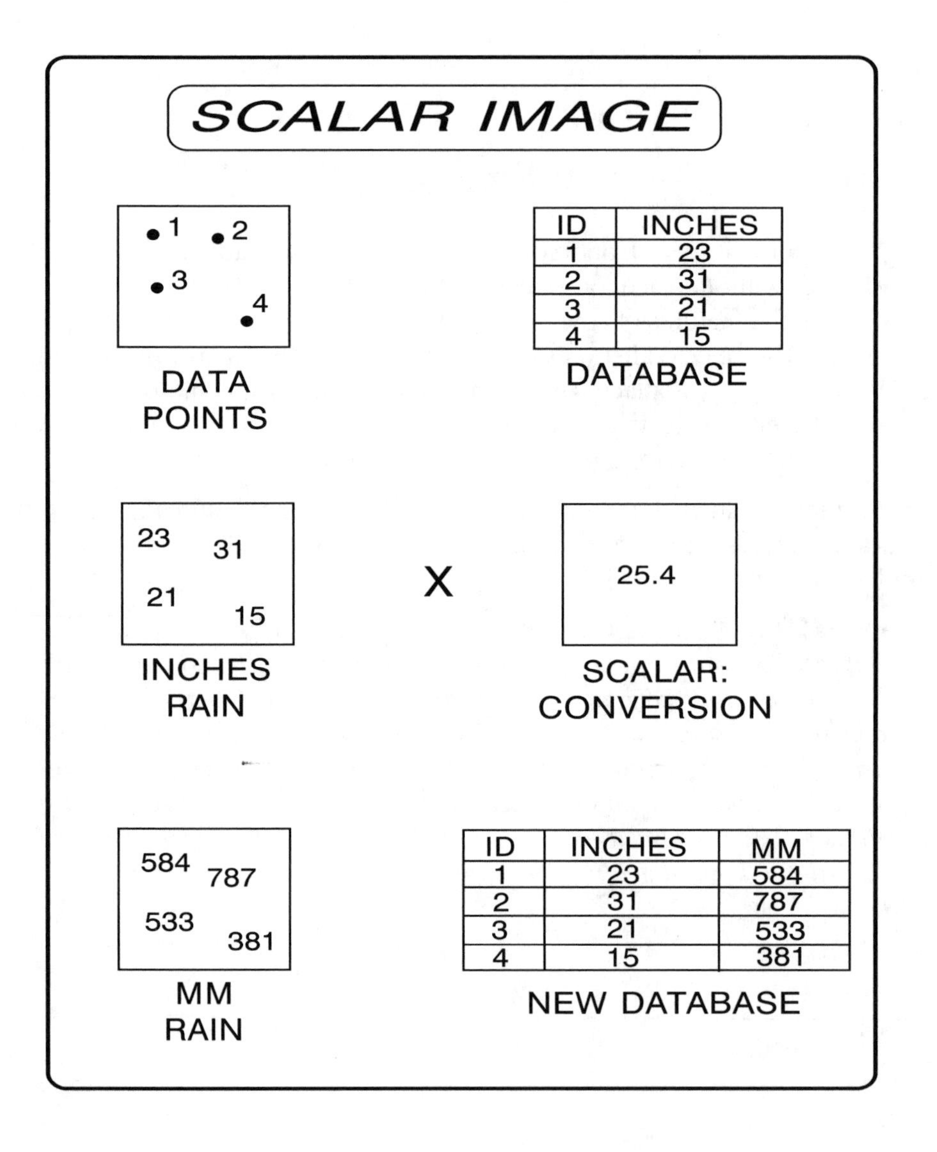
SCALAR IMAGE
1
2
3
4
DATA POINTS
ID | INCHES
1 | 23
2 | 31
3 | 21
4 | 15
DATABASE
23
31
21
15
INCHES RAIN
X
25.4
SCALAR: CONVERSION
584
787
533
381
MM RAIN
ID | INCHES | MM
1 | 23 | 584
2 | 31 | 787
3 | 21 | 533
4 | 15 | 381
NEW DATABASE

Scalar Image

Another overlay procedure is to make a coverage that has a single, uniform value—a scalar image (coverage). The user constructs a scalar coverage simply by giving the value desired. One primary use for the Scalar feature is to transform or convert all features of an existing coverage in order to change all values by a given figure.

The illustration shows a simple coverage of four datapoints that have inches of rain. Datapoints are given at the top, along with the database. At middle left, these points are expressed in inches, but if metric is needed, data conversion will be necessary. Although powerful GISs can convert within the database, some require a Scalar overlay.

A value of 25.4 is assigned to the scalar coverage (inches x 25.4 = millimeters). Then overlay of the scalar image with the Inches Rain coverage using Multiply will produce a coverage in millimeters. The scalar image to be used is at the middle right, created by the user simply as a coverage with a single value of 25.4. The bottom set of figures shows the resultant coverage, expressed in millimeters of rain (rounded values), and the accompanying database that lists both inches and millimeters.

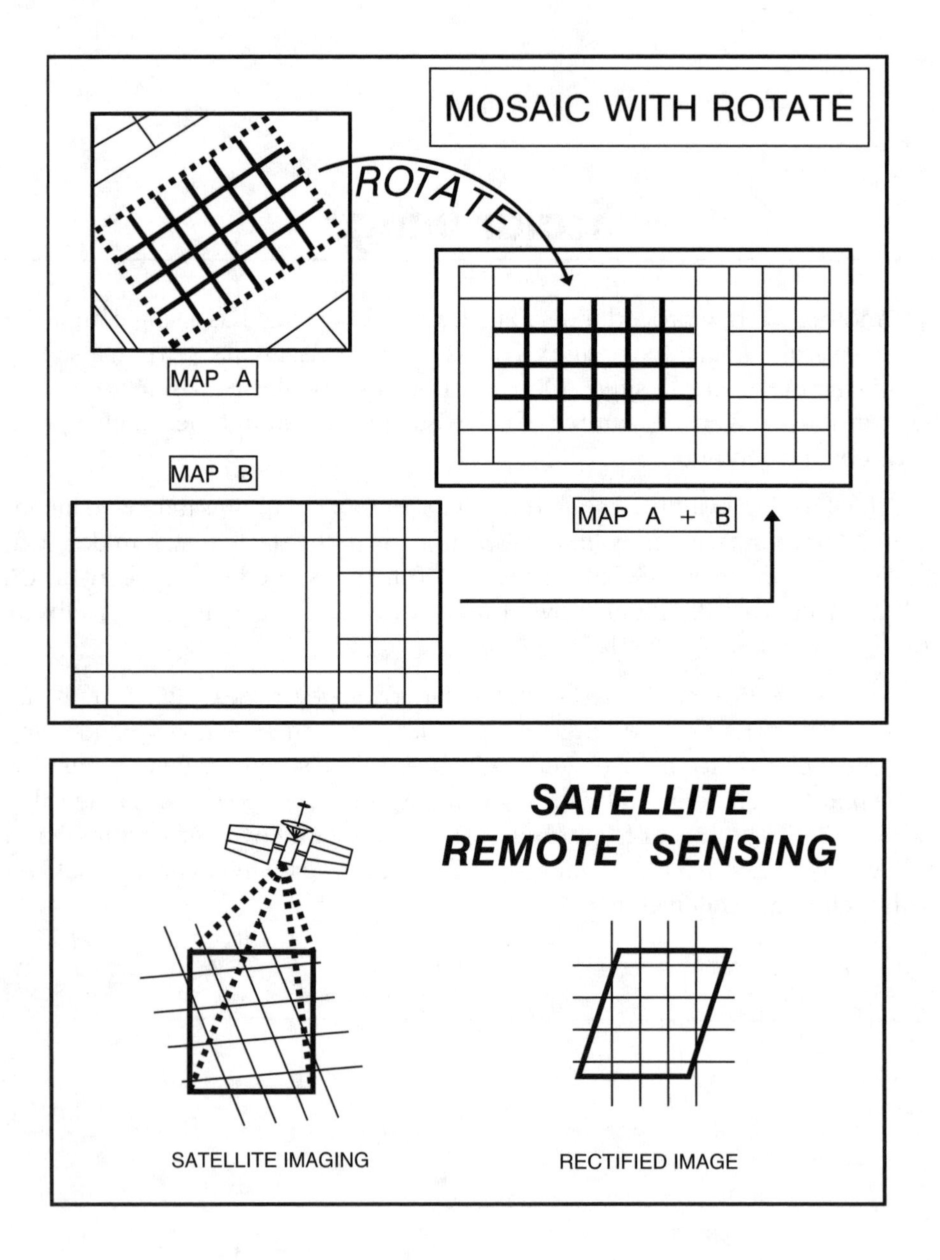
MOSAIC WITH ROTATE
ROTATE
MAP A
MAP B
MAP A + B
SATELLITE
REMOTE SENSING
SATELLITE IMAGING
RECTIFIED IMAGE

Mosaic with Rotate

Another term for overlay often used in GIS is *Mosaic.* Mosaic can also include patching coverages next to each other. The principal point in this illustration is that coverages do not have to be perfectly lined up side by side; they can be turned or mismatched as pairs and still fit together. Remember georeferencing: establishing real-world coordinates on a coverage. When overlaying, many GISs match the coordinate points, not coverage corners or borders. Therefore, if coordinates are accurate in each coverage, the actual shape of the coverage is not important to the computer (though it might look strange to the eye). It is also possible for the user to rotate a coverage manually to make it fit as desired, whether or not georeferencing is involved.

Illustrated at top is a simple example of a planned roads coverage (map A) being merged with an old roads coverage (map B) to show the effects of plans on the current landscape. The Planned Roads coverage was constructed from a different set of coordinates, and perhaps the GIS operator manually turned them to match the existing roads.

The lower illustration shows the effect of turned or twisted coverages. Remote sensing satellites take square images, but their orbital path is usually Northeast to Southwest, making the latitude-longitude system somewhat skewed (distorted) from our normal right-angle perception. Images are "rectified"—rotated and corrected to a normal structure—for better comparison and overlay with existing maps and coverages. The effect on the image is to make a parallelogram, an off-center square that seems unusual in a normal set of coverages.

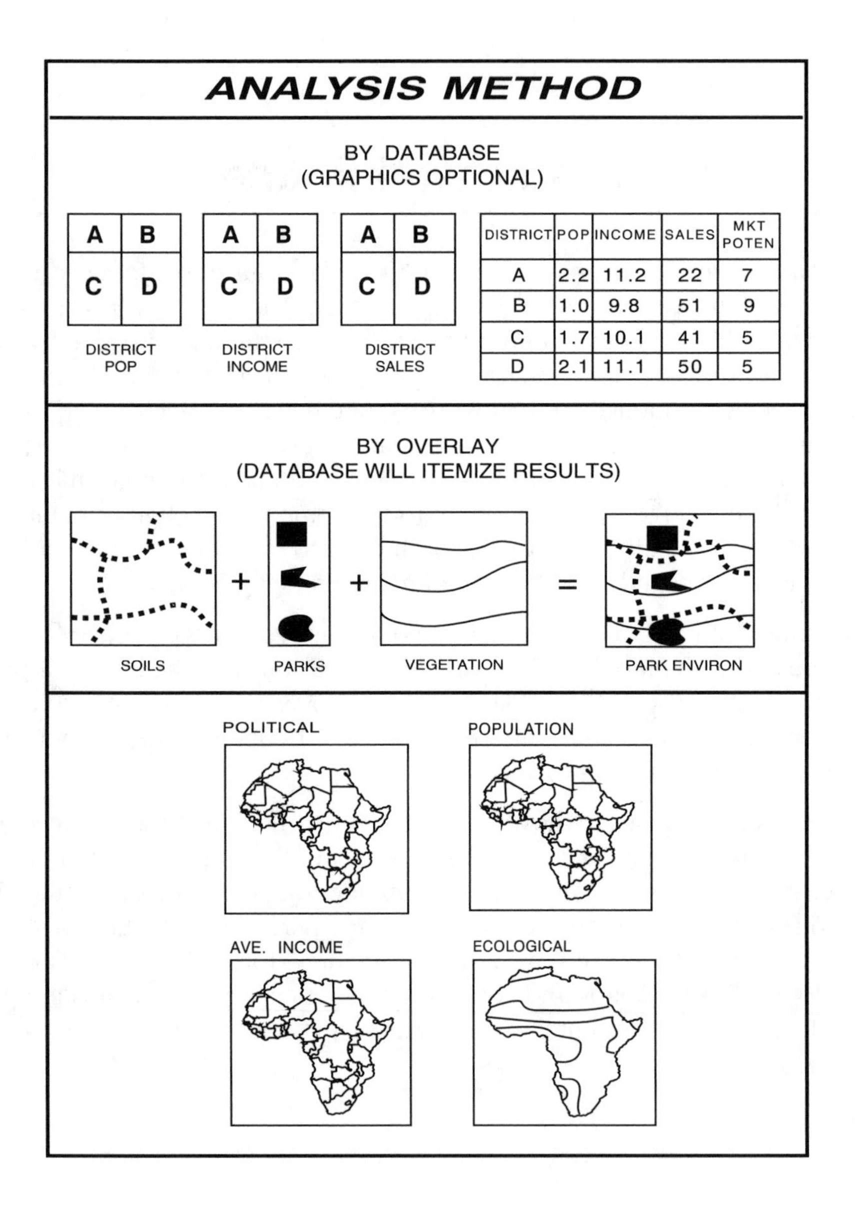
ANALYSIS METHOD
BY DATABASE
(GRAPHICS OPTIONAL)
A B C D
DISTRICT POP
A B C D
DISTRICT INCOME
A B C D
DISTRICT SALES
DISTRICT POP INCOME SALES MKT POTEN
A 2.2 11.2 22 7
B 1.0 9.8 51 9
C 1.7 10.1 41 5
D 2.1 11.1 50 5
BY OVERLAY
(DATABASE WILL ITEMIZE RESULTS)
SOILS
PARKS
VEGETATION
PARK ENVIRON
POLITICAL
POPULATION
AVE. INCOME
ECOLOGICAL

Analysis Method

Two fundamental approaches to basic analysis have been discussed: database and graphic. Low-end GISs usually have only raster structures and weak database capabilities that require manipulation of graphics to generate analytical output.

Powerful GISs have strong database and graphics integration, usually giving the option of taking either a graphics or a database approach to analysis. A useful perspective in making the choice of which to use for a particular application is to regard the nature of the coverages to determine if the database can be used; if not, the graphics method is needed.

If the coverages to be combined have spatially identical features, a database may be best; database operations may be more effective and easier. For example, the top row of the illustration shows that all three coverages use the districts (A through D) as the data feature. They can be combined in the database to generate marketing information. There is no need to use graphics overlays and recoding. A graphics product can be a practical presentation device, but it is not necessary in this situation to perform the analysis. The database has the coverage data, but has added a Market Potential (Mkt Poten) attribute calculated by the other numbers.

If the coverages and features are not similar (second row), the graphics approach is needed, basically because their databases will not match. Shown are soils, a smaller coverage of parks, and vegetation. It would be very difficult, perhaps impossible, to combine these coverages and features for analysis in a database. An overlay of all three coverages is needed to merge and associate the data. Then the database will be able to list attributes, such as area.

The illustrations of Africa at the bottom of the illustration demonstrate a typical set of coverages for a project. Any overlays and analysis involving the Political, Population, and Average Income components can be performed at the database because the features are defined by discrete database records (in this case, by nation). However, to include the Ecological coverage will require graphics overlay because of the nonmatching polygons.

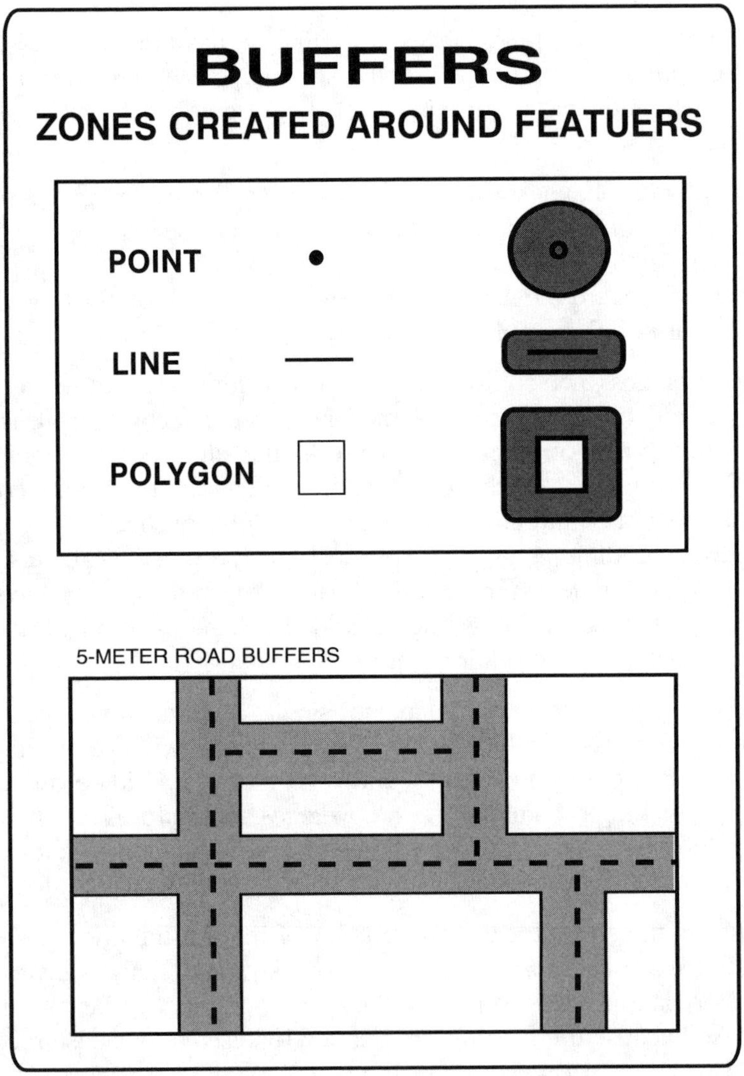

BUFFERS

ZONES CREATED AROUND FEATUERS

POINT

LINE

POLYGON

5-METER ROAD BUFFERS

Buffers

Building zones around features is a very useful and standard GIS capability (and difficult to do manually). Various names are associated with this process, such as *spread, search,* and *corridor;* the common term *buffer* is used here. Illustrated are buffers around each of three data types. The operation can be easy: a desired distance is given and the GIS builds the buffer outward from the selected feature or features.

The example at the bottom of the illustration shows a 5-m buffer around all of the roads. Many GISs permit buffering on selected feature types, such as only on sealed roads.

In some systems, the selected feature must be in a separate coverage before a buffer can be constructed; in other systems, the central feature is simply selected and buffering is performed. In simple raster structures, the number of cells must be given to build the buffer, but most GISs accept real-world distances as standard input. Buffering is a routine operation, along with Recode and Overlay, that often serves as a preparation operation for more advanced analysis.

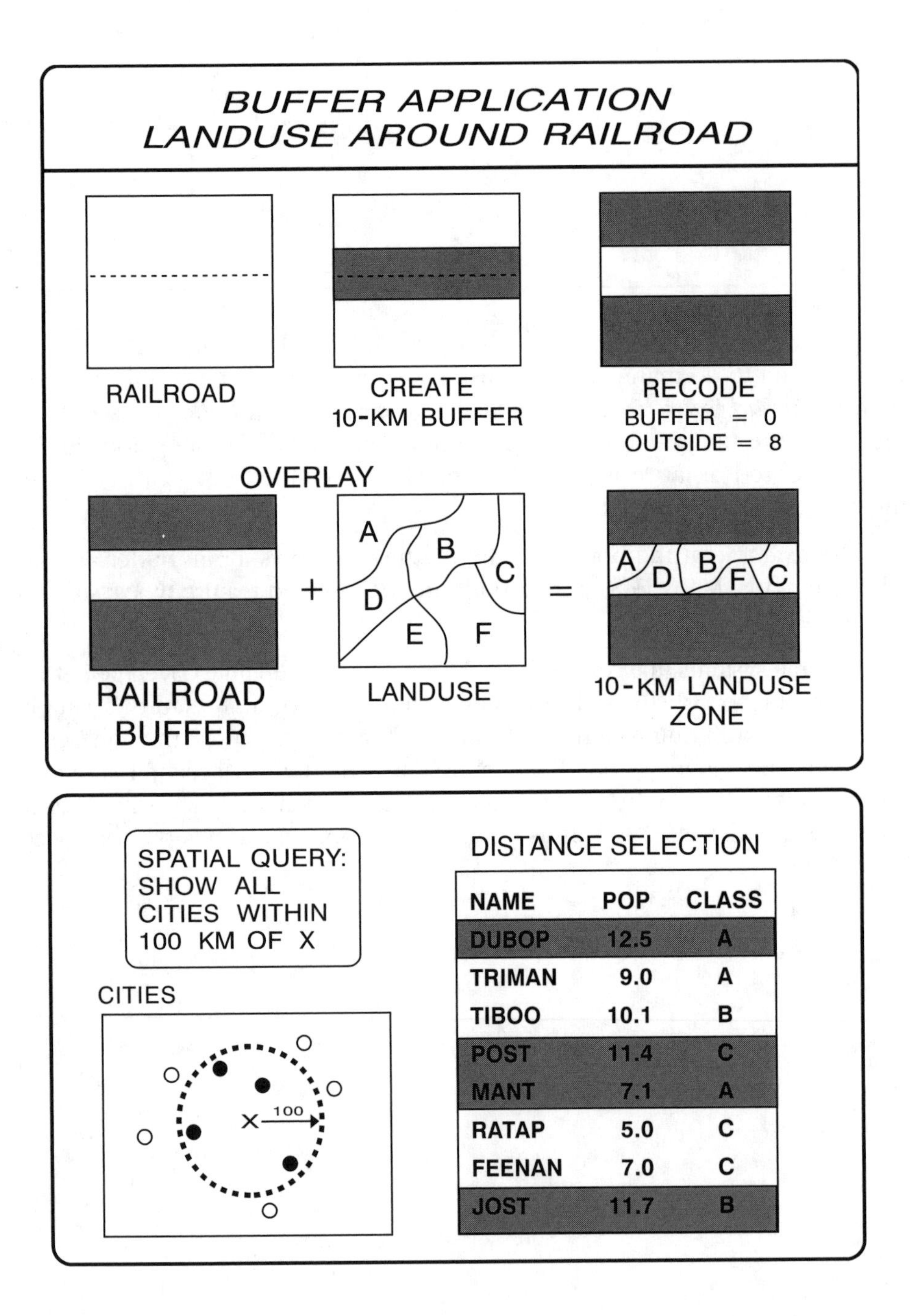

NAME	POP	CLASS
DUBOP	12.5	A
TRIMAN	9.0	A
TIBOO	10.1	B
POST	11.4	C
MANT	7.1	A
RATAP	5.0	C
FEENAN	7.0	C
JOST	11.7	B

Buffer Application: Landuse

The upper illustration shows one application of buffering: determining the landuse 10 km on each side of a railroad. The steps are relatively easy in a raster code system. First, build a 10-km buffer around the railroad. Then a recode is needed: change the buffer zone to 0, or transparent, and the area outside the zone to opaque—a mask. Then overlay with landuse to produce a landuse zone. Follow the illustration and see the relative simplicity of the process.

In a vector topological system, the buffer is created by noting the features and giving the desired distance. The actual buffer is created in a separate coverage, which is then combined (overlaid) with the landuse coverage. As noted, Buffer, Recode, and Overlay are common procedures in many GIS applications.

Distance Selection

The lower illustration shows another useful Buffer application in a query process. All cities within 100 km of point X are selected, possibly for a market analysis or environmental purpose. The database highlights those that occur within the buffer (or perhaps lists them separately), allowing the user to learn more. Many applications continue by making further database queries, such as finding cities with population over 10,000 from the buffered selections. In the illustration, those cities are shaded records in the database.

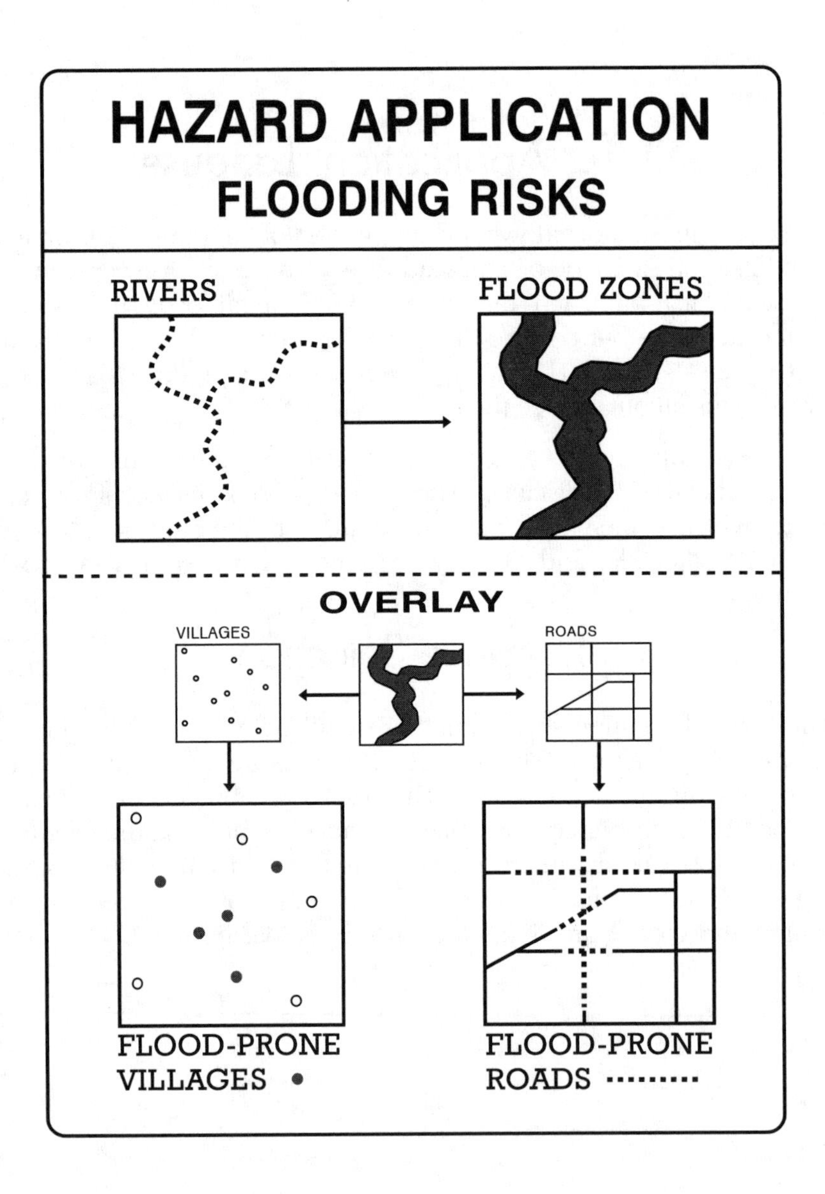
HAZARD APPLICATION
FLOODING RISKS
RIVERS
FLOOD ZONES
OVERLAY
VILLAGES
ROADS
FLOOD-PRONE VILLAGES •
FLOOD-PRONE ROADS ·········

Buffer Application: Flood Analysis

Another useful application of buffering is to analyze risks within a selected distance of some hazard, such as flooding. It is very important that the field manager for disaster relief (or better, disaster prevention) should know which villages and roads are prone to flooding under certain conditions. GIS can assist in two major ways.

First, before the flooding, the manager can run a series of "what-if?" operations, putting in various flood levels to see the results. With that set of data and maps, the manager is ready for a variety of floods and can respond much faster and better.

A second way is to have the landscape coverages ready for analysis when a disaster occurs. The GIS can perform complex analysis for a selected area, large or small, very quickly, permitting the manager to respond rapidly and efficiently. Lives and property can be saved—a wonderful application of GIS technology.

Follow the steps in the illustration. A flood zone buffer is shown at top, and then overlaid (bottom) with Villages to produce a flood-prone coverage and a roads coverage to show which roads are likely to be affected.

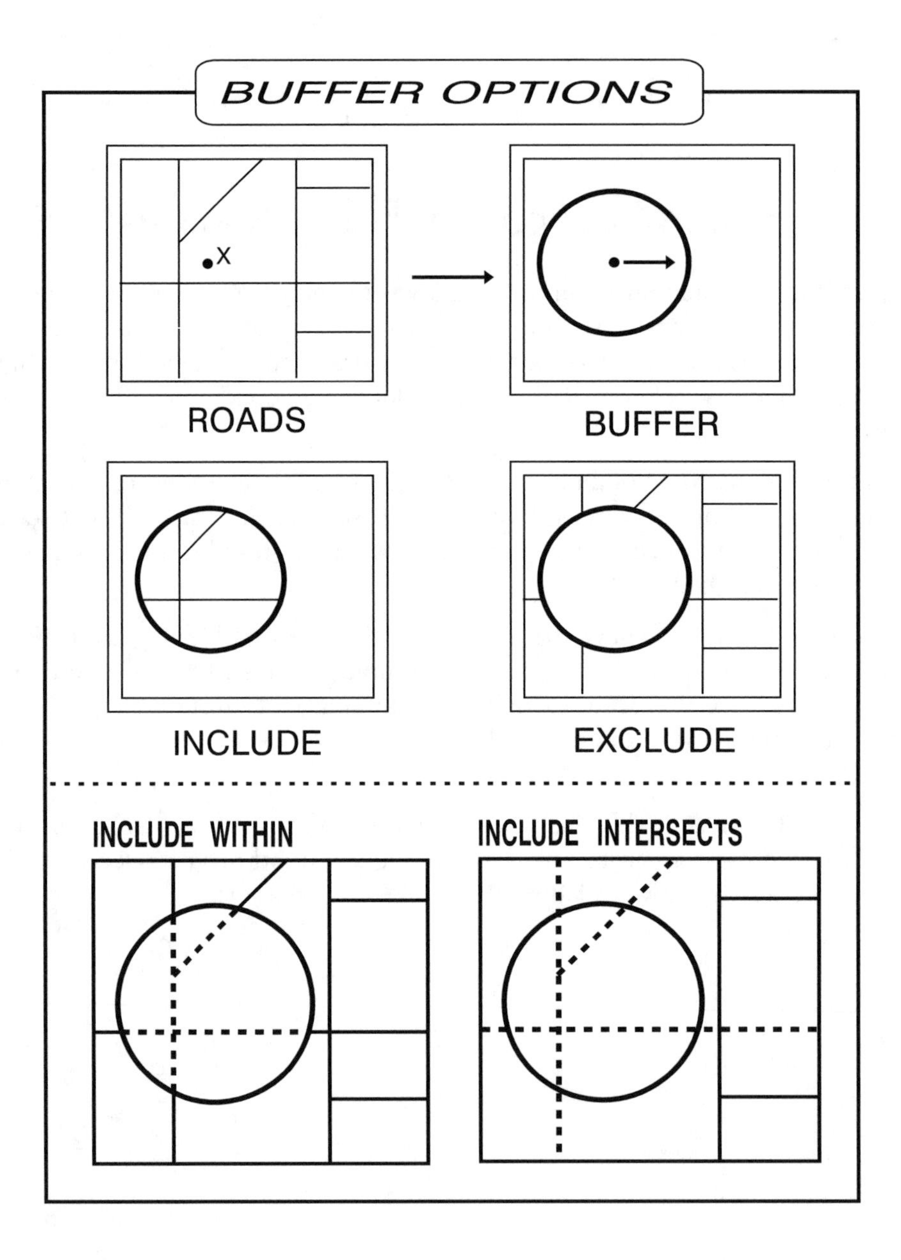
BUFFER OPTIONS
X
ROADS
BUFFER
INCLUDE
EXCLUDE
INCLUDE WITHIN
INCLUDE INTERSECTS

Buffer Options

Buffers can serve several purposes. Once built around a feature, a buffer zone can be a frame or window for including or excluding the features ("contents") within. Include considers only interior features, and Exclude deletes the interior, as shown in the upper four illustrations. Try to imagine some applications for each type.

The two options in the lower illustration show another aspect of Include: incorporating only the parts of features that occur within the buffer, or all of the features that partially occur within the buffer. For example, the illustrated buffer circle is a disaster site impact area and the hazard management team must know the extent of affected roads. Therefore, an Include Within is used to select only the part of the feature existing within the buffer. The sections of roads affected by the disaster are designated.

On the other hand, transportation managers want to know which roads cross into the impact area so that they can establish traffic control all along their routes. Therefore, Include Intersects (an overlay option) is used to select all of the features that may fall only partially within the buffer.

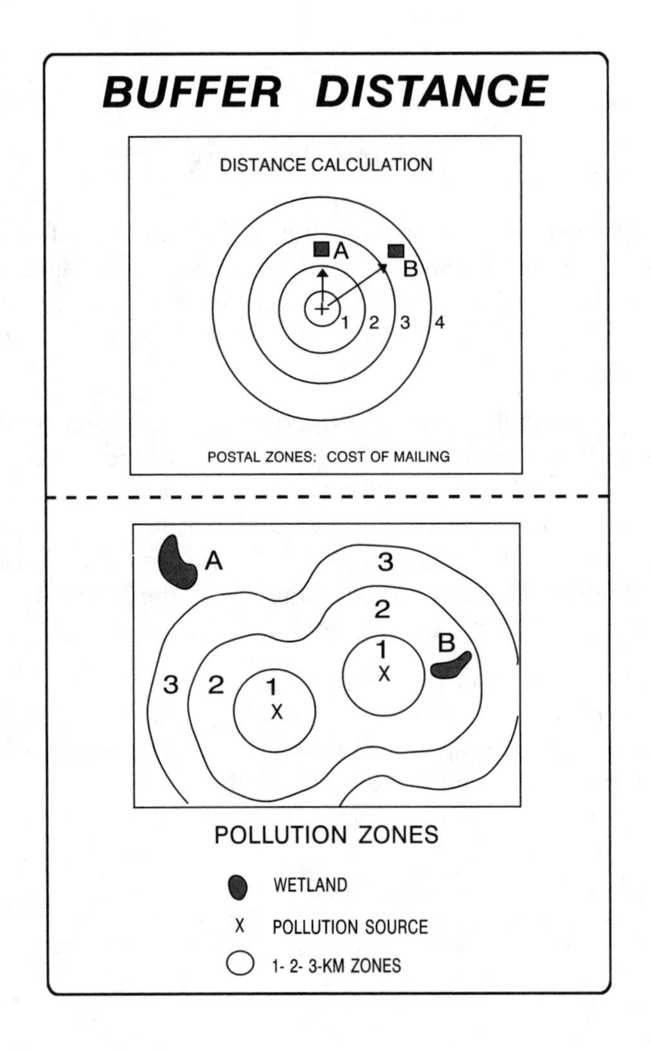
BUFFER DISTANCE
DISTANCE CALCULATION
A
B
1
2
3
4
POSTAL ZONES: COST OF MAILING
A
3
2
1
X
B
3
2
1
X
POLLUTION ZONES
WETLAND
X POLLUTION SOURCE
1- 2- 3-KM ZONES

Buffer Distance

Because buffers are created outward from the central feature or features, it is easy to build concentric zones. In fact, this is how some GISs (particularly raster systems) first construct buffers. When a buffer is commanded, the GIS radiates one-cell increments outward to the edge of the coverage. A recode is necessary to establish the buffer zone. For example, if a 10-km buffer zone is desired from a central point, the GIS first makes the radiating concentric zones to the edge of the coverage. Then the operator recodes all of the cells out to 10 km to the buffer zone code, and all other cells to another, nonbuffer code. This can be tedious work, and is not necessarily efficient.

Some low-end GISs determine distance from one feature to another by building concentric buffer zones between them and then counting the cells to convert to distance (hence the term *Search,* rather than *Buffer,* in these systems). Two practical distance applications for concentric zones are illustrated. The upper illustration shows postal zones from a central site, with the cost of mailing increasing outward in zonal increments. Mailing to destination A (zone 2) will be less expensive than mailing to destination B (zone 3) because of zone distance difference.

The lower illustration shows distributed pollution zones, with progressively less hazardous conditions away from the two sources (X). Wetland B is clearly in a more hazardous situation than wetland A. This is an example of gravity models and "distance decay"—the center has the highest intensity (or magnitude, such as pollution), and the effects decrease outward.

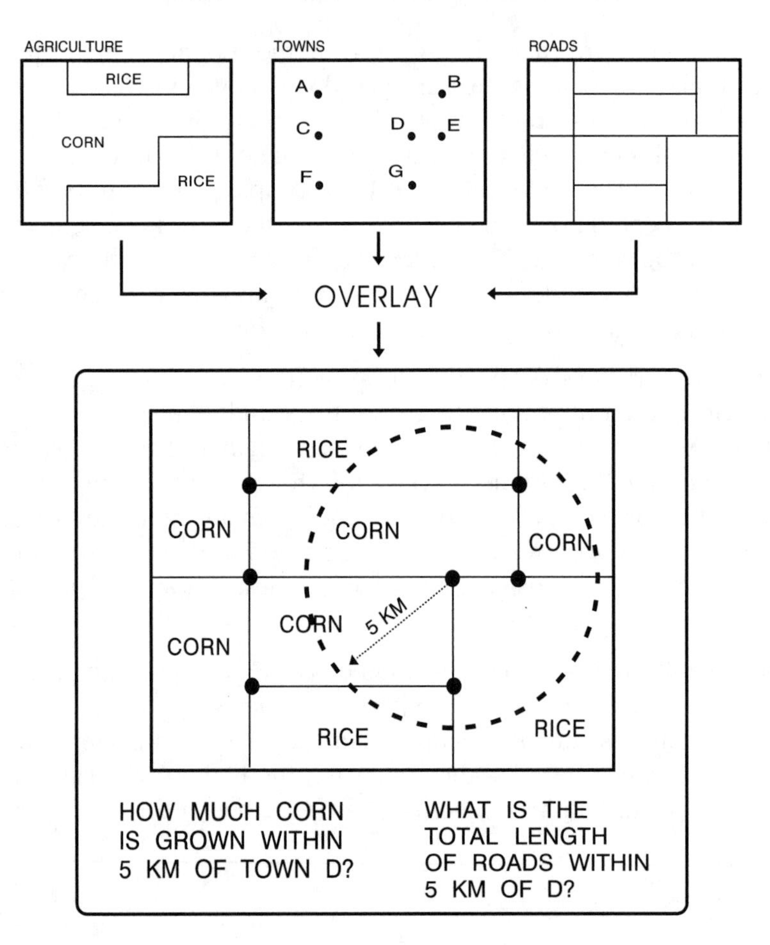
SPATIAL ANALYSIS
AGRICULTURE
RICE
CORN
RICE
TOWNS
A
B
C
D
E
F
G
ROADS
OVERLAY
RICE
CORN
CORN
CORN
CORN
CORN
5 KM
CORN
RICE
RICE
HOW MUCH CORN IS GROWN WITHIN 5 KM OF TOWN D?
WHAT IS THE TOTAL LENGTH OF ROADS WITHIN 5 KM OF D?

Spatial Analysis

There are numerous procedures and techniques using the operations discussed in this section. The illustrated example combines some of these steps in a complex spatial analysis task. First, the important coverage is made from an overlay of Agriculture, Towns, and Roads (which could be the results of recodes from initial data or other data sets). Then a buffer is applied to begin analysis. A 5-km buffer around town D can help to address questions of proximity.

For example, how much corn is within the circular buffer, and what is the total length of all roads within that buffer? An overlay with Include, showing all features within the 5-km buffer, will answer these concerns rather easily. Can you think of a logical application for an Exclude operation?

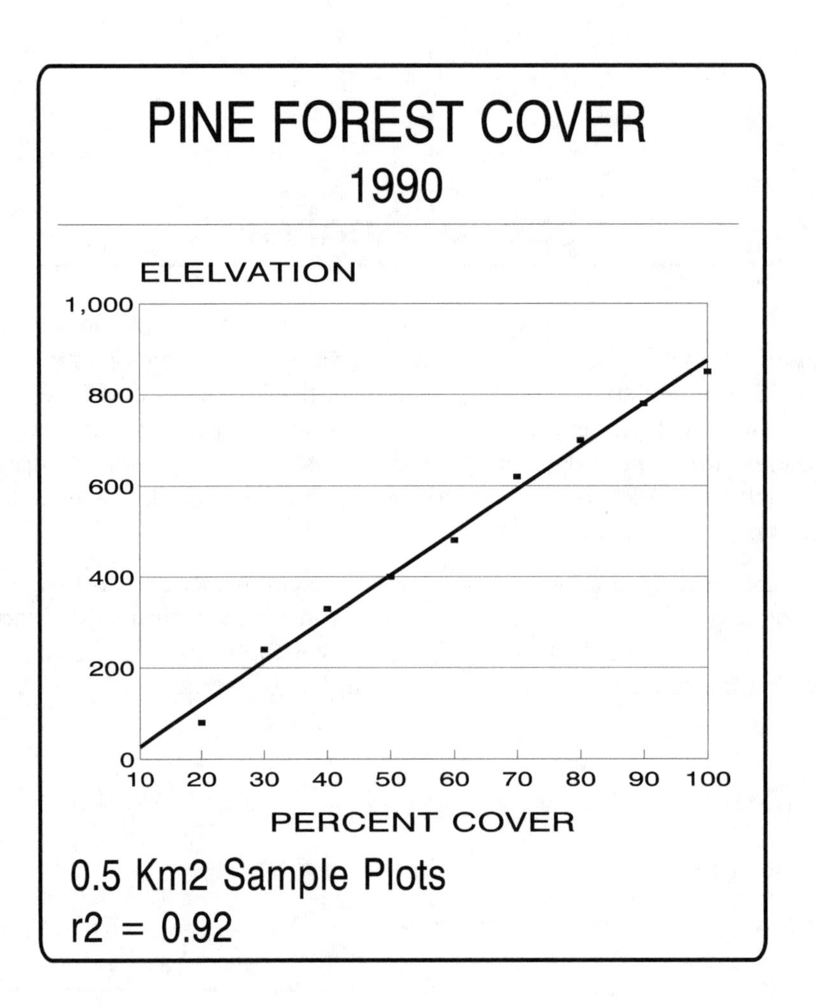
PINE FOREST COVER
1990
ELELVATION
1,000
800
600
400
200
0
10
20
30
40
50
60
70
80
90
100
PERCENT COVER
0.5 Km2 Sample Plots
r2 = 0.92

Statistical Reporting and Graphing

A final basic analysis operation is to analyze data statistically. By generating relatively simple statistical reports and graphs, new information can be appreciated. A listing of the average size of crop fields—with their minimum and maximum yields, along with the range of distances to market—could be very useful in an agricultural survey project.

Illustrated is a graph plotting the percentage of pine forest cover against elevation. It is possible that spatial and ecological relationships may emerge. The points fall very close to the regression line (line of best fit), and the correlation is calculated at 0.92, a very good indicator that density of vegetation increases predictably with elevation. Such a plot is fairly easy to make, and the results can be significant.

Spatial statistical data analysis can be a logical extension of GIS mapping and database analysis. It can add an extra dimension of insight and understanding to the data and analytical results. Again, this shows that GIS is much more than simple mapping software.

Initial Operations in Brief

This chapter has looked at some of the most common, yet highly valuable, GIS operations. They are all used as initial data preparation and first analysis steps. Overlay merges two or more coverages and uses a wealth of "tricks" to produce a useful result. Map Algebra is used to understand and to control overlay data. Several other graphics operations discussed are Clip, Mask, Replace, and Scalar.

Buffering is a highly valuable GIS operation, providing applications in many fields. Buffers may be constructed easily or may involve considerable hands-on work, thought, and analysis.

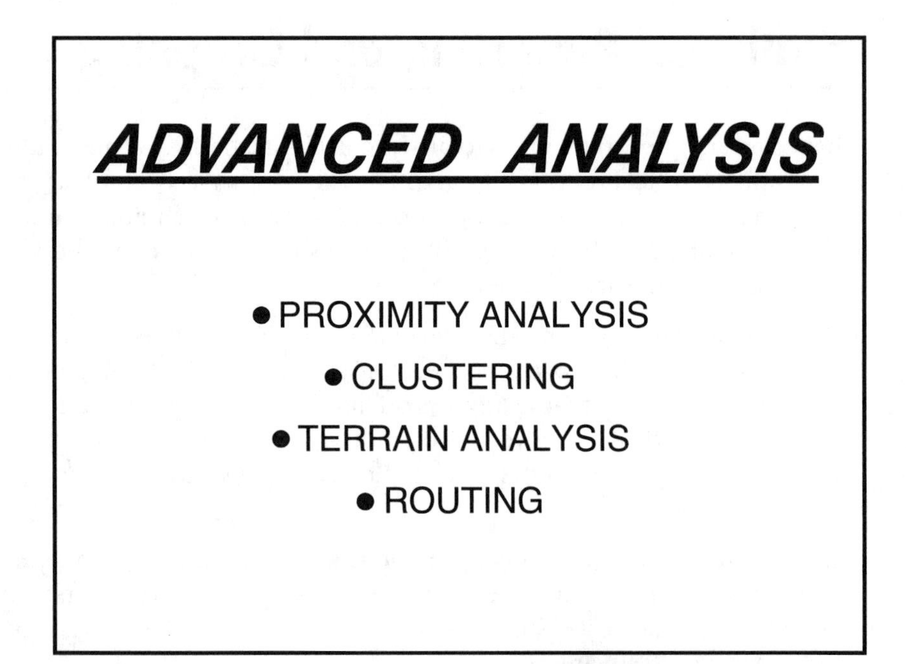

ADVANCED ANALYSIS
• PROXIMITY ANALYSIS
• CLUSTERING
• TERRAIN ANALYSIS
• ROUTING

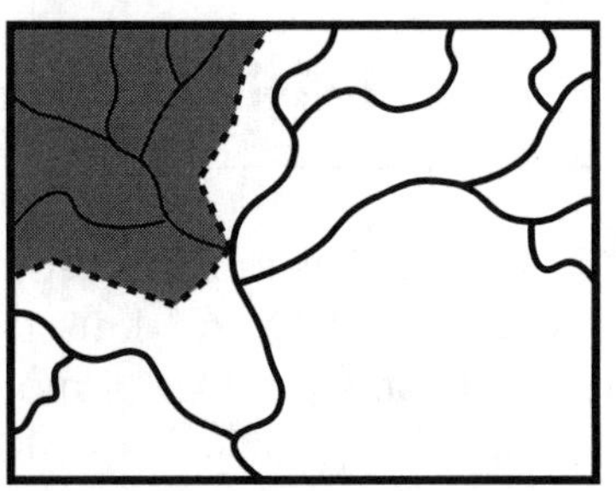

STREAM BASIN DELINEATION

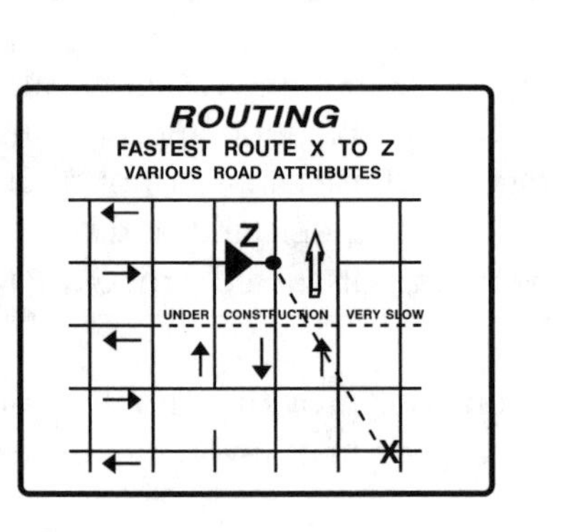

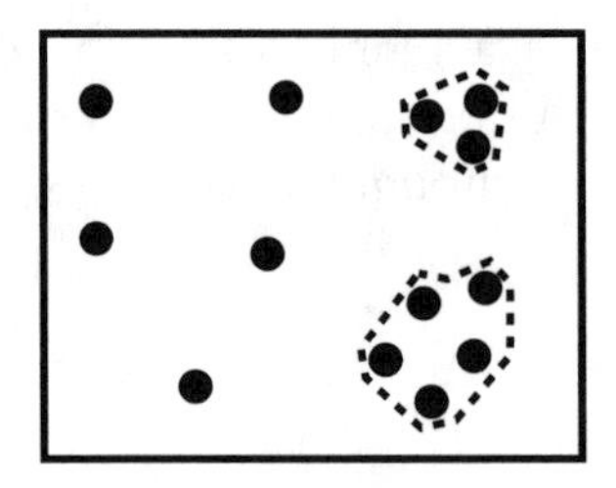

9

Advanced Analysis

Introduction

Advanced analysis operations are procedures that accomplish more sophisticated tasks than those classed as basic analysis here, though the line between the two is not definite. This section presents several types of procedures available in many GISs, but many other advanced operations are not discussed. Here we look at general techniques of proximity analysis, graphics operations, and terrain analysis.

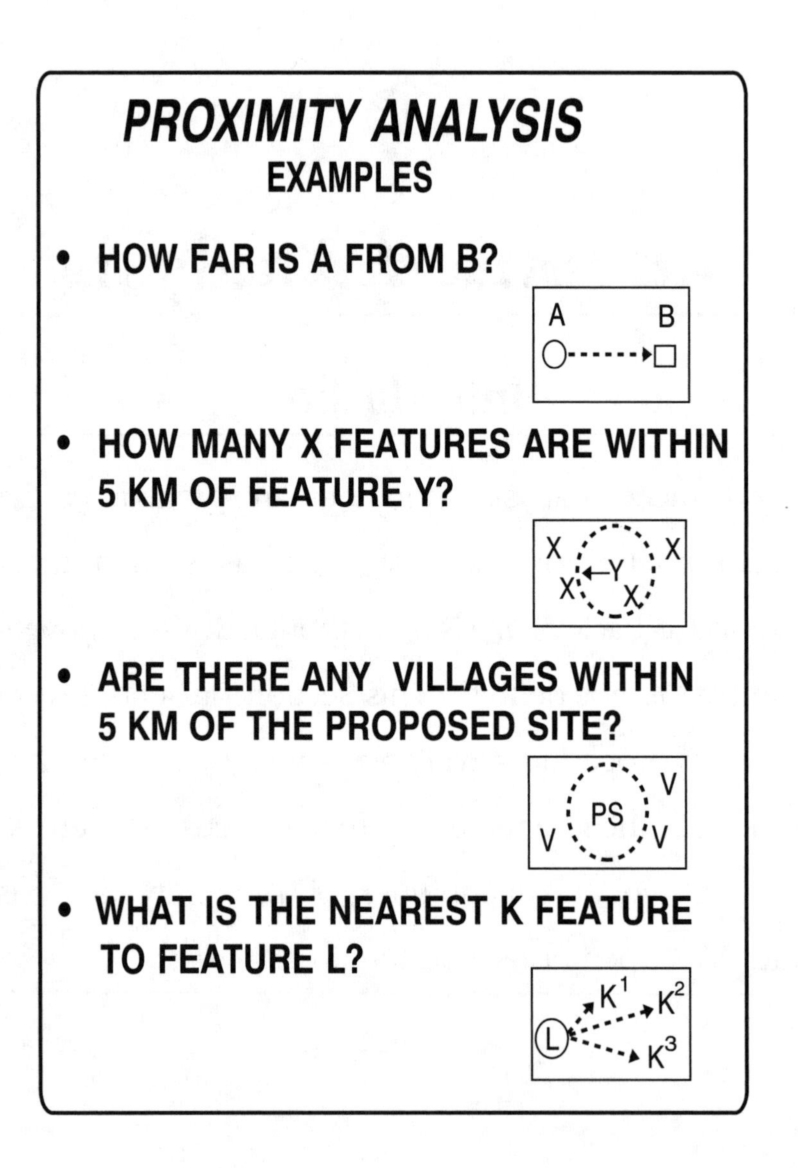
PROXIMITY ANALYSIS
EXAMPLES
• HOW FAR IS A FROM B?
A
B
• HOW MANY X FEATURES ARE WITHIN 5 KM OF FEATURE Y?
X
X
Y
X
X
• ARE THERE ANY VILLAGES WITHIN 5 KM OF THE PROPOSED SITE?
V
PS
V
V
• WHAT IS THE NEAREST K FEATURE TO FEATURE L?
K1
K2
L
K3

Proximity Analysis

The term *proximity analysis* can mean any procedure that performs "neighborhood" or vicinity analysis. Several corresponding examples are shown in the illustration.

- *"How far is A from B?":* Asking the distance between features is a common operation, as has been shown in previous examples. Distance can be expressed in terms of time or cost, in that the conversion of spatial quantities to cost values is easily calculated (dollars per km x distance = total cost of trip). Some GISs allow simple point-and-click selection of features to determine distance. Point to the beginning feature or location, then to the destination.

- *"How many X features are within 5 km of feature Y?":* Getting a count of a certain class of features within a given distance of another feature can be useful, such as the number of communities within the noise buffer.

- *"Are there any villages (V) within 5 km of the proposed site?":* Similarly, searching for specific features within a given distance is possible, such as locating all villages within 5 km of a proposed site (PS). The presence of villages could exclude the proposed site from further consideration. With a good database system, the query could be: "Are there any villages with populations over 500 within 5 km of the proposed site?"

- *"What is the nearest K feature to feature L?":* The last example combines distance and identification in a search for the nearest feature of a given class, a rather easy task. First the program locates all K features and then calculates their distance from L, reporting the one with the shortest distance. The next illustration is a good example.

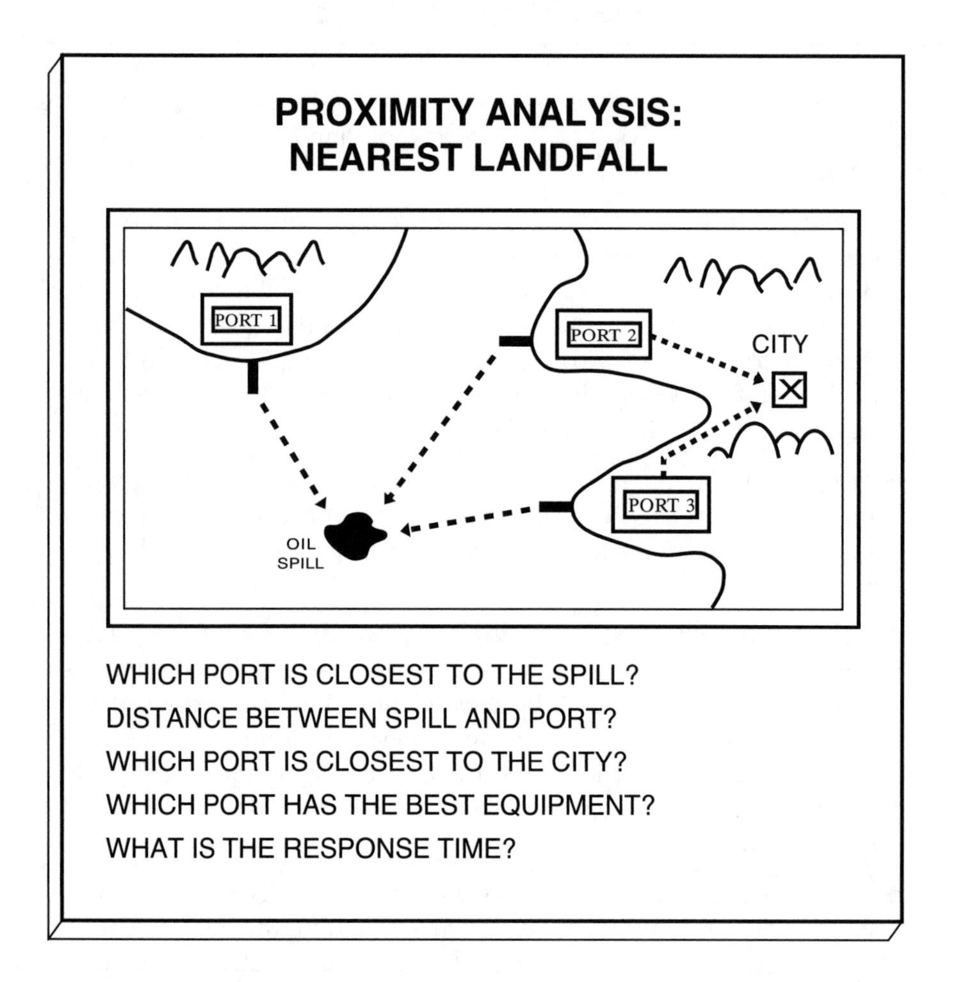
PROXIMITY ANALYSIS:
NEAREST LANDFALL
PORT 1
PORT 2
PORT 3
CITY
OIL
SPILL
WHICH PORT IS CLOSEST TO THE SPILL?
DISTANCE BETWEEN SPILL AND PORT?
WHICH PORT IS CLOSEST TO THE CITY?
WHICH PORT HAS THE BEST EQUIPMENT?
WHAT IS THE RESPONSE TIME?

Nearest Landfall

Calculating distance and attaching database attributes in a search is a GIS strength. One valuable application of proximity analysis is the determination of the nearest or best site of emergency assistance, such as the closest port to a marine oil spill. Response time is critical, and a rapid selection of the best port can be significant. As noted, distance can be expressed as time, helping to determine response time to an emergency. In the illustration, port 1 seems to be the closest.

Distance may be a major factor, but other attributes (from the database) might also be important. For example, of the three closest ports in the illustration, which one has available equipment or people? Which one has the best access to land-based facilities (such as storage of equipment) that are situated in city X? The closest port to the spill may be port 1, but the best one may be port 3.

Any number of attributes could be stored and used. Port 3 may be the best as determined by location, but its boats could be very slow. Boats in port 2 are very fast and can cover the longer distance in less time than the port-3 boats can cover the distance they have to travel. Such calculations are relatively easy in most GISs, giving the emergency manager a powerful tool to aid in response time and conditions.

The manager decides the important factors of a particular event and then chooses database attributes accordingly, which are then applied to GIS analysis. In the end, graphics (coverage and map display) and database information are produced.

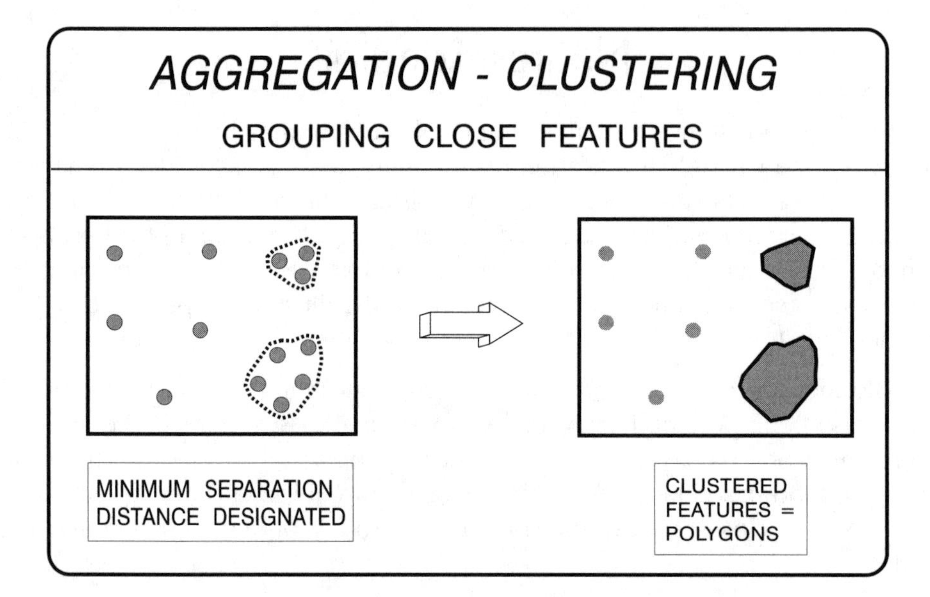
AGGREGATION - CLUSTERING
GROUPING CLOSE FEATURES
MINIMUM SEPARATION DISTANCE DESIGNATED
CLUSTERED FEATURES = POLYGONS

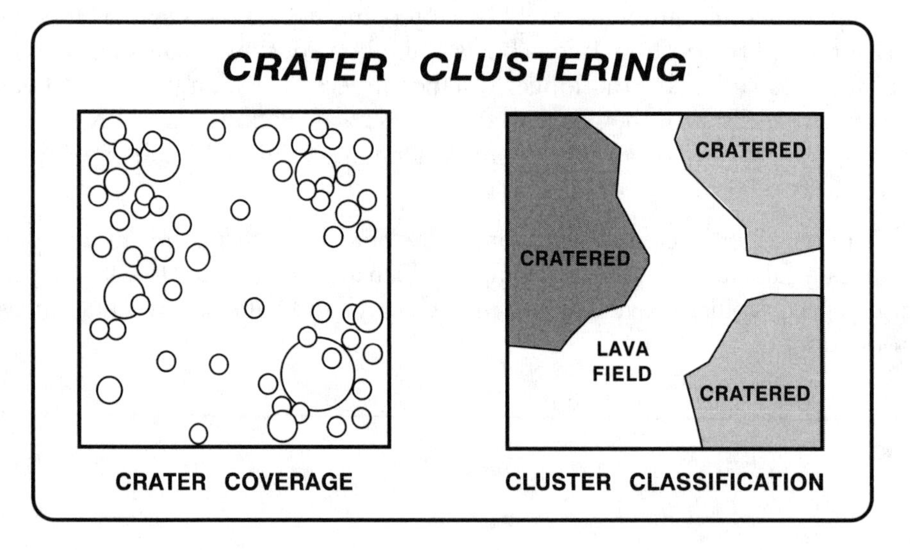
CRATER CLUSTERING
CRATERED
CRATERED
LAVA FIELD
CRATERED
CRATER COVERAGE
CLUSTER CLASSIFICATION

Aggregation

Another proximity analysis operation groups features that are very close to each other. Aggregation looks at how close fragments of features are, and if they are within a specified distance, it will combine them into a single feature. The upper illustration shows small, circular features (could be agricultural fields) scattered around an area. Those on the right side are so close together that they should be aggregated (clustered) into a single polygon. The user gives a minimum separation distance, and identical features within that distance are grouped into polygons.

A typical reason to aggregate is to eliminate irrelevant, confusing, or unimportant space or features in between important ones. For example, a large agricultural unit may appear to be separate fields because of numerous small roads. Also, small patches of natural vegetation around the fields may be insignificant to the larger interpretation of agriculture, so they should be ignored.

Crater Clustering

Some landscapes are best defined by many small, discontinuous features that give the appearance of a speckled (spotted) terrain, such as craters on the moon or karst (irregular limestone region) sinkholes—limestone solutions—on Earth. The craters are not important individually, but when clustered they characterize the unit of land. These land units are also known as "photomorphic regions" because they are defined by the principle morphology (appearance). The lower illustration shows a portion of the moon, where aggregation of craters (within a certain distance of each other) designate the classification. There is cratered terrain and smoother, lava field terrain. The clustered coverage is easier to include with overlay operations, whereas the crater coverage remains as detailed, primary data.

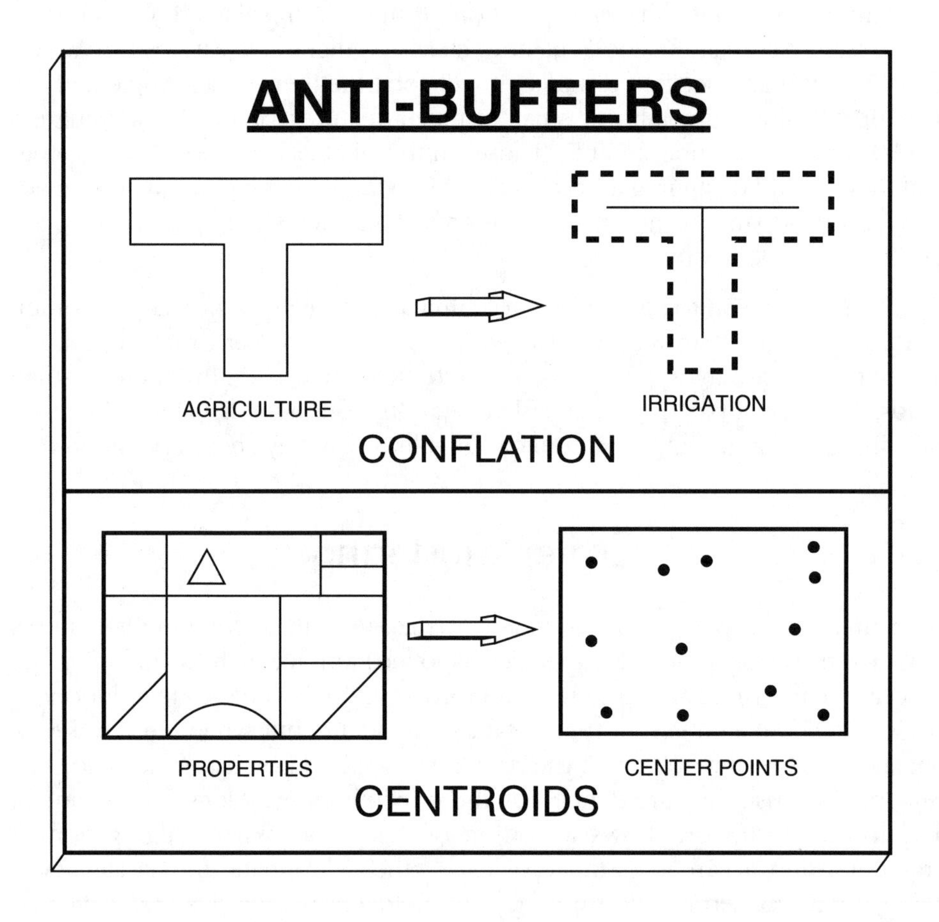
ANTI-BUFFERS
AGRICULTURE
IRRIGATION
CONFLATION
PROPERTIES
CENTER POINTS
CENTROIDS

Graphics Operations

Anti-buffers

The opposite of buffers is the constriction of features, usually polygons. The process of reducing a feature to a simpler element is called *conflation* (upper illustration). In the illustration, a T-shaped agricultural field is minimized to a T-line, the smallest feature that represents the original field. One application could be to develop the most efficient irrigation system.

Polygons can also be represented by their center point, called a *centroid* (lower illustration). The center of areas is determined and a single point replaces the entire polygon. One use could be the simplification of a mixed-size and mixed-shape polygon area to a more efficient visual and database structure (where the spatial characteristics of polygons are not important, only their presence). Obtaining a count is easier, for example, and using points rather than polygons for some analytical operations may be more convenient. Lines could be reduced to their center points, also. Can you think of an application where centroids are just as useful as their polygons?

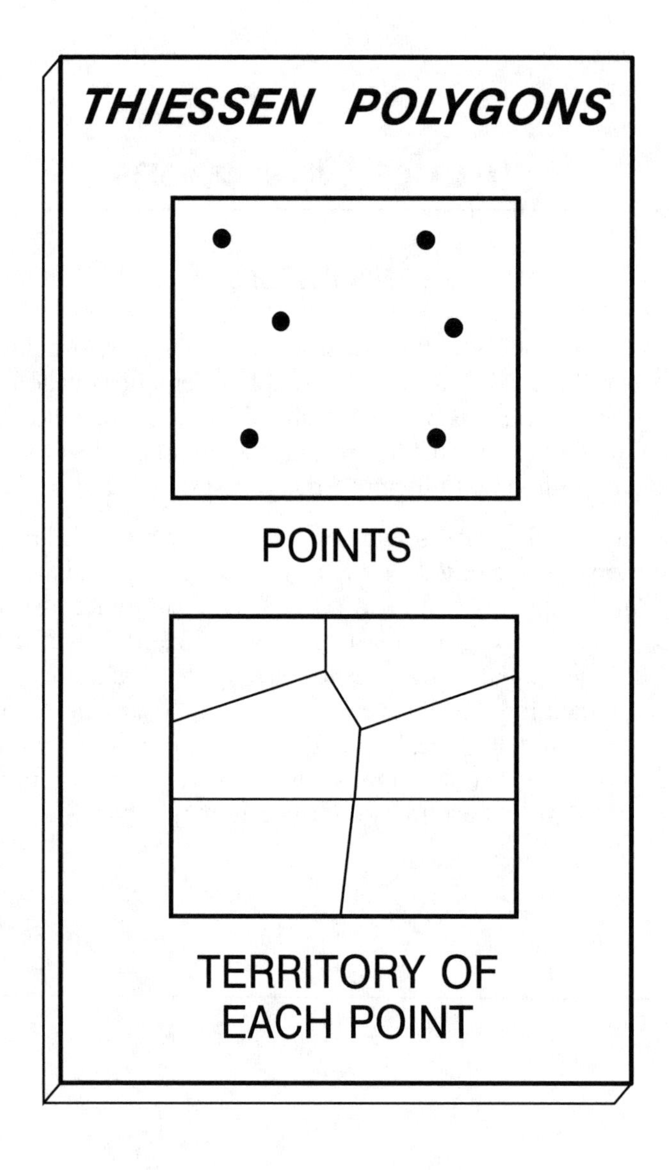
THIESSEN POLYGONS
POINTS
TERRITORY OF
EACH POINT

Thiessen Polygons

The opposite of centroids are equal areas around points, termed *Thiessen polygons.* They are the "territories" of points. The program expands each point's area until it meets the next one coming from a neighbor point or until it runs into a coverage edge.

The illustrated points coverage at top is converted to Thiessen polygons of the point features (bottom). The points can be cities, and the Thiessen polygons define their "hinterlands"—areas of influence or service. Applications include determination of market area of stores and calculation of areas served by rain gauges.

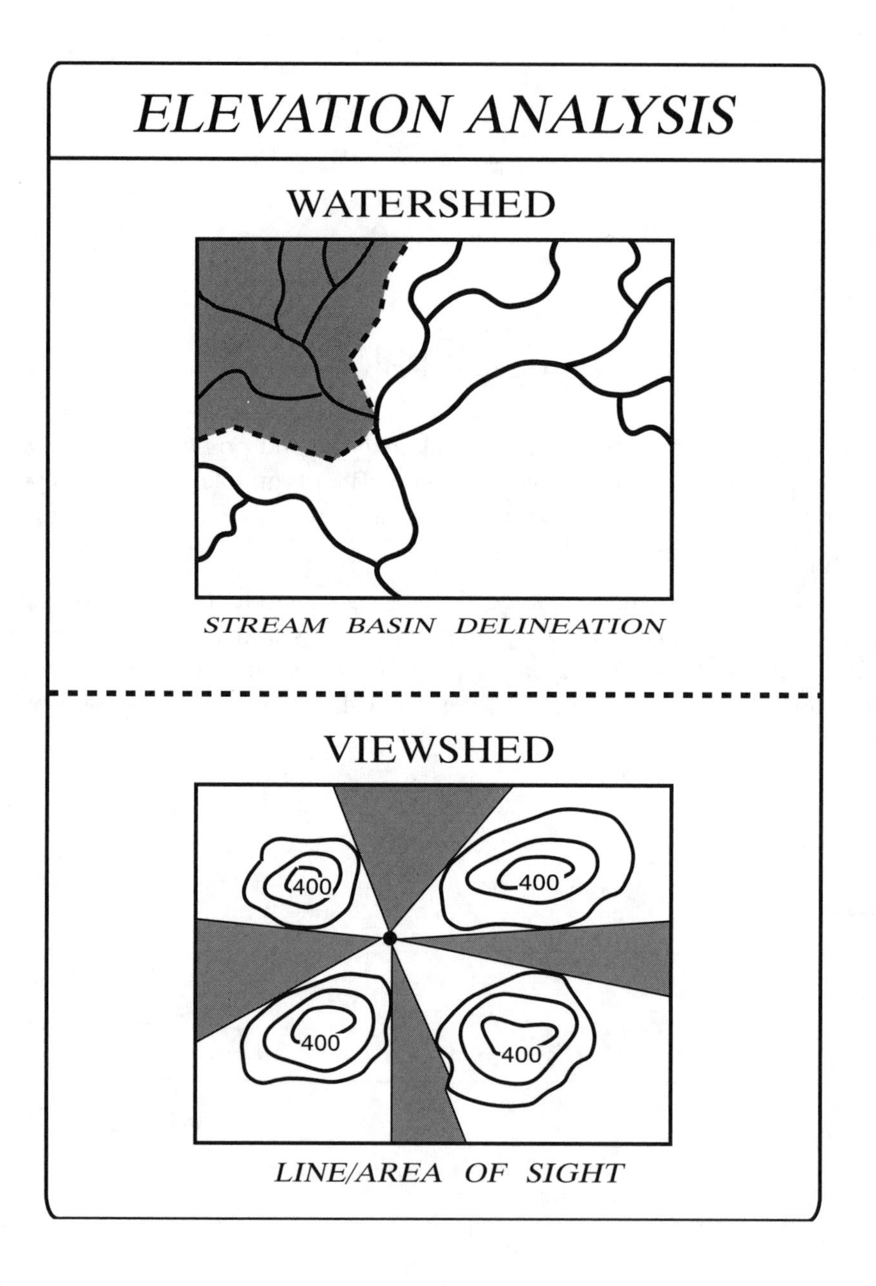
ELEVATION ANALYSIS
WATERSHED
STREAM BASIN DELINEATION
VIEWSHED
400
400
400
400
LINE/AREA OF SIGHT

Terrain Analysis

Elevation Analysis

Whereas Thiessen polygons are the territories of points, the area drained by a stream system is defined by its highest surrounding elevation, its "rim." A stream's *watershed* (basin, catchment) is the area it drains; all water in that area will eventually exit into the stream. Many GISs have operations that effectively spread a stream's watershed area uphill until the highest elevation is reached, which is the divide for the neighboring stream's watershed.

In the top illustration, the watershed has been drawn for the stream in the upper left of the watershed map. Note that its tributaries have been included because they also drain into the main stream. Each stream on the coverage, regardless of size, has a natural watershed, and a good GIS could determine all of them.

A *viewshed* is the area that can be viewed or seen from a particular location, given the surrounding topography or other obstacles in the line of sight. It is important, for example, to know how far and what can be seen from a fire observation tower. Where is the best location to maximize the view? Height of vegetation, eye level (height of the observation deck), and other features can be used in the operation. The bottom illustration shows a sighting position restricted by surrounding topography—hills higher than the viewing elevation.

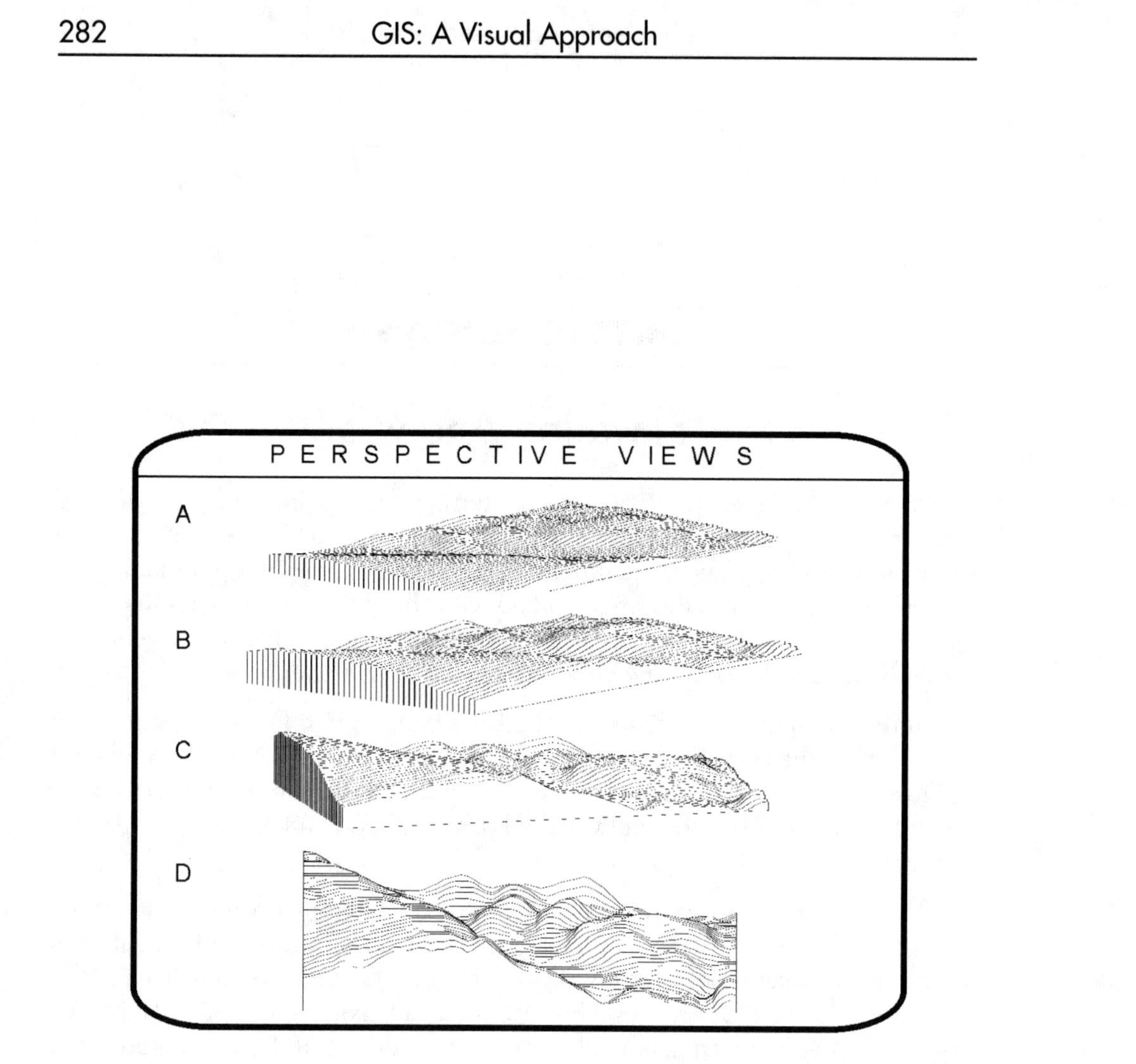
PERSPECTIVE VIEWS
A
B
C
D

Perspective View

When basic elevation has been determined, there are various terrain analysis procedures that can be applied. A common technique is to draw the landscape in 3D to present a realistic appearance of topography. This is often termed a *perspective view* (or isometric model). The user normally has control over the azimuth (rotation) of the view, height and angle of the view, and scale of landscape exaggeration (to enhance subtle topographic differences).

The four examples are elevation perspectives of a small section of the Dirol Plain in Africa. A small valley extends form the northwest corner to bottom center, with low hills on each side. The top display (A) is turned about 45 degrees counterclockwise from north, with a viewing angle of 45 degrees above the horizon (90 would be directly above; 45 gives a good perspective), and no vertical exaggeration. Perhaps this is a realistic perspective of the terrain. Of course, turning at various angles would give different views of the landscape.

Example B has the same viewing angles, but with twice the vertical exaggeration. Topography is artificially enhanced; the low hills are more pronounced. This can be useful for landscape planners.

Example C has even more vertical exaggeration and is turned differently, about 30 degrees from north. Part of the valley is seen better. The bottom view, D, is highly exaggerated (about 3x from the original data), has a lower viewing angle (30 degrees), and is turned to 10 degrees. The valley floor is visible and the terrain differences are conspicuous (visibly significant).

There are numerous other possible views that can be used. Not shown are "drapes" of other data onto the landscape, such as landuse or rainfall amounts, allowing visual analysis of relationships between elevations and other data. Also, other types of data can be put into perspective views, such as rainfall. Consider a 3D rainfall coverage with a natural vegetation overlay to show the effect of precipitation on the ecosystem. Perspective views are flexible and valuable operations.

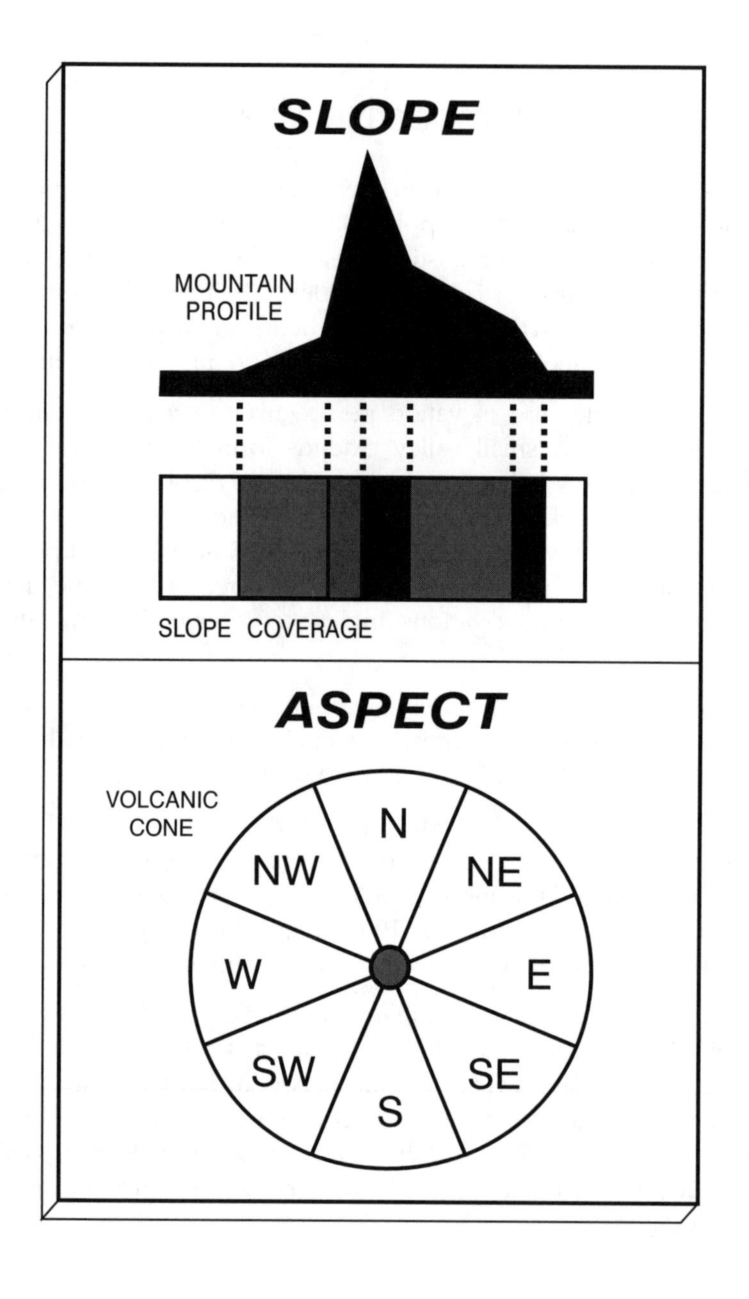
SLOPE
MOUNTAIN
PROFILE
SLOPE COVERAGE
ASPECT
VOLCANIC
CONE
N
NE
E
SE
S
SW
W
NW

Slope and Aspect

Other terrain or elevational analysis techniques include calculation of a profile (cross-section) and determination of slope, the change in elevation, and aspect, the direction a slope faces. A profile is created by drawing a line across an elevation coverage display, reading elevations along the line, and then viewing the "slice" from a horizontal perspective. Slope involves calculation of topographic changes on a landscape. In a raster system, slope is calculated by the change in elevation from one cell and its uphill neighbor. A greater difference in elevation between two points signifies greater slope.

The upper illustration shows a mountain profile indicating various slopes, from level to very steep. Slope coverages are made by reading elevations all over the coverage and converting them to slope classes. Below the profile is a simplified rectangular slope coverage, although most coverages are much more complex. A profile could also be constructed from a slope coverage.

Aspect is the direction a slope faces, usually expressed in cardinal compass terms (North, Northeast) or degrees. GISs can determine aspect of a slope by considering its direction of downslope. The lower illustration shows a symmetrical volcanic cone with equal amounts of aspect, except the top crater, of course. Most topography has much more complicated aspect categories.

Why is aspect significant? Solar insolation differs on northern and southern aspects of slopes, for example, and there are temperature and moisture variations. These factors influence vegetation and agricultural growth. Consequently, aspect can be an important ecological and planning consideration.

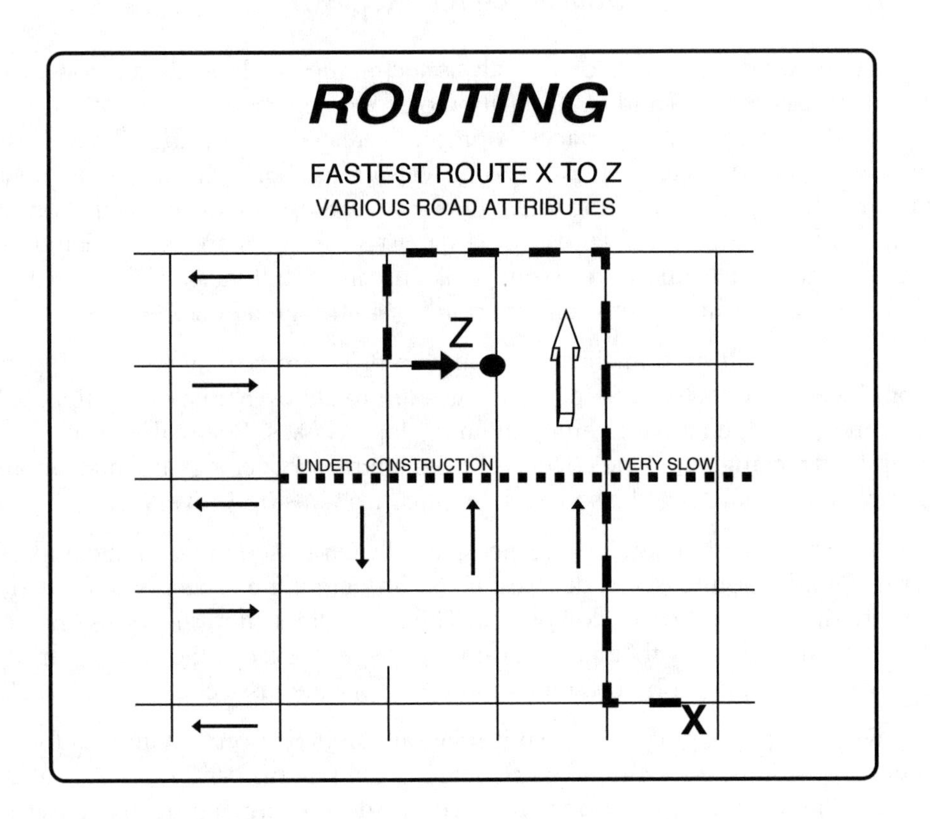
ROUTING
FASTEST ROUTE X TO Z
VARIOUS ROAD ATTRIBUTES
Z
UNDER CONSTRUCTION
VERY SLOW
X

Network Operations

Routing

A road or river network consists of multiple, connected lines. One of GIS's major strengths is its ability to perform numerous operations on networks. Finding the shortest path from one location to another involves tracking all possible routes and presenting the one having the shortest distance.

It is possible to attach attribute information to each line or network segment, such as rate of travel (speed limit), road condition, one-way streets, or road name. Therefore, given the distance and travel rate, it is rather easy for the program to determine which route is fastest (which may not be the shortest).

The illustration shows a road system, of which most streets are one-way. The fastest route between X and Z is calculated. The shortest route would be to take the "under construction" road, but it would be slower than the chosen route.

There are numerous applications of network GIS. Efficient routing is a major requirement for many businesses and services, especially for emergency responses. It is often said that "time is money"; therefore, for some enterprises, GIS can help save financial resources. In fact, there is a separate subdivision of GIS devoted to transportation, sometimes called Transportation GIS or GIST.

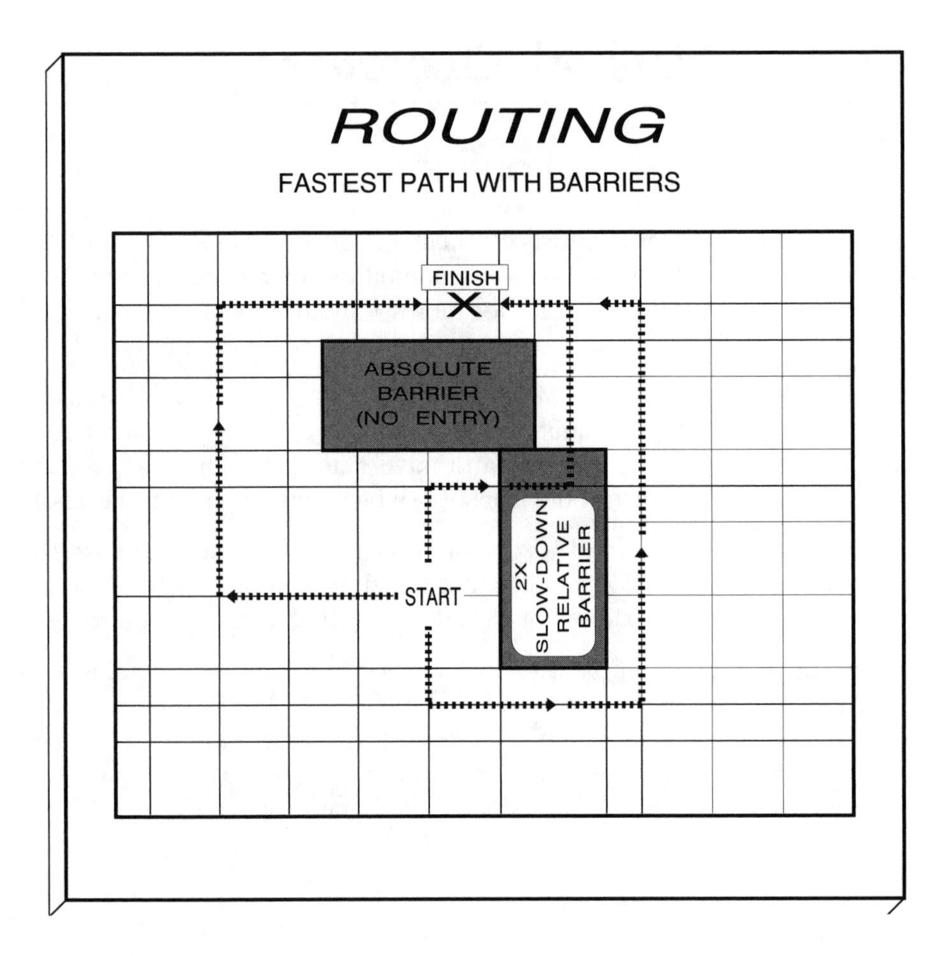
ROUTING
FASTEST PATH WITH BARRIERS
FINISH
ABSOLUTE BARRIER (NO ENTRY)
2X SLOW-DOWN RELATIVE BARRIER
START

Routing with Barriers

Travel speeds are usually not consistent over a network; many factors retard invariable and fast movement. "Friction surfaces" define barriers that slow movement. For example, travel over a rough road may slow down the rate by a factor of two. The "under construction" road of the previous illustration is an example of a friction surface. An absolute barrier, such as a road closing, stops travel altogether.

Illustrated is analysis of a street pattern with both a friction surface (relative barrier) and an absolute barrier. Various paths from Start to Finish are considered. It may be that travel around a 2x friction surface is faster than the shorter route through it. Various speeds can be used to test various scenarios and conditions. GIS is very good at making network analysis efficient. As noted in the oil spill illustration, emergency response applications make use of these types of operations.

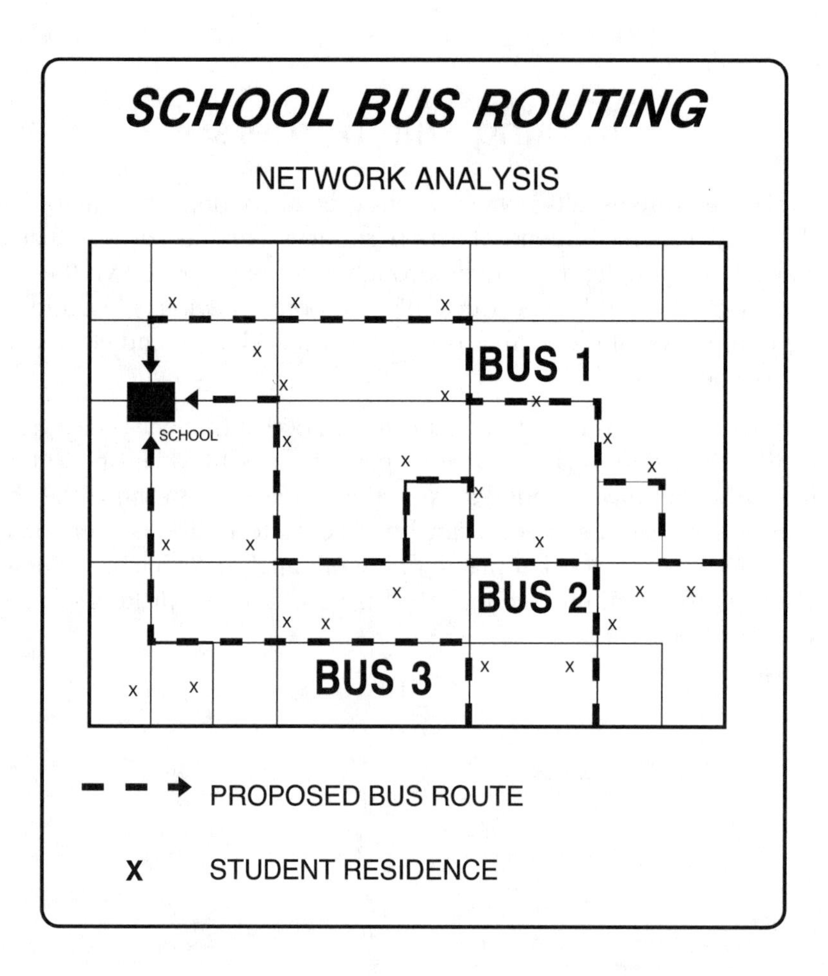
SCHOOL BUS ROUTING
NETWORK ANALYSIS
BUS 1
BUS 2
BUS 3
SCHOOL
PROPOSED BUS ROUTE
X STUDENT RESIDENCE

Network Analysis: School Bus Routing

An example of *network analysis* is determination of the most efficient routing of school busses. Students live all over, but the transportation manager must decide the best routes for each bus in order to maximize available resources. Time and fuel are major concerns, as is proper service for the students. In the illustration, it was decided that no student should have to walk more than one-half block to a bus stop. The three assigned busses were given the routes as noted by the heavy, dashed lines.

GIS can help solve these problems by building in criteria such as distribution of students, maximum walk distance, shortest route, and other considerations. Whereas the small illustrated example is easily solved, most school systems have very complex situations and therefore find GIS very useful.

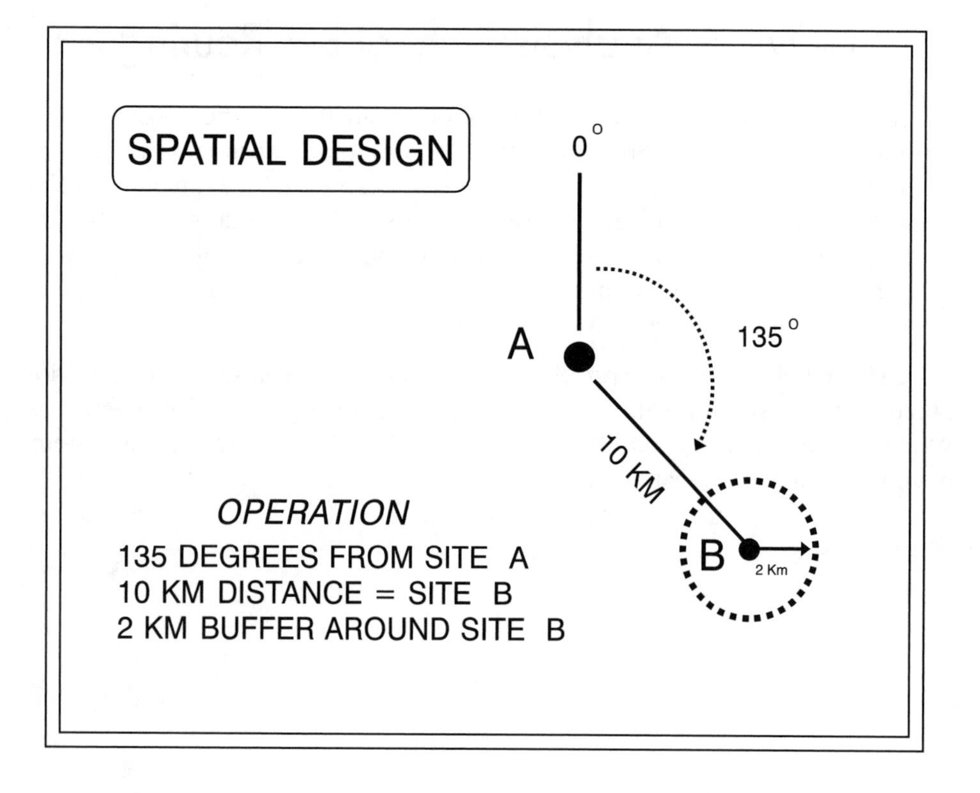
SPATIAL DESIGN
0°
135°
A
10 KM
B
2 Km
OPERATION
135 DEGREES FROM SITE A
10 KM DISTANCE = SITE B
2 KM BUFFER AROUND SITE B

Spatial Design

A powerful GIS can combine procedures to offer various design and analysis capabilities. In this illustration, for example, the user wants to draw a 10-km road exactly 135 compass degrees from point A, then establish a 2-km buffer around the end point (B). These types of design techniques are useful to city engineers and planners who develop utility networks (sewer, electricity, and so on). They are also very useful in the final design of maps.

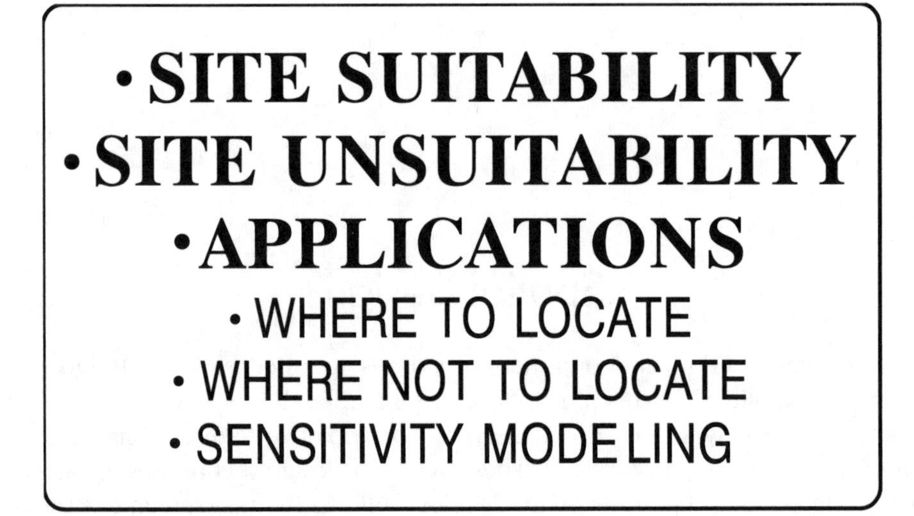
•SITE SUITABILITY
•SITE UNSUITABILITY
•APPLICATIONS
• WHERE TO LOCATE
• WHERE NOT TO LOCATE
• SENSITIVITY MODELING

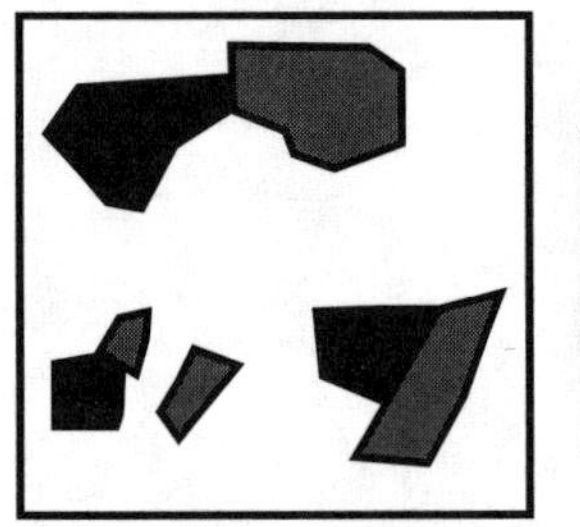
SITE SUITABILITY
BEST SITES
SUITABLE SITES
UNSUITABLE SITES

SENSITIVITY MODEL
VERY SENSITIVE
MODERATELY SENSITIVE
NOT SENSITIVE

10

Site Suitability and Models

Introduction

The variety of operations discussed in previous sections are used for many purposes. One of the major applications that uses a mix of procedures is *site suitability*—finding the most suitable site according to specific spatial and attribute conditions. A typical application would be locating a new industry, where several factors have to be considered, such as proximity to transportation, landuse, topography, land values, and others.

Site Suitability

Site suitability may include locating the most unsuitable site (or sites), as well as where *not* to locate (site unsuitability). Where is the least desirable place to locate something? This is still referred to as a site suitability procedure.

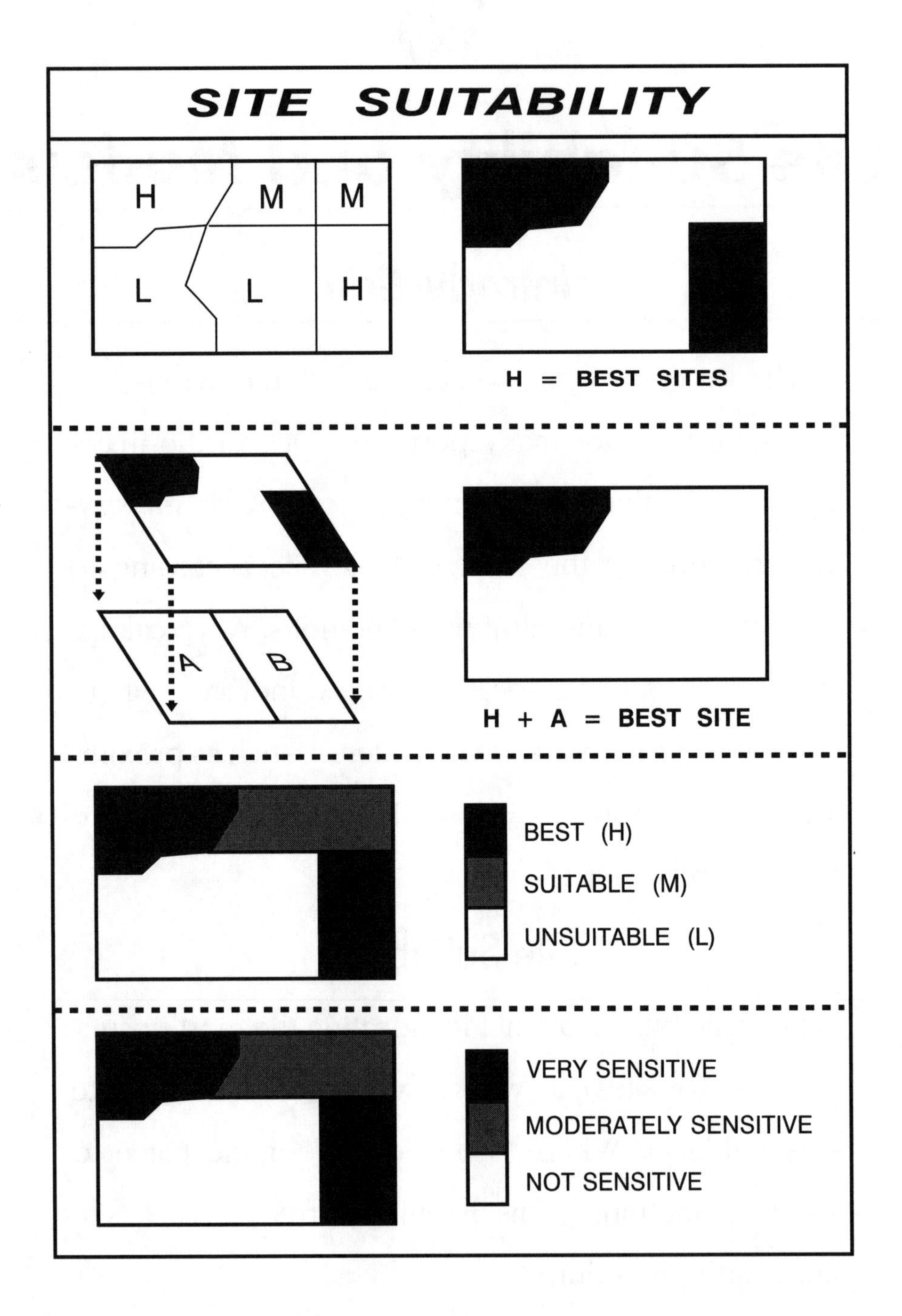
SITE SUITABILITY
H
M
M
L
L
H
H = BEST SITES
A
B
H + A = BEST SITE
BEST (H)
SUITABLE (M)
UNSUITABLE (L)
VERY SENSITIVE
MODERATELY SENSITIVE
NOT SENSITIVE

Site suitability analysis typically involves sifting through one or more data files to select the best attributes defined by specific directions. The upper illustration shows a rectangular area with soil quality attributes of H (high), M (medium), and L (low). The best sites are defined as those having high soil quality, and the resultant site suitability coverage presents a simple display.

Often site suitability analysis involves more than one coverage, perhaps overlaying them graphically or combining databases to locate where the best possible spatial or attribute conditions exist. The second set of illustrations shows the Best Sites coverage from the upper illustration overlaid with a district coverage (districts A and B). The best site is defined as the high soil quality areas in district A. Therefore, only one H site is selected.

It is only slightly more difficult to define site suitability in degrees or levels of acceptability, such as Best, Suitable (acceptable), and Unsuitable. The Soil Quality coverage at top has been analyzed as High = Best, Medium = Suitable, and Low = Unsuitable. The resultant site suitability coverage (third illustration) is close to the soil quality landscape (not quite—find the differences) but is now a redefined data set.

Often the site suitability application is to locate sites that are "sensitive" to given conditions. For example, environmental management often needs to include ecological sites that may be damaged by a proposed project, and site suitability procedures are used to locate them. It is even possible to find sites of various sensitivity measures, such as primary and secondary risk locations. The basic process is referred to as "sensitivity modeling." The lower illustration shows the three-level coverage just above, but the categories are redefined in ecological classes.

SITE SUITABILITY DATABASE APPROACH

PROPERTY NUMBER	AREA (SQ M)	OWNER	TAX CODE	SOIL QUALITY
1	100,000	TULATU	B	HIGH
2	50,100	BRAUDO	A	MEDIUM
3	90,900	BRAUDO	B	MEDIUM
4	40,800	ANUNKU	A	LOW
5	30,200	ANUNKU	A	LOW
6	120,200	SILIMA	B	HIGH

SELECTION CRITERIA:

The best property has:

Area: >40,000 sq m
Owner: not Silima
Tax Code: B
Soil Quality: High

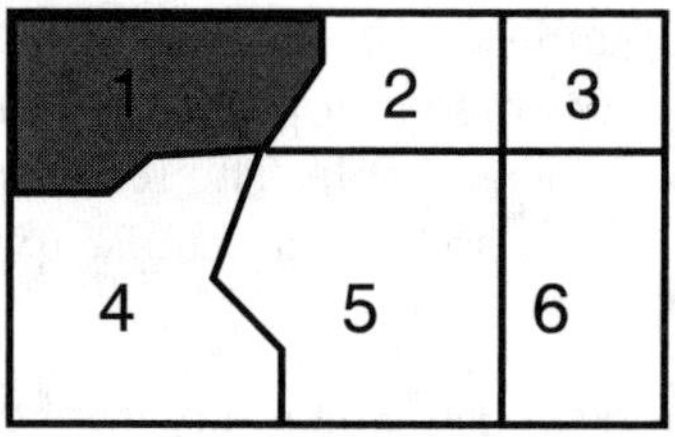

Answer: Property number 1

Database Approach

One of the simplest approaches to finding best sites is to sift through a database to select the features that meet all or most criteria. The illustration shows a query to locate properties having the attributes listed ("Selection Criteria:"). As detailed in the discussion concerning relational databases, the program begins at the first field to select the relevant properties, then moves on to subsequent fields, arriving (it is hoped) at one or more records that meet all conditions. A graphical display of coverages may not be necessary except for visualization of results, as shown in the map with property number 1 shaded. The selected site may be used in additional GIS operations. Nonetheless, often the site must be found using both database and coverage work.

SITE SUITABILITY
RUBBER INDUSTRY CRITERIA
Southern Thailand
• Must be >100 Ha
• Access to Rubber Plantation
Best: within 5 km
Acceptable: 5-10 km
• Access to Sealed Roads
Best: within 5 km
Acceptable: 5-10 km
• Proximity to River/Large Stream
Best: within 3 km
Acceptable: 3-7 km
• Within Phatthalung District

Site Suitability Example

A typical site suitability construction gives the basic purposes of a goal and sets specific criteria. Then it is the GIS operator's responsibility to determine how to achieve the preparation and final product. The illustration shows a standard set of conditions that are to be met in an industrial siting proposal in southern Thailand.

Obviously, some recoding and buffering will be needed. Weights could be assigned to the important factors, such as access to rubber plantations. Once the input coverages and structure have been prepared, they will be combined, probably using several overlays to achieve a final product from which to make site suitability assessment. Think how coverages will be constructed for each criterion. The next illustration presents an example of an operations flow chart.

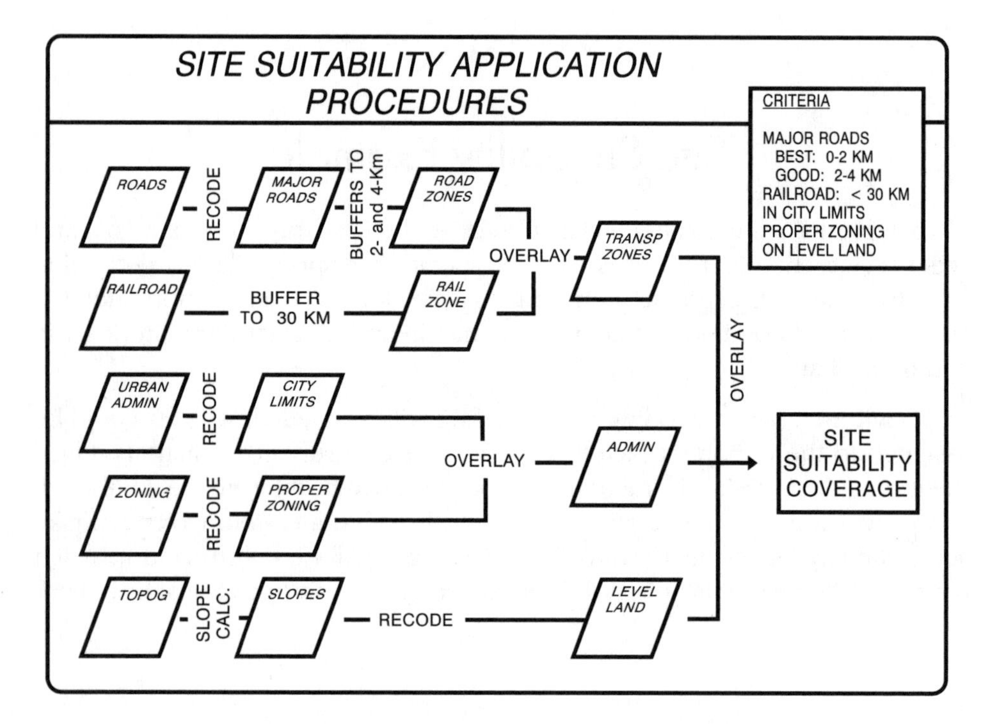
SITE SUITABILITY APPLICATION PROCEDURES
CRITERIA
MAJOR ROADS
BEST: 0-2 KM
GOOD: 2-4 KM
RAILROAD: < 30 KM
IN CITY LIMITS
PROPER ZONING
ON LEVEL LAND
ROADS
RECODE
MAJOR ROADS
BUFFERS TO 2- and 4-Km
ROAD ZONES
OVERLAY
TRANSP ZONES
RAILROAD
BUFFER TO 30 KM
RAIL ZONE
OVERLAY
URBAN ADMIN
RECODE
CITY LIMITS
OVERLAY
ADMIN
SITE SUITABILITY COVERAGE
ZONING
RECODE
PROPER ZONING
TOPOG
SLOPE CALC.
SLOPES
RECODE
LEVEL LAND

Site Suitability Flow Chart

Although the concept is easy, sometimes site suitability can be a complicated process and a bit confusing, particularly when multiple coverages are used. Therefore, good preparation and organization are needed. The illustration shows a typical flow chart of the steps to be taken to prepare input coverages and achieve a final product. A list of criteria is in the upper right-hand box. There are probably other ways to achieve some of these steps, but this arrangement and its operations seem logical.

Note the organization. Roads and Railroad coverages are worked in adjacent steps, first individually (making buffers) and then in an overlay giving transportation zones. The administration coverages are treated similarly. A logical structure makes the process more efficient. Follow each step to understand the process.

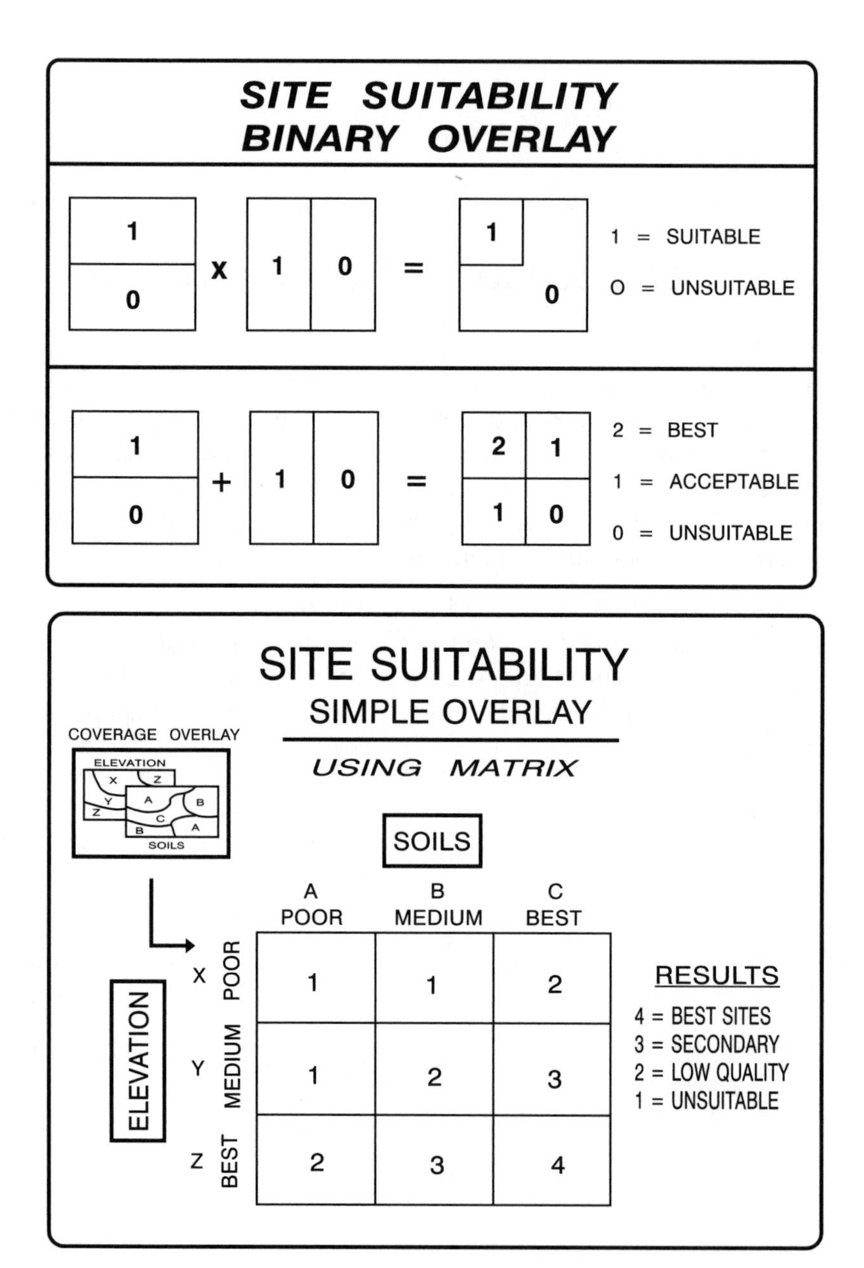
SITE SUITABILITY
BINARY OVERLAY
1
0
x
1
0
=
1
0
1 = SUITABLE
O = UNSUITABLE
1
0
+
1
0
=
2
1
1
0
2 = BEST
1 = ACCEPTABLE
0 = UNSUITABLE
SITE SUITABILITY
SIMPLE OVERLAY
USING MATRIX
COVERAGE OVERLAY
ELEVATION
X
Z
Y
A
B
Z
C
B
A
SOILS
SOILS
A
POOR
B
MEDIUM
C
BEST
ELEVATION
X POOR
Y MEDIUM
Z BEST
1
1
2
1
2
3
2
3
4
RESULTS
4 = BEST SITES
3 = SECONDARY
2 = LOW QUALITY
1 = UNSUITABLE

Overlay

Binary Overlay

Overlay of two or more coverages is a common site suitability operation. The upper illustration is a simple procedure of combining two binary coverages: 1 = Suitable, 0 = Unsuitable. Overlay with Multiply will result in Suitable where 1 and 1 occur; otherwise, any site with Unsuitable will be unacceptable in the end.

To get a *rating model,* overlay the same coverages using Add. The "best" sites are where two Suitables occur (1 + 1 = 2). Acceptable is where one Suitable exists (1 + 0 = 1), and of course two Unsuitables are clearly unacceptable (Unsuitable). These are two simple yet very useful approaches to site suitability.

Matrix Overlay

Another approach to obtaining suitability ratings is to use the matrix to calculate results. Although a bit more difficult, the matrix offers better control over data. In the lower illustration, Elevation and Soils are paired and entered as rows and columns on the matrix. Both coverages have been recoded to Best, Medium, and Poor attributes, and output values will be assigned by the user. Four levels are desired: Best (4), Secondary (3), Low Quality (2), and Unsuitable (1).

Read the matrix carefully. Best sites consist of Best Soil (class C) and Best Elevation (class Z). Secondary sites, coded 3, have one Best class from either the Elevation or Soil coverage and a Medium (soil B or elevation Y). The other categories, Low Quality (2) and Unsuitable (1), can be read in the matrix.

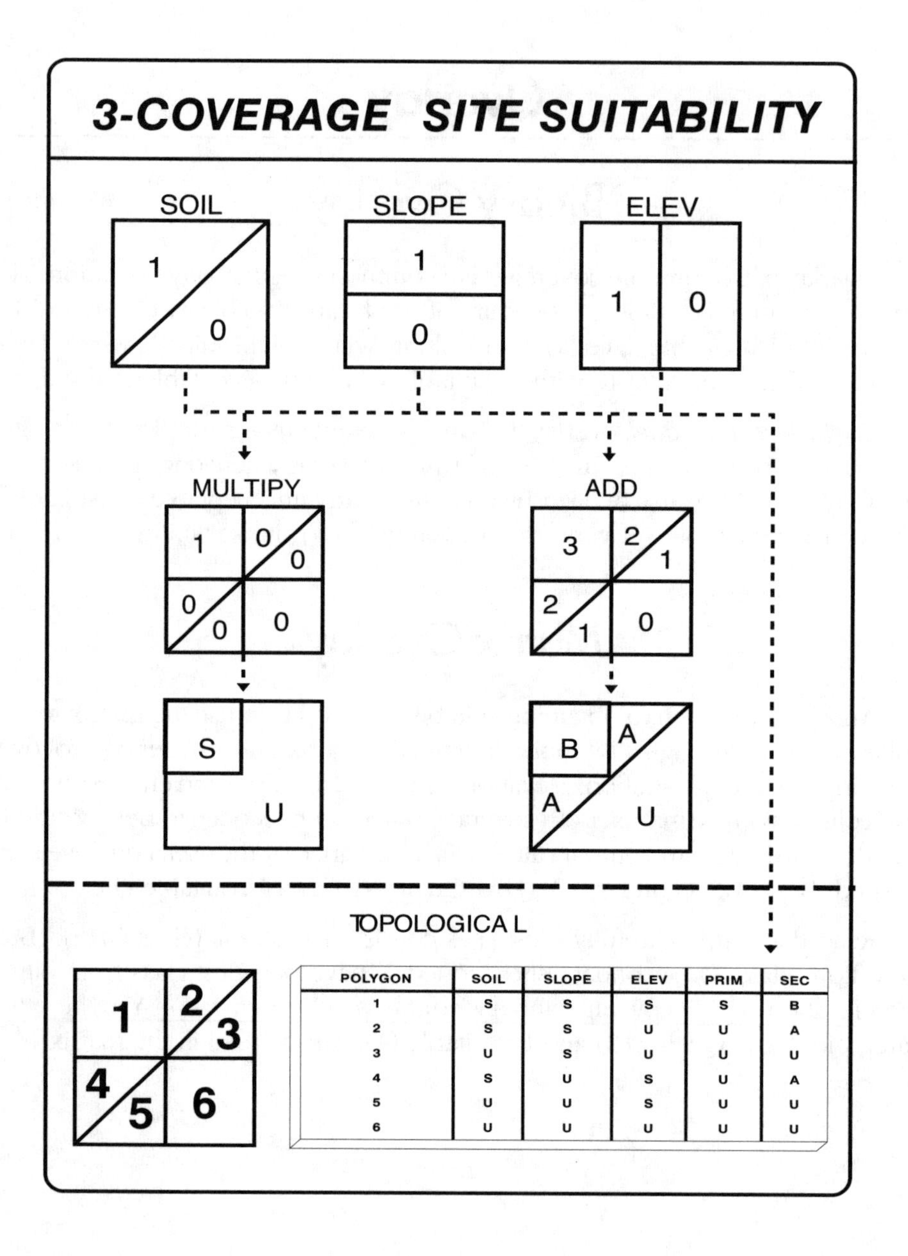

POLYGON	SOIL	SLOPE	ELEV	PRIM	SEC
1	S	S	S	S	B
2	S	S	U	U	A
3	U	S	U	U	U
4	S	U	S	U	A
5	U	U	S	U	U
6	U	U	U	U	U

Three-coverage Overlay

Overlaying three coverages for site suitability is conceptually easy, requiring only careful preparation of input coverage data. The illustration shows three coverages—Soils, Slope, and Elevation—to determine the suitable sites for some activity. The coverages have binary codes of 1 for Suitable and 0 for Unsuitable. As in the previous example, we can use an overlay with Multiply to get absolute or binary suitability, or overlay with Add to get ratings (levels of suitability).

The overlay with Multiply will result in suitable sites with value 1 and all others having value 0. Easy: 1 x 1 x 1 = 1; any combination of 1s and 0 = 0 (1 x 1 x 0). This is no different than two cover suitability operations (or four or more covers).

Overlay using Add gives three levels of ratings. All suitable categories (1 + 1 + 1) will sum to 3, Best (B). Acceptable (A) has two Suitables and one Unsuitable, summing to 2 (1 + 1 + 0 in any combination). Two or more Unsuitables are unsuitable (1 + 0 + 0 or 0 + 0 + 0). Study this example and the logic becomes obvious.

Despite the need for coding and recoding in most raster systems, the use of numbers in site suitability analysis has advantages. For example, a variety of formulas can be used in the overlay process to test "what-if" scenarios or theories of influence, such as weighing Elevation over Soils and Slope. (Modeling will be discussed.) Raster coding does have advantages.

Topological

The lower illustration shows the overlay results, with each polygon's identity number (1 through 6, which are *not* suitability values). The database gives each polygon's attributes according to the coverages. With a good relational database and topological system, coding of values is not needed.

The database also includes two fields to present suitability results. PRIM is Primary Suitability that responds like the overlay with Multiply (only all suitables make a final Suitable, or S); any polygon with an unsuitable classification results in an Unsuitable (U). The Secondary Suitability column (SEC) is like overlay with Add, giving B (Best), A (Acceptable), and U (Unsuitable).

The decision to use either raster or vector site suitability analysis is typically determined first by the available system format. When there is a choice, project considerations are used. If the final product is simple, as in the illustration, perhaps the vector system is better, provided the more complicated database structure doesn't create problems. If further analysis and modeling will be used, the raster format may be better. Experience is the best guide.

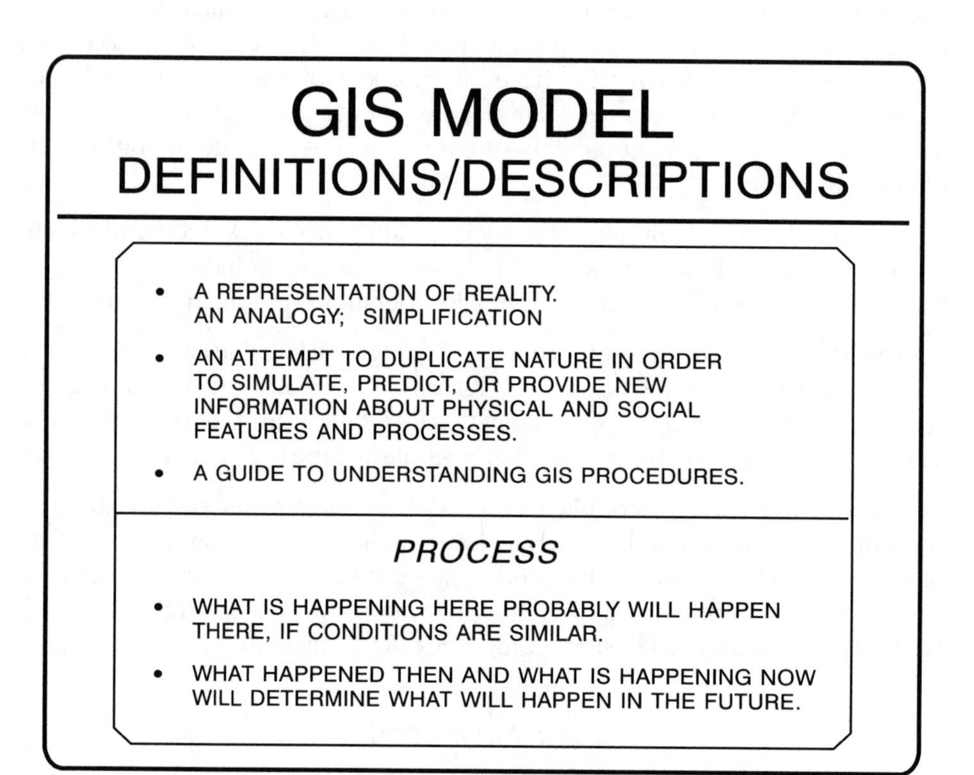
GIS MODEL
DEFINITIONS/DESCRIPTIONS
• A REPRESENTATION OF REALITY. AN ANALOGY; SIMPLIFICATION
• AN ATTEMPT TO DUPLICATE NATURE IN ORDER TO SIMULATE, PREDICT, OR PROVIDE NEW INFORMATION ABOUT PHYSICAL AND SOCIAL FEATURES AND PROCESSES.
• A GUIDE TO UNDERSTANDING GIS PROCEDURES.
PROCESS
• WHAT IS HAPPENING HERE PROBABLY WILL HAPPEN THERE, IF CONDITIONS ARE SIMILAR.
• WHAT HAPPENED THEN AND WHAT IS HAPPENING NOW WILL DETERMINE WHAT WILL HAPPEN IN THE FUTURE.

Models

GIS Model

In GIS applications, work done and results achieved can be for a single goal, but we often hope they are useful for more than one site or task. When the procedures are found successful, it is nice when they can be applied to other sites involving similar tasks. Therefore, we not only provide achievement for a specific objective but make a "model" that may be valuable for other applications.

A model in GIS is a generalized set of conditions or procedures. We use the term rather loosely, but it means basically that what we are doing is not unique and can be used for other areas, conditions, or circumstances. Although there are numerous definitions and uses of the term *GIS model,* the following points are presented as working concepts. Read the list in the illustration and return to it for review after the following section.

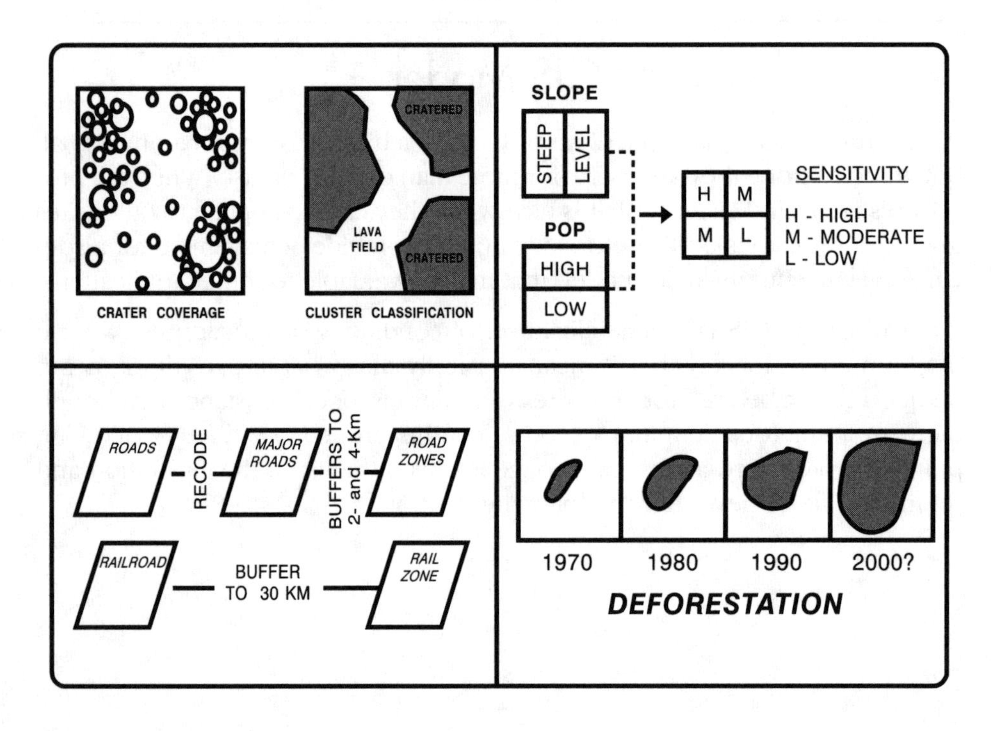
CRATERED
LAVA FIELD
CRATERED
CRATER COVERAGE
CLUSTER CLASSIFICATION
SLOPE
STEEP
LEVEL
POP
HIGH
LOW
H
M
M
L
SENSITIVITY
H - HIGH
M - MODERATE
L - LOW
ROADS
RECODE
MAJOR ROADS
BUFFERS TO 2- and 4-Km
ROAD ZONES
RAILROAD
BUFFER TO 30 KM
RAIL ZONE
1970
1980
1990
2000?
DEFORESTATION

The Nature of a GIS Model

A GIS model:

◆ Is only a representation of reality; it is a generalization. It does not include everything on the landscape but uses only those items that illustrate or are important to the task or goal at hand. The model reduces complexity and confusion. A model is an analogy, a simplification of the real world. The first image in the illustration presents a cratered lunar landscape, with the clustered generalization serving as a model for the terrain—showing reduced detail but remaining a reasonable representation.

◆ Attempts to duplicate nature in order to simulate, predict, or provide new information about physical or social features and processes. It illustrates how natural or human processes work. The next illustration shows how GIS presents the relationship between slope and population density, and the resultant sensitivity effects. This is a good model because it explains in a simple manner but does not attempt to show the actual landscape.

◆ Is a set of GIS procedures to accomplish a given task. This type of model shows which GIS steps are needed to complete a general or specific goal. The image in the lower left of the illustration is a flow diagram showing GIS steps to accomplish some task. The operator can use this as a "cookbook" menu procedure.

One important function of a model is to demonstrate what is going on, the actual process, in a simple and understandable manner. A model states:

◆ What is happening here (or in one place) will probably happen there (or another place) if conditions are similar (unless there is evidence to suggest a different process).

◆ What happened then (in the past) and what is happening now will determine or influence what will happen later (called predictability). Trend analysis attempts to determine trends and then project into the future or past. The lower right illustration shows 1970 to 1990 deforestation, with a projection to the year 2000. Data exist for the years 1970 to 1990, but the forecast is only a prediction based on existing data.

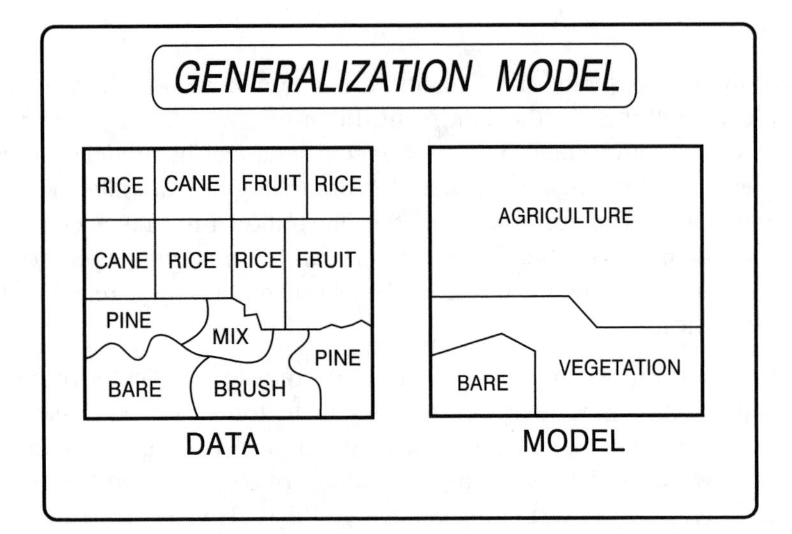
GENERALIZATION MODEL
RICE
CANE
FRUIT
RICE
CANE
RICE
RICE
FRUIT
PINE
MIX
PINE
BARE
BRUSH
DATA
AGRICULTURE
BARE
VEGETATION
MODEL

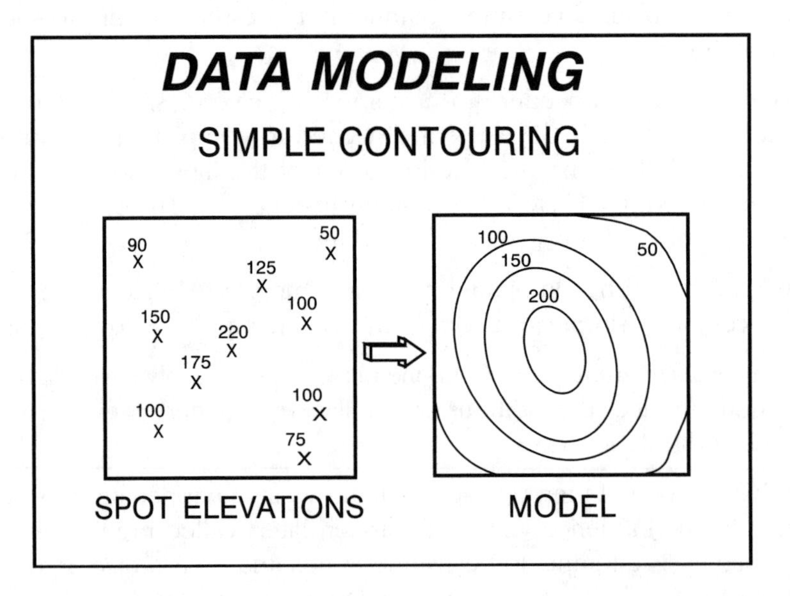
DATA MODELING
SIMPLE CONTOURING
90
50
125
100
150
220
175
100
100
75
SPOT ELEVATIONS
100
50
150
200
MODEL

Generalization Model

Actually, any map is a model of reality because only selected features are shown; a photograph shows everything. The upper illustration shows a simple model. The Data coverage, made from original field or image data, is modeled to a more general structure. The model reduces the landscape from detailed to more general categories, from field data to thematic coverage. Note that the shapes have also been simplified. Whereas the landscape map is a model of the real world, the thematic coverage is a model of both the real world and the detailed landscape map.

Data Modeling: Contours

A simple model can take available data and generalize them for a larger area (see lower illustration). Without additional information, the model "assumes" that what is seen or known at specific sites determines or influences what happens in the area overall. Elevations can be determined at specific data sites by surveying, and then a contouring program can extrapolate (estimate) those points as relevant to the area. The points are believed to be representative of the entire area. Elevation can be represented by contour lines—lines on which a given elevation is constant (the line has that specific elevation everywhere it occurs).

The principal idea is that elevations at any given point can be inferred by the contour lines, but in reality there may be some inaccuracies and distortions. A contour map is only a generalization of the real world; it is a model.

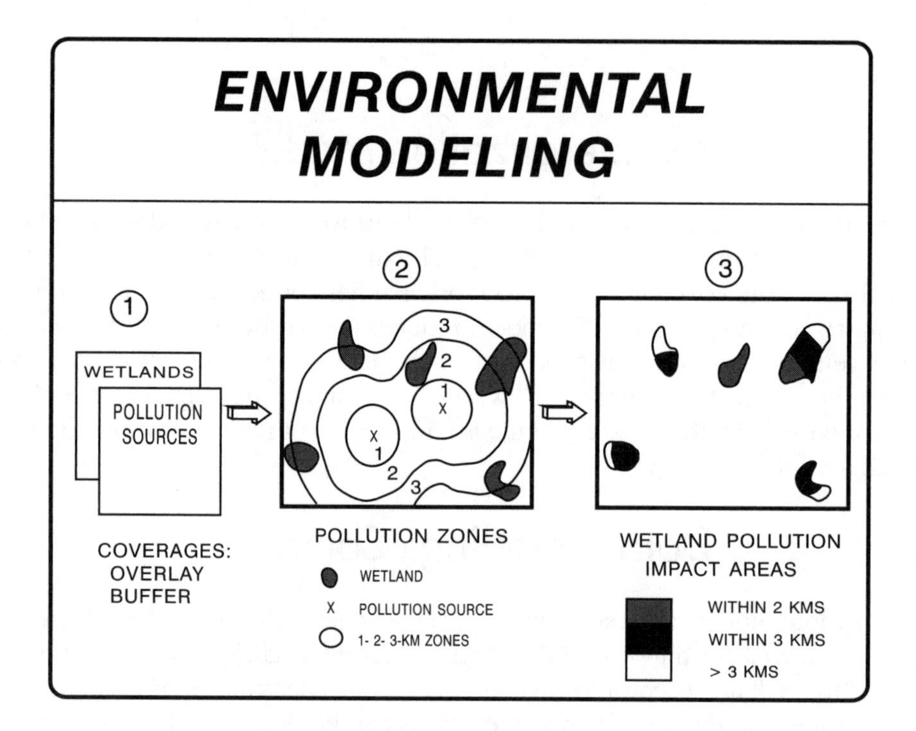
ENVIRONMENTAL
MODELING
1
2
3
WETLANDS
POLLUTION
SOURCES
COVERAGES:
OVERLAY
BUFFER
POLLUTION ZONES
WETLAND
X POLLUTION SOURCE
1- 2- 3-KM ZONES
WETLAND POLLUTION
IMPACT AREAS
WITHIN 2 KMS
WITHIN 3 KMS
> 3 KMS

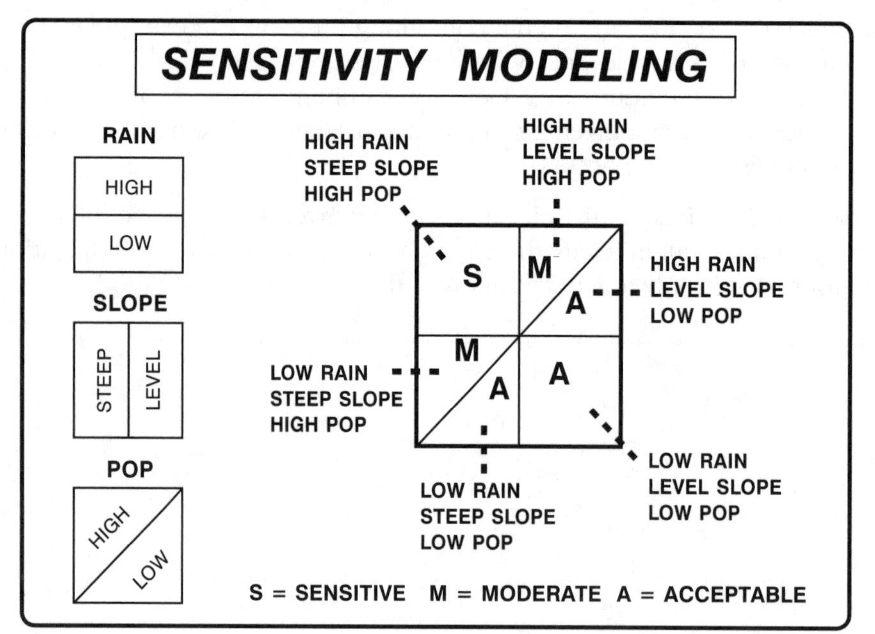
SENSITIVITY MODELING
RAIN
HIGH
LOW
SLOPE
STEEP
LEVEL
POP
HIGH
LOW
HIGH RAIN
STEEP SLOPE
HIGH POP
HIGH RAIN
LEVEL SLOPE
HIGH POP
HIGH RAIN
LEVEL SLOPE
LOW POP
LOW RAIN
STEEP SLOPE
HIGH POP
LOW RAIN
STEEP SLOPE
LOW POP
LOW RAIN
LEVEL SLOPE
LOW POP
S
M
A
M
A
A
S = SENSITIVE M = MODERATE A = ACCEPTABLE

Environmental Modeling

Environmental landscapes and processes are usually complex, and modeling ecosystems for environmental analysis is a valuable application of GIS. It is easy, for example, to change one of the parameters (such as pollution zone sizes) or to add new sources to test alternative situations ("what-if" analysis). The data may be somewhat generalized, but the basic elements are believable. The upper illustration shows both the process and landscape of a wetlands (environmentally sensitive ecosystems) pollution project. Steps 1 through 3 (circled) guide the GIS procedure, and landscape data show expected results.

To model the potential pollution of wetland areas, specific data have been combined and buffered (1) in the illustration to make a Pollution Zones coverage (2)—a model of pollution transportation from the X site pollution sources. Zones of distance from the pollution sources are based on the logical assumption that wetlands closer to a pollution site are in higher danger of ecosystem damage.

The wetlands have been zoned into impact areas (3) so that proper management procedures can be started. Pollution may or may not occur as predicted due to many variables, but this is a generalized idea based on limited available data. The model may show where additional analysis is needed for better results.

Sensitivity Modeling

A basic model shows what will happen under given data or circumstances. Before applying GIS operations to a set of coverages for environmental analysis, the user must understand potential results and their meanings. The model in the lower illustration gives a general structure of merging rain, slope, and population density to test which combinations may be sensitive and which may not.

The model coverage shows Sensitive (S) sites having high rain, steep slopes, and high population density. Moderate (M) have two highs and one lesser attribute. Read the other measures and note how the model shows all possibilities. Reality will be different, of course, but this may be a good guide.

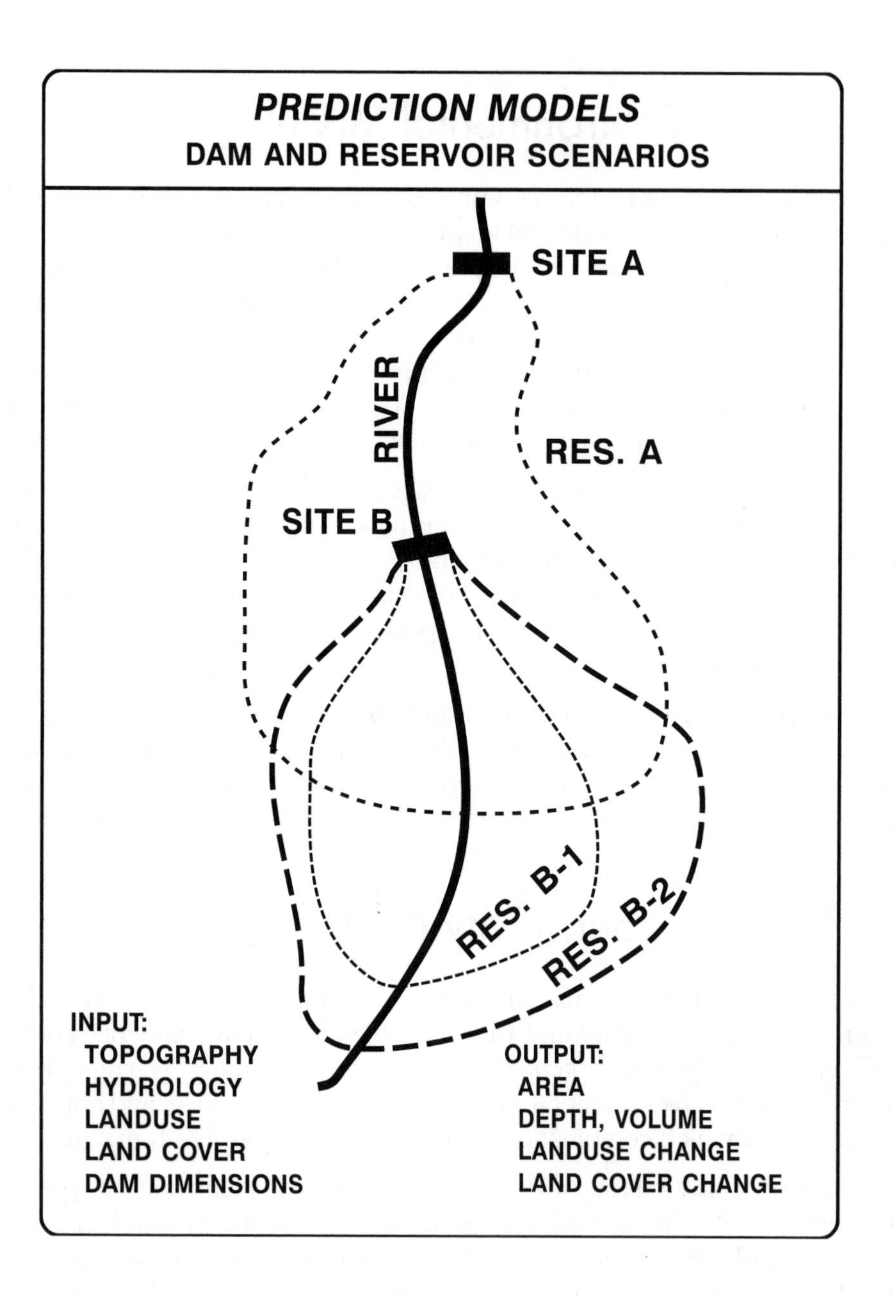
PREDICTION MODELS
DAM AND RESERVOIR SCENARIOS
SITE A
RIVER
RES. A
SITE B
RES. B-1
RES. B-2
INPUT:
TOPOGRAPHY
HYDROLOGY
LANDUSE
LAND COVER
DAM DIMENSIONS
OUTPUT:
AREA
DEPTH, VOLUME
LANDUSE CHANGE
LAND COVER CHANGE

Prediction Models

GIS models are used to test various scenarios ("what-if" cases) of proposed spatially related projects or processes. They are especially useful in predicting impacts of various features, such as a proposed dam. The location and dam attributes can be changed to test options.

Illustrated are two possible dam sites (A and B) and predicted results of two dam configurations at site B. Input data consist of several factors that directly affect reservoir characteristics, such as topography and elevation, hydrology, and dam dimensions (particularly height). Landuse and Land Cover (not shown) could be used to gauge environmental impacts. Population distribution could be added.

Output includes the probable spatial extent of the reservoirs (Res. A, Res. B-1, and Res. B-2), depth and volume of each, and the changes (losses) in given landuses and land cover. Maps showing old and new landuse and land cover would be helpful.

Some GISs offer "goal" objectives: a way to define what is wanted and then have the GIS help to achieve that goal. For instance, in the dam scenario in the illustration, a nearby historic site must not be flooded by the reservoir. The location and elevation of the site is put into the calculation and the program helps to configure the reservoir to protect the site. Water level would be a major factor, which is determined by the dam. In the end, perhaps several reservoir/dam possibilities may be produced and the user continues the process of environmental analysis.

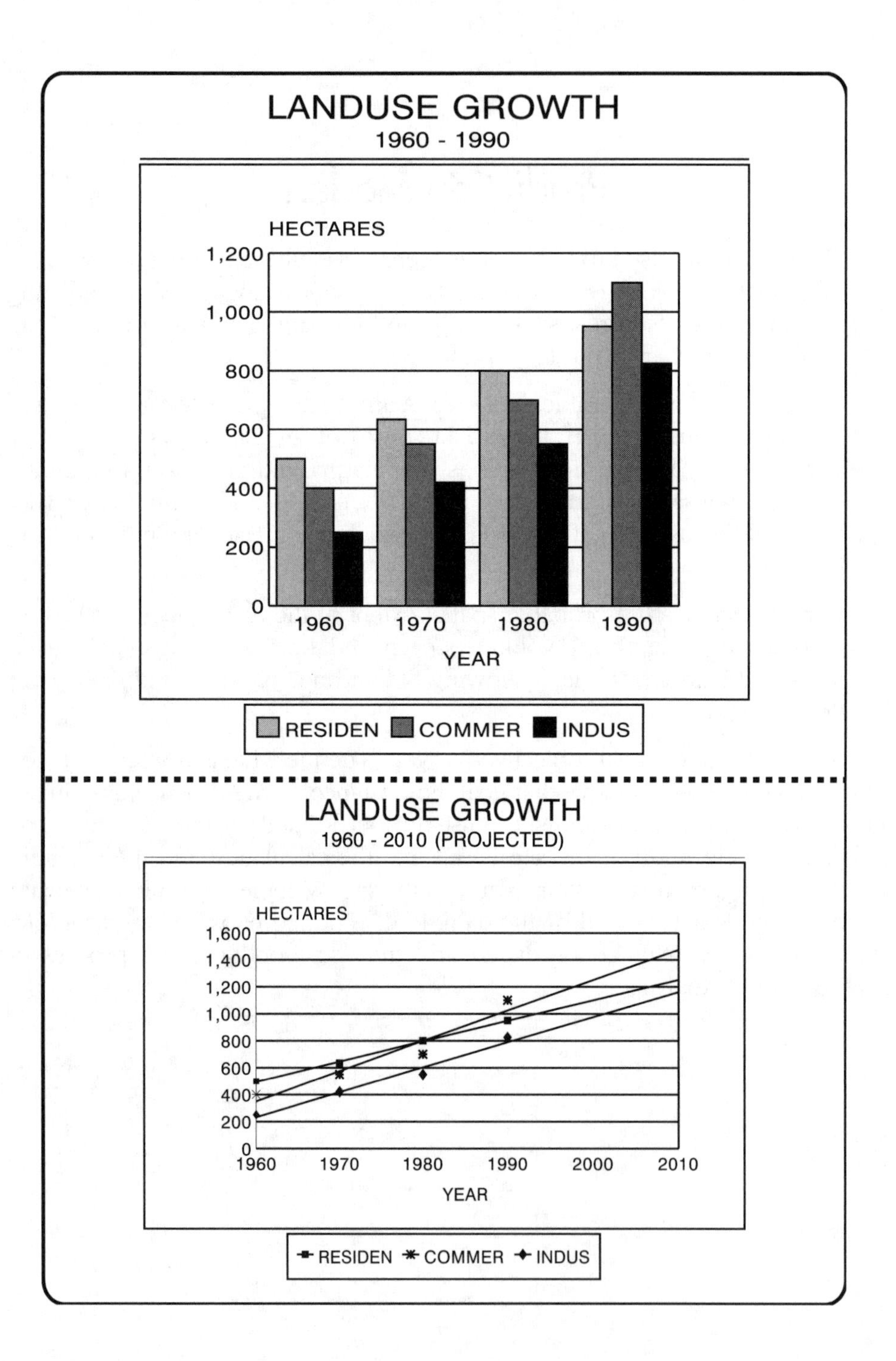
LANDUSE GROWTH
1960 - 1990
HECTARES
1,200
1,000
800
600
400
200
0
1960
1970
1980
1990
YEAR
RESIDEN
COMMER
INDUS
LANDUSE GROWTH
1960 - 2010 (PROJECTED)
HECTARES
1,600
1,400
1,200
1,000
800
600
400
200
0
1960
1970
1980
1990
2000
2010
YEAR
RESIDEN
COMMER
INDUS

Statistical Models

There are various forms of quantitative models. One example has been mentioned: Soil + Slope x 2.5 = Erosion Potential. Statistical techniques model data by giving measures that are representative. The average of a set of numbers may not actually exist as a datum, but it is one way of describing the entire data list. The illustration shows two ways to represent GIS data other than as coverages or databases.

The upper graph is a simple bar chart, showing three types of landuse during each year measured. Comparisons of landuses within a given year are easy to see, and growth from one period to another is possible. It is a general view of existing landuse and gives an idea of growth over 30 years.

The lower line graph is another method of describing the growth. The point symbols represent each landuse measure at each time period, but the important feature is the line. A line connecting each point would show no more than the bar graph of the upper illustration, but this line is a "trend line"—the line of best fit or average straight line that runs through the data points for each landuse.

A trend line is a generalization of the landuse quantities. Each true point may occur above or below the line, but the line is a representation over time.

Equally important is that the trend line can be extended to the future (or to the past), giving an indication of what will happen if current trends continue. In this case, landuse sizes will continue to increase. However, reality may be different, given numerous possible variables. This is a good start at understanding the area's land use change, but it is only a start.

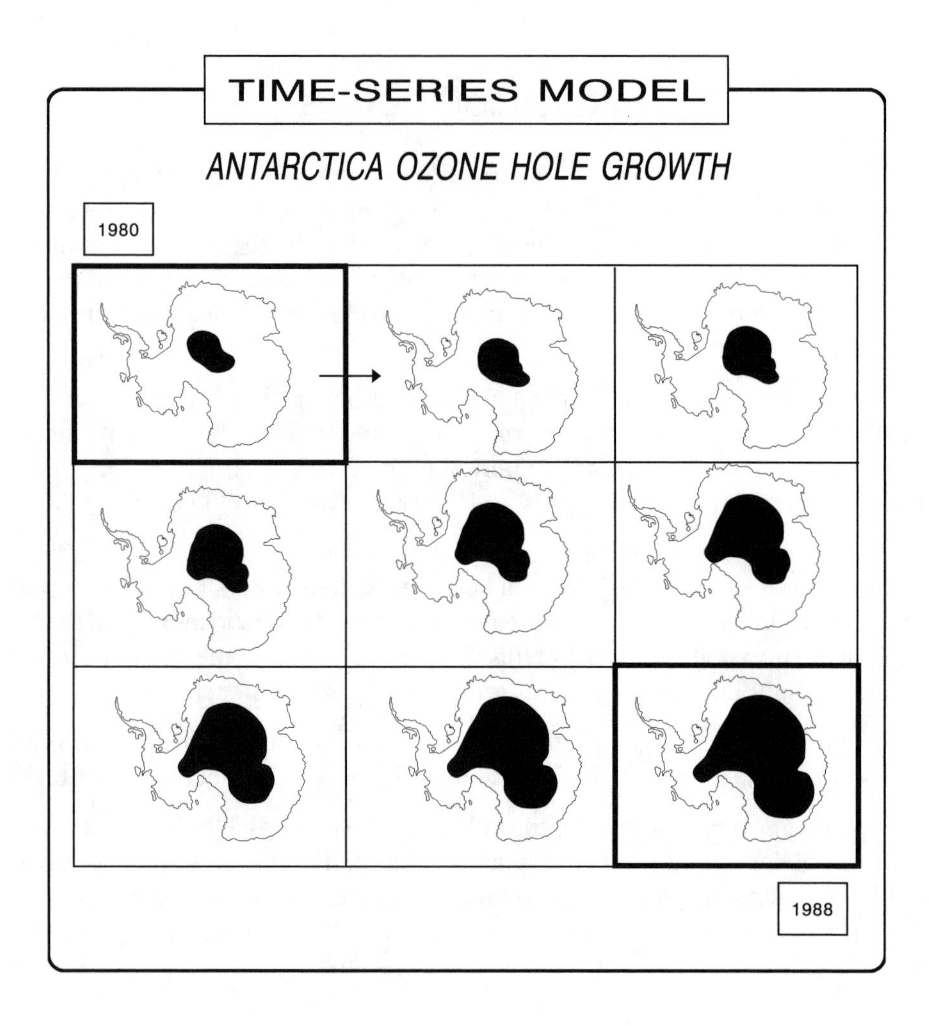
TIME-SERIES MODEL
ANTARCTICA OZONE HOLE GROWTH
1980
1988

Time-Series Model

Models are simplifications of reality. They represent what is happening generally. When applied to time, models can help understand what is happening over a period of years. They can estimate what probably happened or will happen based on data provided for other times. These can be called time-series models.

When conditions for two times are given, a model may project backward, forward, or in between to extrapolate conditions. The illustration presents a hypothetical case in which the 1980 and 1988 Antarctica ozone hole configurations have been provided and the time-series model estimates changes during the intervening years (1981 to 1987).

These conditions may not have actually occurred, and the model should not be thought of as providing specific, conclusive data, but a credible set of scenarios have been created from which an idea of the changes can be considered. A model can do no more than that.

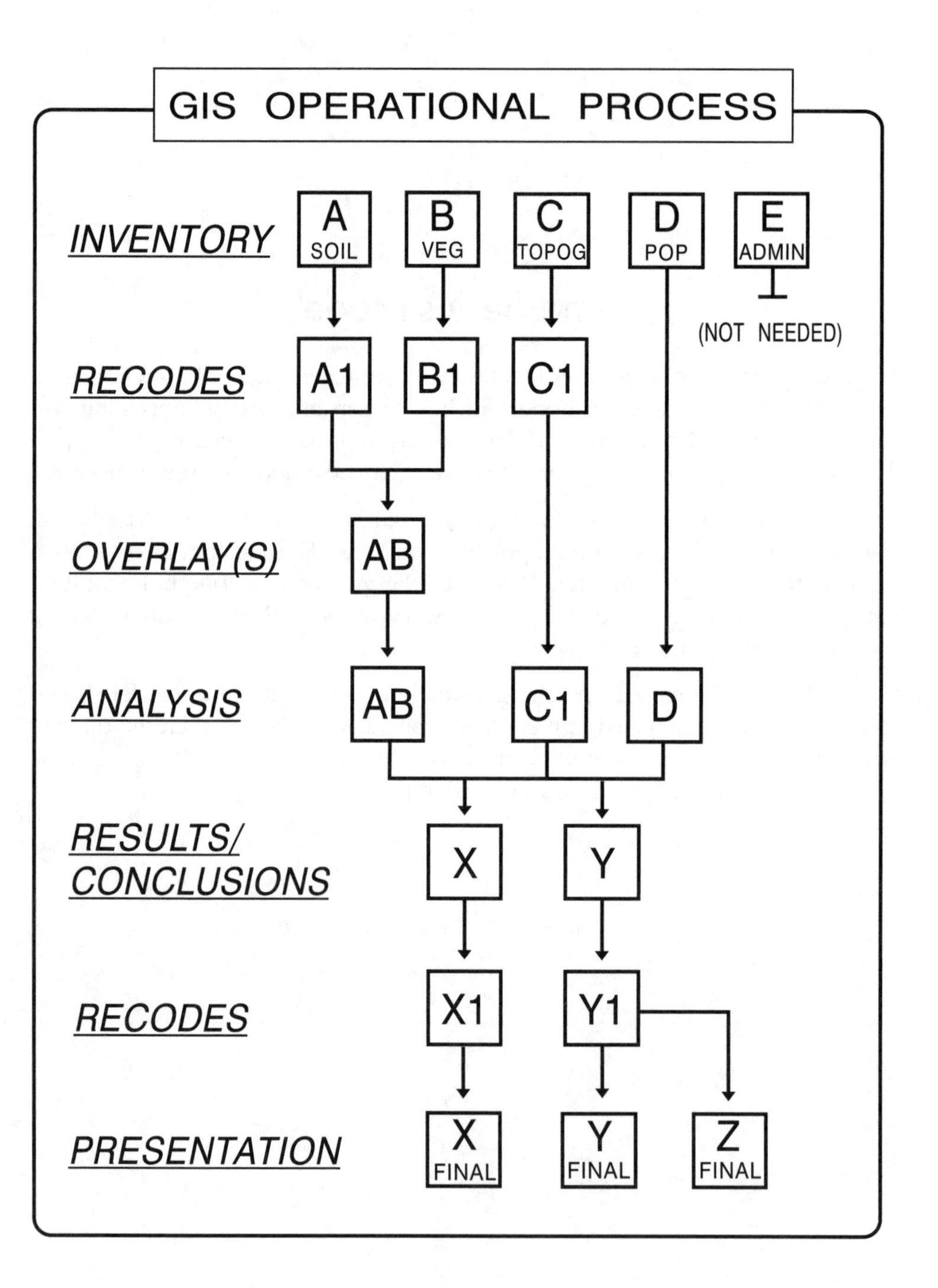
GIS OPERATIONAL PROCESS
INVENTORY
A
SOIL
B
VEG
C
TOPOG
D
POP
E
ADMIN
(NOT NEEDED)
RECODES
A1
B1
C1
OVERLAY(S)
AB
ANALYSIS
AB
C1
D
RESULTS/
CONCLUSIONS
X
Y
RECODES
X1
Y1
PRESENTATION
X
FINAL
Y
FINAL
Z
FINAL

GIS Operational Process

An operational model provides specific steps to be followed in order to achieve a specified goal. It shows the procedures to be used—the process flow chart. Such a model is not data specific. The process should work with different data sets or in a particular region (within a certain range).

In the illustration, an ecological analysis is outlined, beginning with the input coverages specified (from any part of the region under study). Each step is a certain procedure that leads to the next step. Presentation of results is the final step. The process is the model. Some users refer to this as the "cookbook" approach.

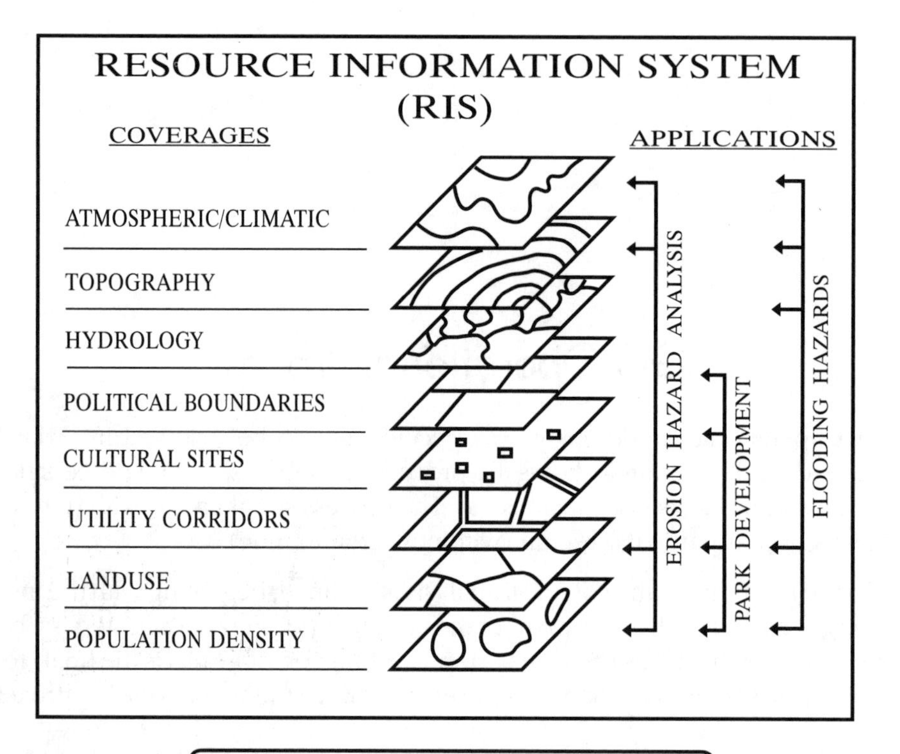
RESOURCE INFORMATION SYSTEM
(RIS)
COVERAGES
APPLICATIONS
ATMOSPHERIC/CLIMATIC
TOPOGRAPHY
HYDROLOGY
POLITICAL BOUNDARIES
CULTURAL SITES
UTILITY CORRIDORS
LANDUSE
POPULATION DENSITY
EROSION HAZARD ANALYSIS
PARK DEVELOPMENT
FLOODING HAZARDS

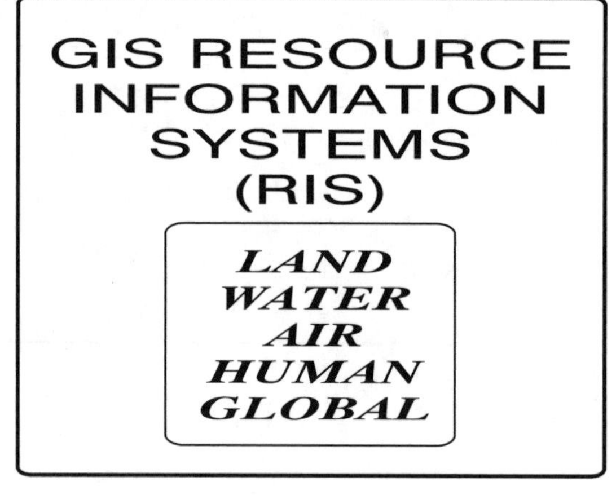
GIS RESOURCE
INFORMATION
SYSTEMS
(RIS)
LAND
WATER
AIR
HUMAN
GLOBAL

Application Planning

Resource Information System

Most of the previous steps have discussed individual operations, but a GIS usually has larger applications. Illustrated at top is a Resource Information System (RIS) that provides a user, or probably many users, with a set of data and a flexible means to achieve numerous goals.

The coverages typically come from different sources, such as remote sensing imagery, digitized maps, databases, and the like. They are stored so that users can extract various data sets for diverse applications. For example, the erosion hazard analysis people need atmospheric data, topography, landuse, and population density. The park development project uses political boundaries, cultural sites, landuse, and population density. The flooding hazard analyst requires the five coverages pointed to in the illustration.

An RIS is one of the major uses of GIS and is particularly valuable in that there can be many applications. The cost of maintaining a central database and GIS is far less than each application organization having its own. Also, there will be consistency in regional data, and therefore a higher-quality database.

GIS RISs can exist for all natural and social environments. The lower illustration lists Land, Water, Air, Human, and Global. Most GIS applications today are land-oriented, but developing technology is permitting better capabilities. For example, air and water are 3D environments, not a single surface, as is land. 3D GIS requires powerful computers and large data sets. The human environment uses both the natural and the human world, and there is no problem in making GIS dedicated to the cultural landscape. Global data sets are usually very large and cover either all or large parts of the Earth.

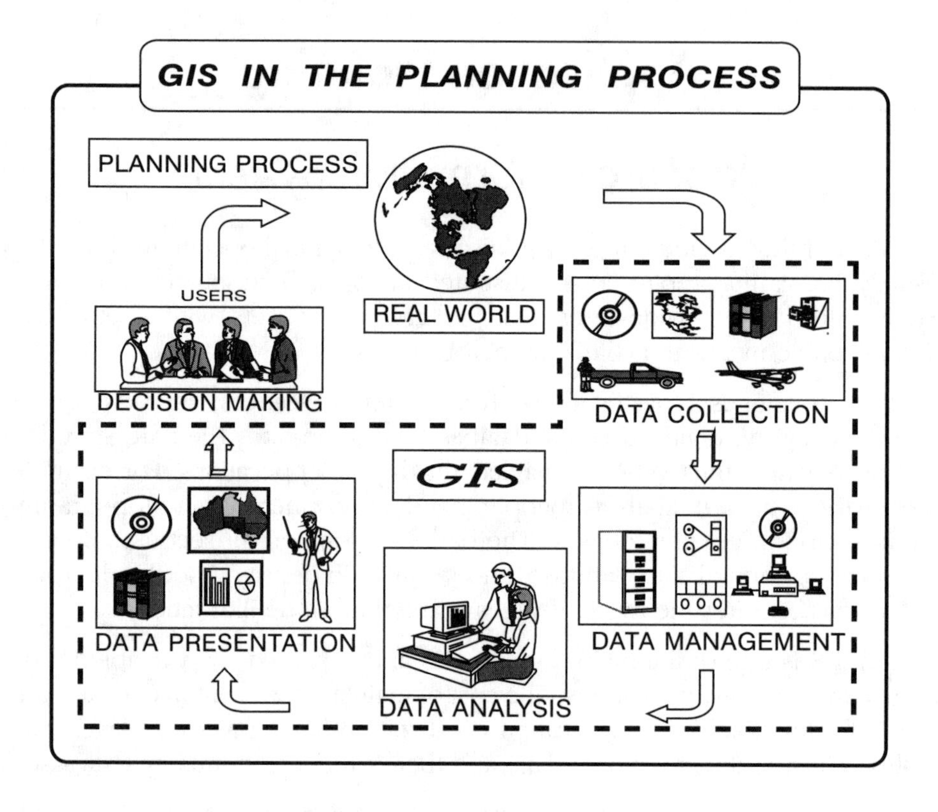

Planning Process

The basic planning process for most applications follows a general outline. The "real world" possesses a need for action, for some change or special maintenance. It contains much of the necessary data related to that need. Thus, as shown in the illustration, the process for achieving the best management and decisions is the following.

◆ *Data Collection:* Data come from many sources, including CD-ROMs, tapes, maps, reports, tabular data, field collection, and remote sensing.

◆ *Data Management:* Data are then stored, managed, and prepared for use.

◆ *Data Analysis:* This is the real work, where research is carried out and conclusions made. This is where GIS operations occur.

◆ *Data Presentation:* Data must be presented in some way, perhaps stored on CDs for later use, made into maps and reports, summarized into statistics, or displayed.

◆ *Decision Making:* When all of the data have been processed, the users go through the output to make decisions in constructing the best response to a given need.

◆ *Dissemination:* Decisions are passed to the "real world" for action.

The dotted line surrounds the steps performed by GIS—most of the planning process. GIS has become the major tool for a large range of applications. It's importance is obvious.

Site Suitability and Models Recap

This chapter discussed the very important GIS paradigms of site suitability and models. Site suitability is really a process of numerous GIS operations achieving the most suitable site for a particular goal or set of objectives. Site unsuitability is the identical process of finding the least desirable location. Site sensitivity considers various levels of grades or magnitudes of acceptability.

GIS models provide generalizations of the real world, simplified for easy understanding and use. They show what is happening in a location to assist in given goals, such as environmental management. Prediction of processes is a major use of models.

A GIS procedure model is a flow chart of steps taken to achieve certain applications or operations. Construction of a Resource Information System (RIS) is a primary application of such steps. For many objectives, these are highly valuable in the planning process.

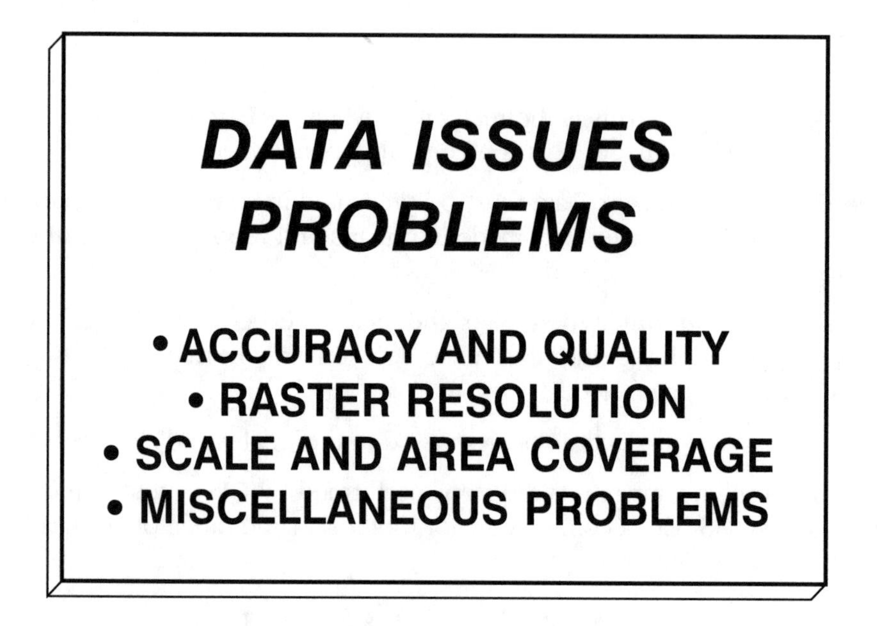
DATA ISSUES
PROBLEMS
• ACCURACY AND QUALITY
• RASTER RESOLUTION
• SCALE AND AREA COVERAGE
• MISCELLANEOUS PROBLEMS

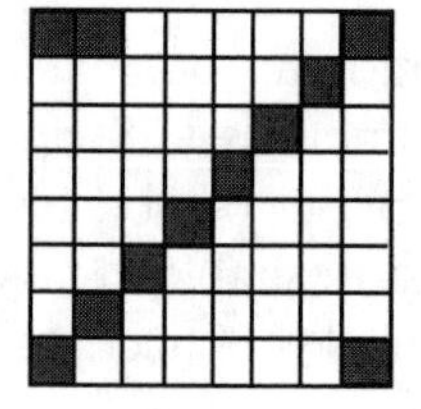
HIGH RESOLUTION GRID

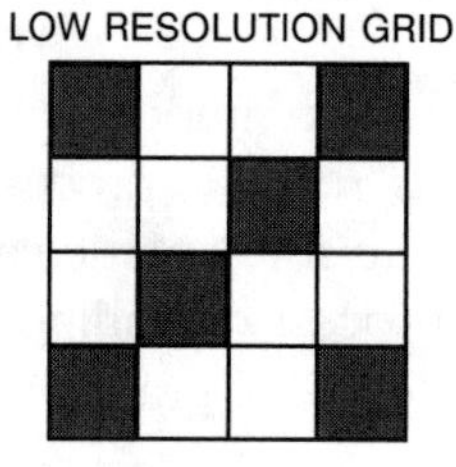
LOW RESOLUTION GRID

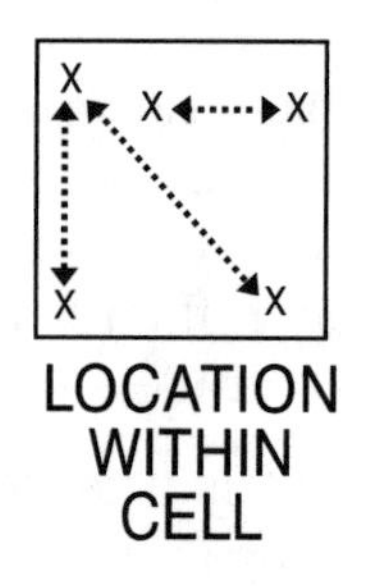
X
X
X
X
X
LOCATION
WITHIN
CELL

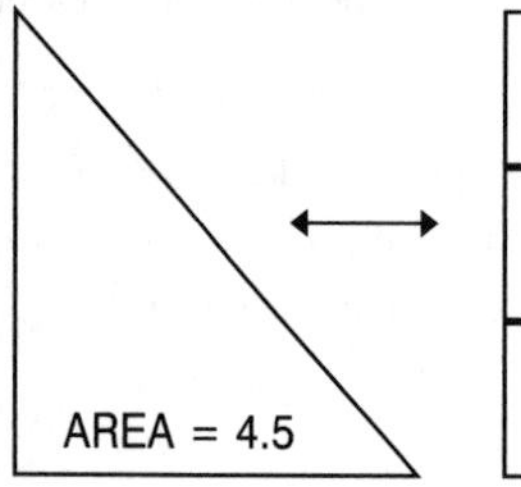
AREA = 4.5

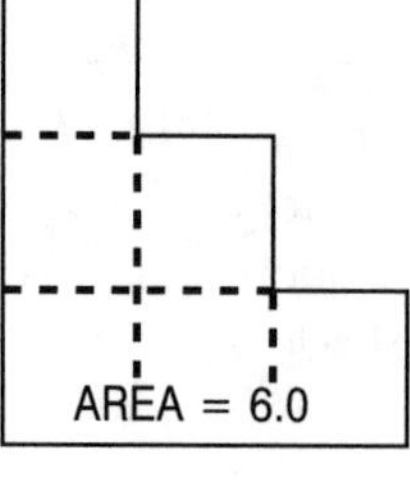
AREA = 6.0

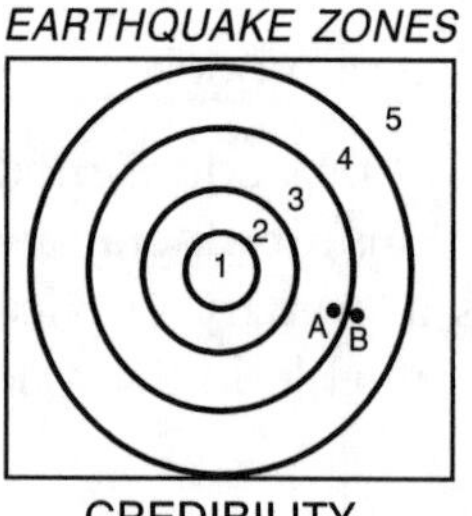
EARTHQUAKE ZONES
5
4
3
2
1
A
B
CREDIBILITY

11

Data Issues and Problems

Introduction

Data is the most important part of GIS. If we do not take care of data, everything else is in danger. This chapter looks at some of the more important data issues in GIS operations and applications. Several data issues and problems have been examined, which are useful considerations in all aspects of GIS.

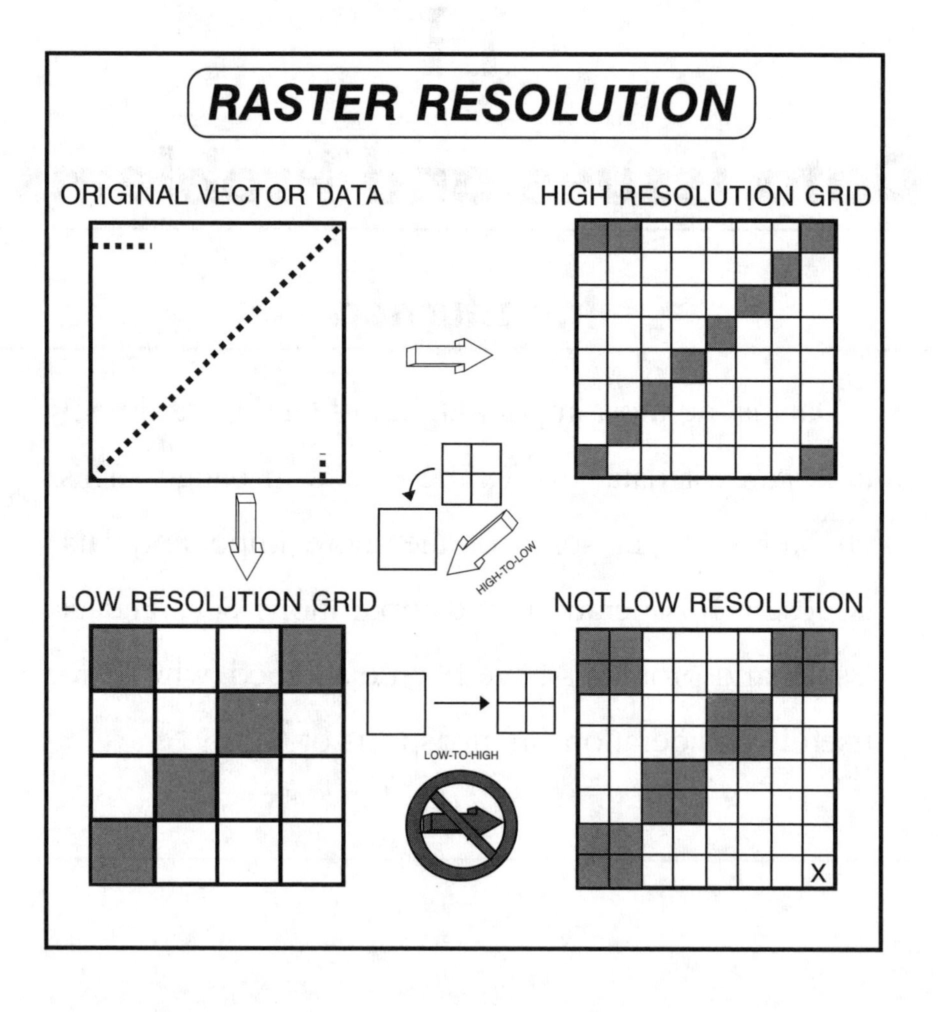
RASTER RESOLUTION
ORIGINAL VECTOR DATA
HIGH RESOLUTION GRID
HIGH-TO-LOW
LOW RESOLUTION GRID
NOT LOW RESOLUTION
LOW-TO-HIGH
X

Raster Problems

Raster Resolution

Let's take another look at raster formats. We've seen how the raster format generalizes data and even creates unusual shapes for diagonal features. The Original Vector Data box (upper left in the illustration) is a simple digitized file awaiting transformation into the raster GIS. There are three line features: a short one at the upper left, the long corner-to-corner diagonal, and a very short one at the lower right. It is possible to use a low-resolution (lower left) or a high-resolution grid (to the right) for rasterizing. Of course, the low-resolution grid is very general (16 cells), whereas the higher resolution (64 cells) is more detailed. Note the effects of resolution on the digitized features.

NOTE: *Normal GIS coverages use many more grid cells than shown in the illustration, of course, which is an exaggeration or magnification of a real example. The concept is the important issue.*

It is possible to change the grid from high to low resolution after initial rasterization. The small diagram in the center, with the arrow pointing from the high to low grid, shows four high-resolution cells being generalized to a single low-resolution cell. The operation may be a simple command to re-rasterize the coverage from an 8 x 8 to 4 x 4 cell structure.

It is possible that the change will produce worse results than would have occurred under direct "vector to low-resolution" gridding. For example, a single high-resolution cell representing the lower right short line could be eliminated rather than kept. It is only a single cell, but when the surrounding block of four cells are changed to one low-resolution cell in the transformation, the program could decide that three blank, high-resolution cells determine what goes into the single large, low-resolution cell.

Although high to low resolution may be acceptable, it is not appropriate to change from low to high (bottom row). It is technically possible, but the results will be misleading. For example, the four new high-resolution cells made from the former large, low-resolution, single cell give the impression that there is a high-resolution cell data resolution, but in truth there is not. Remember that the low-resolution cell was the minimum mapping unit (the maximum resolution for data accuracy) and that there is no improvement simply by dividing it into four parts. Accuracy doesn't change just because the cell is now smaller. Compare the two high-resolution versions on the right. The lower one is considerably different from the more accurate grid at top. For example, the single very short line is missing in the lower right image. Misleading data from resolution changes can be a serious, and often undetected, problem.

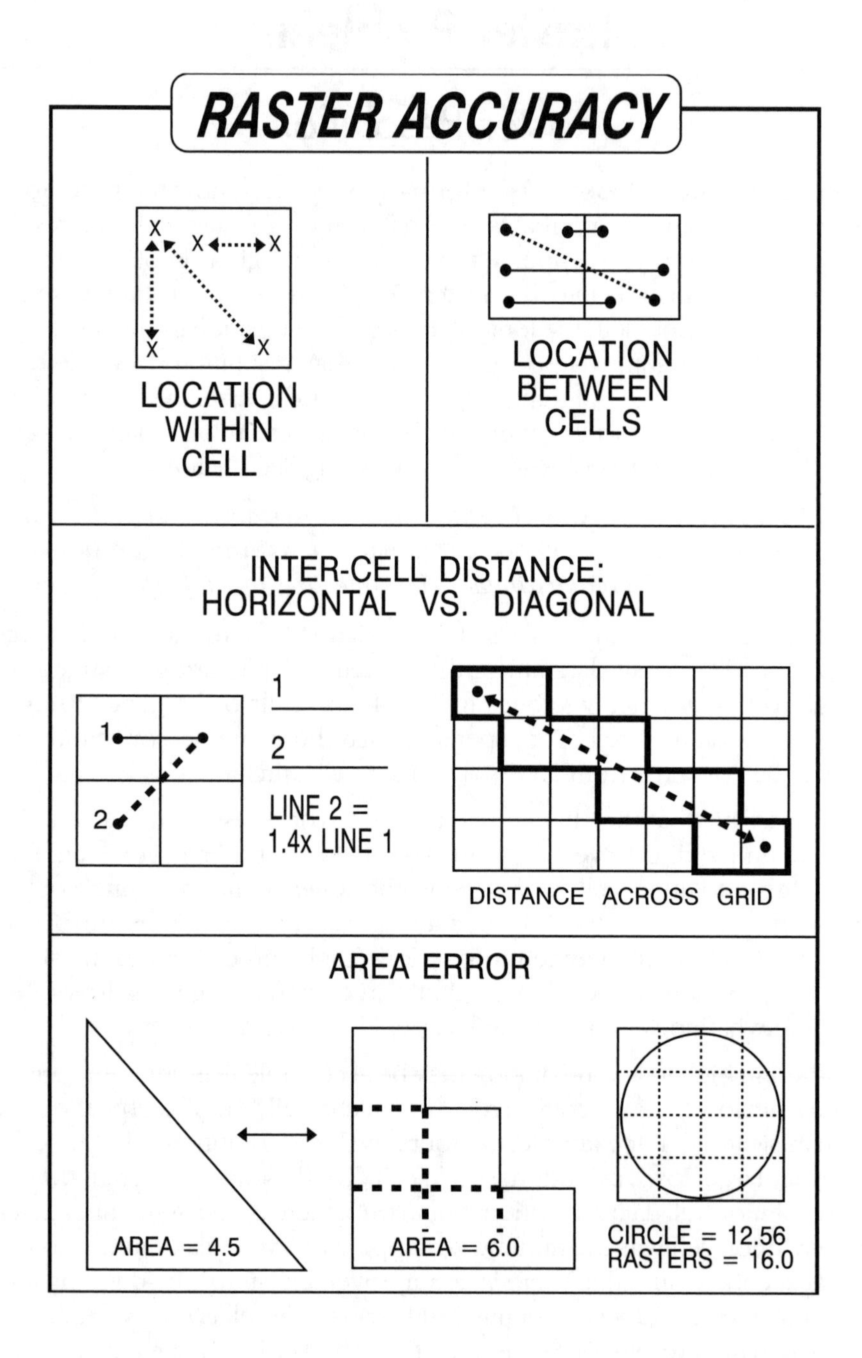
RASTER ACCURACY
X
X
X
X
X
LOCATION WITHIN CELL
LOCATION BETWEEN CELLS
INTER-CELL DISTANCE: HORIZONTAL VS. DIAGONAL
1
2
1
2
LINE 2 = 1.4x LINE 1
DISTANCE ACROSS GRID
AREA ERROR
AREA = 4.5
AREA = 6.0
CIRCLE = 12.56
RASTERS = 16.0

Raster Accuracy and Precision

Questions of raster data precision (the exact location) and accuracy (maximum spatial truth) are often a problem. Recall that the cell is the maximum resolution—the minimum, smallest mapping unit—and there is no way to know exactly where any small feature occurs within the cell. The first box (upper left) in the illustration shows that an X feature can be located anywhere within the cell area, but the location according to the raster format is simply the entire cell (or basically, anywhere within the cell). This may not be very important for some applications, but locational uncertainty can be significant for others.

NOTE: *Obviously, the cell size becomes important when there is need for high spatial accuracy. A higher grid number results in greater accuracy.*

Uncertainty becomes greater when measuring across cells. The actual real-world distance between two features in adjacent cells can vary considerably (second top box)—either very close together (inside edges), to the outside edges of the cell, or even to opposite diagonal corners. Nonetheless, the raster distance is two cells because the cell is the minimum measuring unit.

The left middle box shows the problem of distance between adjacent horizontal and diagonal cells. Actual distance between the centers of diagonal cells is 1.4x that of horizontal or vertical cells. But raster measurements are made according to cell count; thus, the distance either horizontally or diagonally is reported simply as two cells. If the cell is 1 km on a side, the distance is noted as 2 km, but the ground distance really might be 2.8 km (corner to corner).

The problem is compounded when measuring across a grid (middle box, right). The straight distance between the corner cells (dotted line) measures 6.9 cell-width units, whereas the cell count is 7. The extra tenth of a cell width seems small, but over a normal grid of several thousand cells the difference can be significant. Measurements become inconsistent and inaccurate.

Area measurements (bottom box) are also generalized. The triangle measures 4.5 cell units, but requires six cells in a raster format. The circle is 12.56 cell units, but takes 16 raster cells to depict.

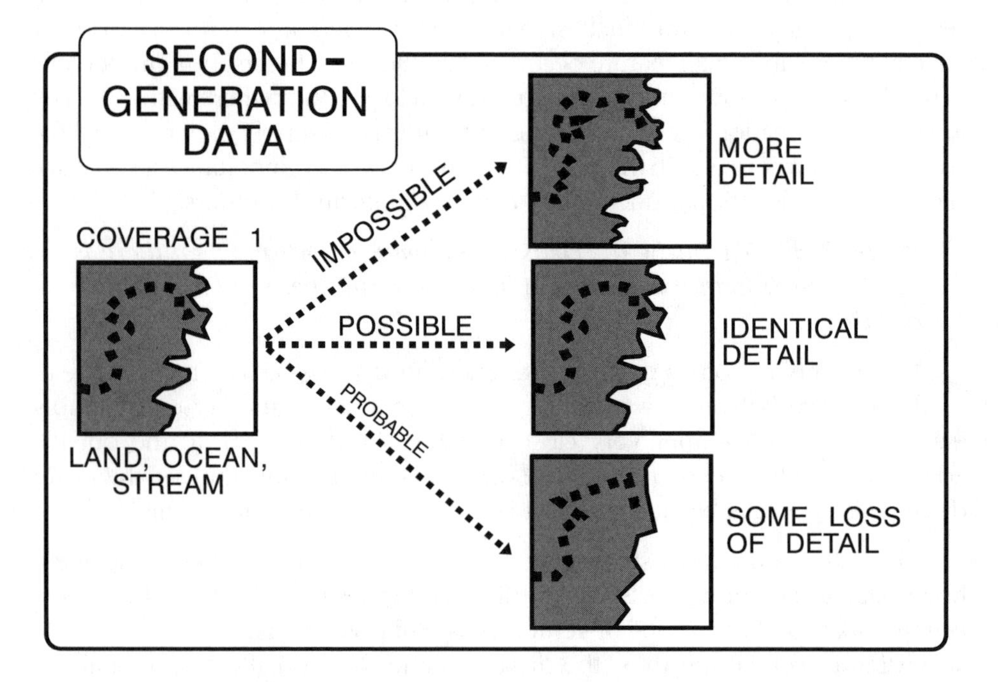
SECOND-GENERATION DATA
COVERAGE 1
LAND, OCEAN, STREAM
IMPOSSIBLE
POSSIBLE
PROBABLE
MORE DETAIL
IDENTICAL DETAIL
SOME LOSS OF DETAIL

Second-generation Data

Original data accuracy cannot be improved—there is no way to verify adjustments unless new data are collected. This means that the second generation of data can be no more accurate than the first generation, and is usually worse because of various copying, handling, or transformation problems. The illustration shows that more or better detail for the second generation is impossible (upper right). Identical detail is possible if a perfect copy is made (middle), but this typically involves some loss of detail (bottom).

It may be possible to make the second generation "appear" better, but that would not be due to increased resolution, only through visualization tricks. Even then, the accuracy is not improved.

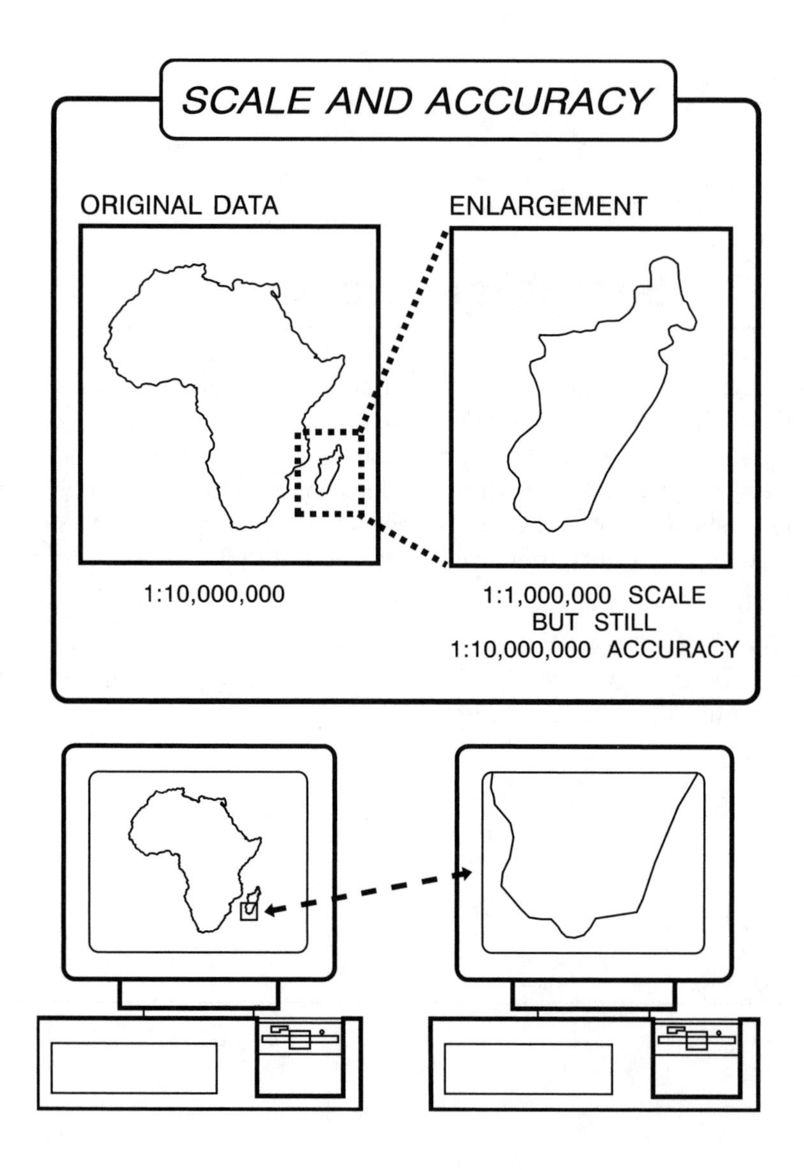
SCALE AND ACCURACY
ORIGINAL DATA
ENLARGEMENT
1:10,000,000
1:1,000,000 SCALE
BUT STILL
1:10,000,000 ACCURACY

Scale

Scale and Accuracy

The scale of original data can be enlarged, but accuracy is not improved. Shown is an original Africa map at 1:10,000,000 scale, with an enlargement of Madagascar to 1:1,000,000. Despite the slightly better appearance of the enlargement, there is really no more detail because there is no better data accuracy than the original scale.

Sometimes the appearance of an enlargement is an improvement over the original presentation because there is more space to present all of the available detail; there is too much detail to show at the original smaller scale. The lower illustration displays the same 1:10,000,000-scale Africa on a monitor at very small visual scale. It appears to be a low-resolution coverage because the view is restricted by computer presentation limitations. Nonetheless, it is still the same data and has the same accuracy quality as both upper illustrations.

Of course, in GIS the coverage can be enlarged. The view at right shows the southern part of Madagascar. Note that the detail is no better than the corresponding version above—we apparently have magnified the data to see all of the details. It is still the same scale accuracy. This is a "data versus display" difference in apparent accuracy.

Accuracy is often overlooked in GIS data. Each source of data has accuracy limits, but we rarely know what they are. Vector point data, for example, can be very precise (very exact location), but we cannot be sure how accurate that precision really is (truthful placement of nodes and vertices). Therefore, the quality of the data may be questionable, especially if information on the original data is not available. This can be a serious concern for GIS users, though typically there is little that can be done.

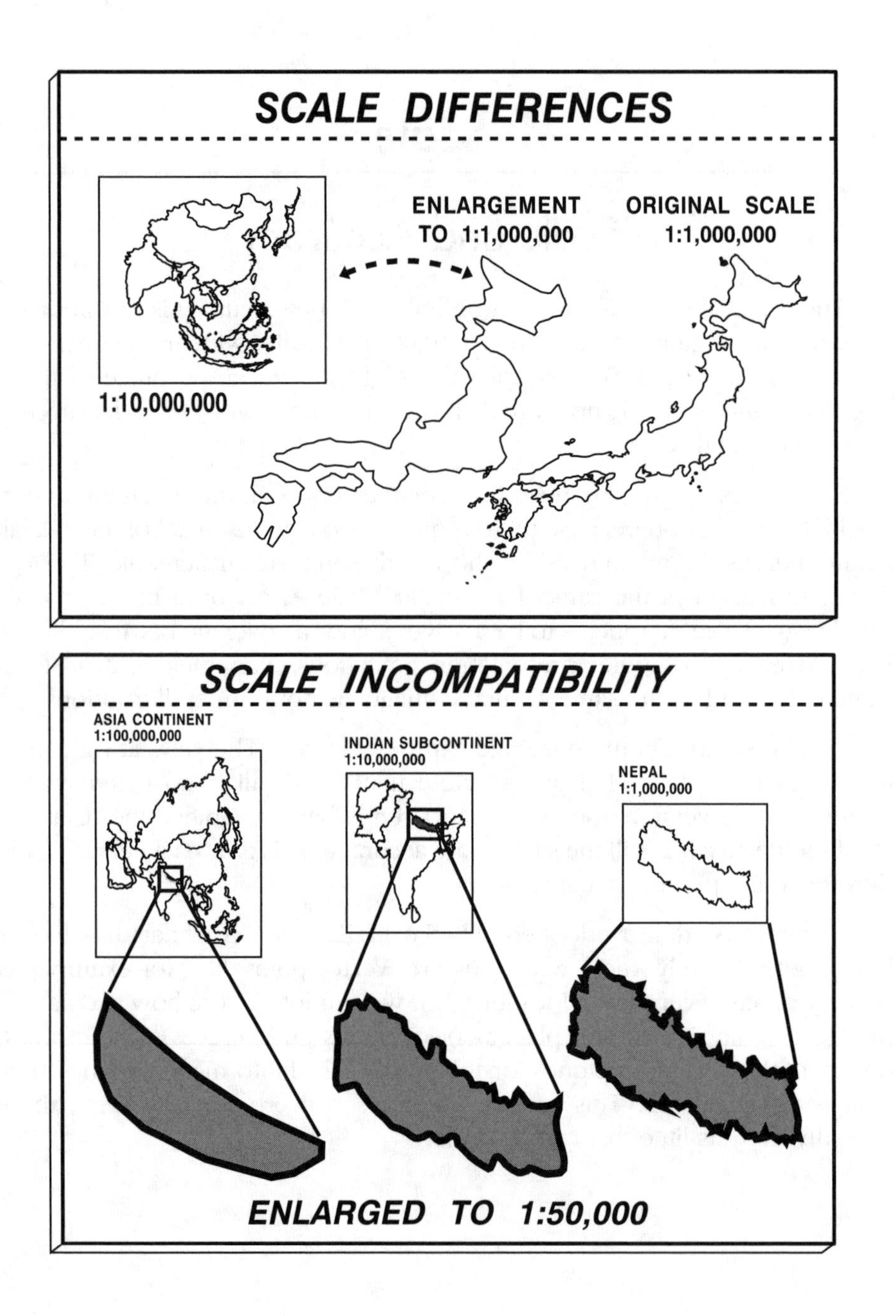
SCALE DIFFERENCES
ENLARGEMENT
TO 1:1,000,000
ORIGINAL SCALE
1:1,000,000
1:10,000,000
SCALE INCOMPATIBILITY
ASIA CONTINENT
1:100,000,000
INDIAN SUBCONTINENT
1:10,000,000
NEPAL
1:1,000,000
ENLARGED TO 1:50,000

Scale Differences

The accuracy issue can be important when changing or using various scales. It is possible to include coverage features that came from different scales, so there would be confusing questions of spatial accuracy. At the top of the illustration are two images of Japan: one as a 1:1,000,000 enlargement from a 1:10,000,000 scale map, and the other from original 1:1,000,000 data (these could be coverages). The differences in data quality are obvious. Again, simple magnification of a feature or coverage does not improve data accuracy.

Scale Incompatibility

The lower illustration shows three original scale maps or coverages (boxes), with Nepal clipped and enlarged to a common scale. The resultant scale is the same, but the detail is considerably different because the original map scales were different. Accuracy may be equally variable, also, but there is no reasonable way of knowing the limits of precision and location of features.

It seems from these illustrations that the largest scale has the best resolution and data quality. That may be true for many applications, but the larger area covered by small-scale maps may be more important.

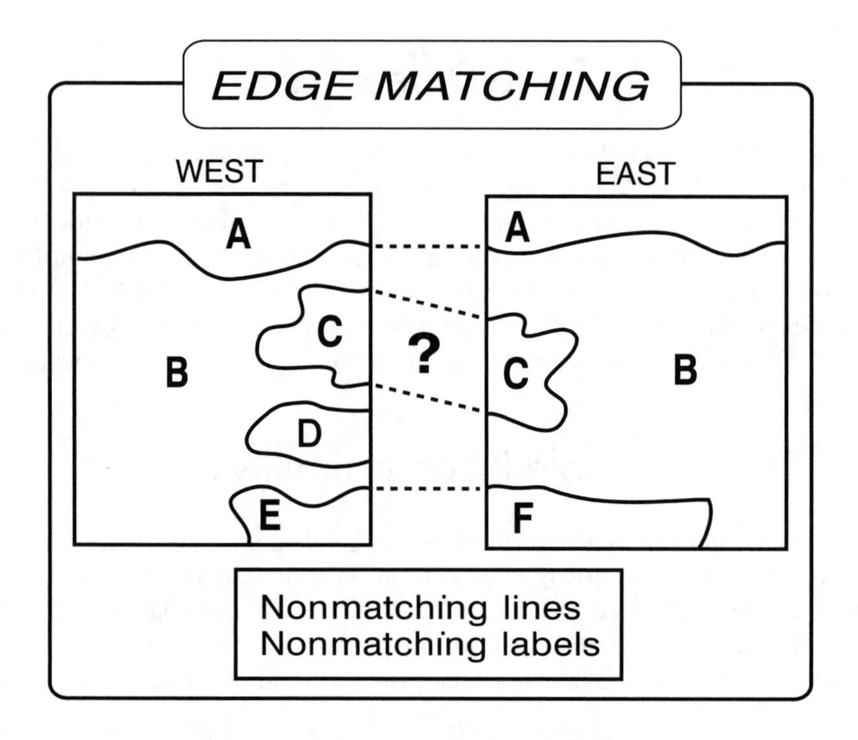
EDGE MATCHING
WEST
EAST
A
B
C
D
E
?
A
C
B
F
Nonmatching lines
Nonmatching labels

Edge Matching

One major problem in GIS is that adjacent coverages may not match properly when joined. Attaching coverages to make large-area data files is a common GIS operation, but when edge features do not align, the operator must make decisions.

Illustrated are two simple coverages, perhaps agricultural areas. Feature A attaches properly and it presents no problem. Feature C, however, does not match and must be repaired. Unless there is some evidence of which feature is in the correct position, some relocation will be needed. But what does the operator do? Lower the feature on West, raise it on East, or do both? In any case, there is the probability that an unavoidable mistake is being made and that feature credibility and quality will be compromised.

Feature D exists on the West coverage, but is absent from East. Is this a mistake? It appears that the feature extends into the East area, although the actual border could be the edge of the West coverage. This could be true if each coverage is defined by administrative boundaries—many features have sharp borders along administrative lines. Perhaps East was constructed before the D feature was made—a data quality issue based on time. What does the operator do in this case? Leave it as shown and make sure data dates are noted? Find another data source showing the true spatial extent of D and add it to East (but giving that coverage different feature dates)? It's a problem.

The bottom feature is labeled E on West but F on East. This seems to be a labeling mistake, but which one is correct? Perhaps it is an interpretation incompatibility. The data crew in East assigned the feature according to its criteria, whereas the West workers used a different means of deciding feature names. Both are correct, but how should the feature be labeled here?

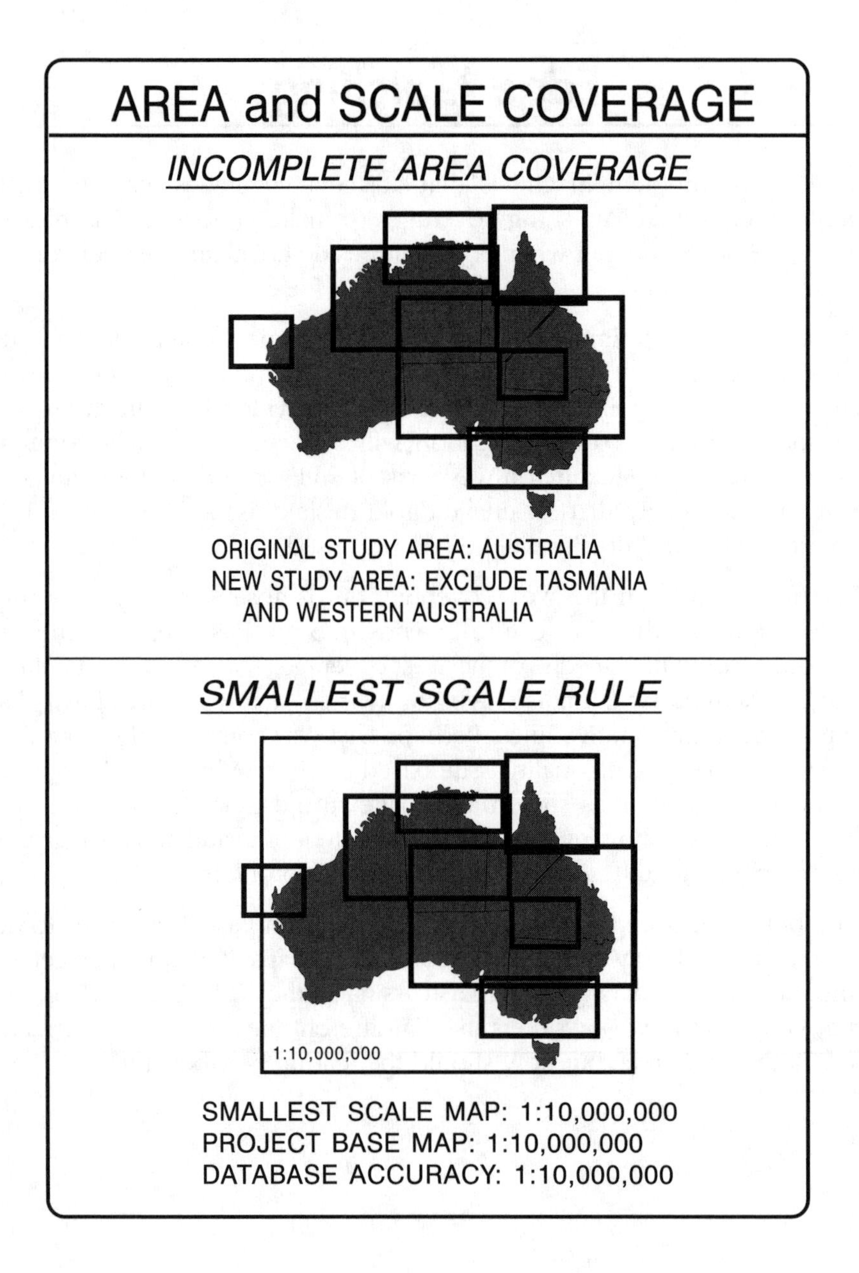
AREA and SCALE COVERAGE
INCOMPLETE AREA COVERAGE
ORIGINAL STUDY AREA: AUSTRALIA
NEW STUDY AREA: EXCLUDE TASMANIA AND WESTERN AUSTRALIA
SMALLEST SCALE RULE
1:10,000,000
SMALLEST SCALE MAP: 1:10,000,000
PROJECT BASE MAP: 1:10,000,000
DATABASE ACCURACY: 1:10,000,000

Area and Scale Coverage

Incomplete Area Coverage

A major practical problem in GIS is getting complete data of the study area. Typically, there are gaps in area coverage (especially for large areas and regions), which forces either extra work locating the missing data or redesign of the project. The upper illustration shows map and data coverage of Australia for a particular project, but there are missing parts. The project must be redesigned to exclude Western Australia and Tasmania (island to the south) or to include the additional time, money, and effort necessary to incorporate the missing areas. [The diagonal lines in Queensland (northeast state) are an error of data digitizing and should be ignored.]

Smallest-scale Rule

There is a rule that states that regardless of area covered, the smallest-scale data used determines the accuracy of all data. In regard to the lower illustration, suppose all of Australia is covered this time because a 1:10,000,000 map was found to correct the data shortage previously mentioned. If all of Australia is used, accuracy limitation is at the smallest original map scale—1:10,000,000. Higher accuracies from larger scales will not improve the lowest-accuracy data. This is a GIS equivalent of the old saying that a chain is only as strong as its weakest link; a data set is only as accurate as its least accurate data. Therefore, the base scale and data quality of this GIS project are from the 1:10,000,000 map. Remember, it is not possible to improve original data.

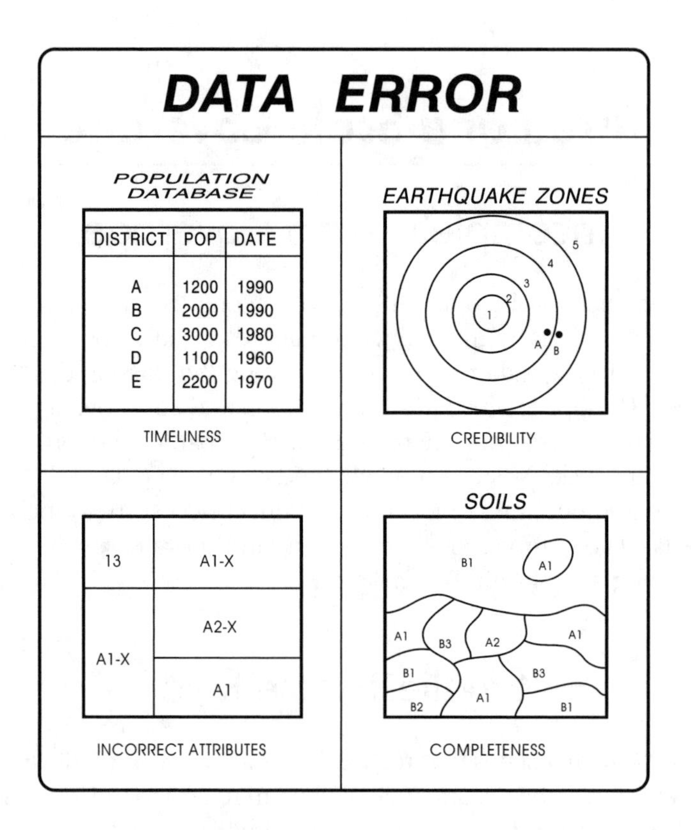

Data Problems

Errors

We tend to believe what we see on maps—that everything is true and accurate. However, maps and coverages can have significant error from numerous sources and the user should be aware of how sensitive data are to various influences. Some of the typical error sources not yet discussed are:

◆ *Timeliness or Age of Data:* The date of a coverage's original data can be critical, say, in population statistics. The date for any type of theme or feature that can change through time should be considered carefully to ensure accuracy or relevance. A city's utility links (e.g., phone, electricity, water) can change, and the date is important.

Certainly the date of the original data should always be given. Old data may compromise overall data quality.

- Illustrated at the upper left is a small database that has mixed dates for five populations. The 30-year difference makes the population figures incompatible and almost useless for comparative analysis.

◆ *Too Much Credibility:* Some data can seem to be more accurate or important than they really are. By nature, radiating phenomena such as earthquake zones, temperature regions, and rainfall patterns are transitional and seldom is there a definite boundary between designated measures. Sharp divisions are drawn out of necessity and convenience (transitions are difficult to present), but the user should not infer precision.

- Although sites A and B in the illustration are in different earthquake intensity zones, they will probably experience little difference in impact. In fact, the nature of earthquakes is inconsistent and the next quake will likely have different zones anyway. This means that the illustrated map cannot be used for very credible scientific data or prediction but only as a very generalized indicator of earthquake impacts—nothing more than historic data. Soil types are also typically transitional, but defined boundaries imply sharp changes.

◆ *Incorrect Attributes:* We normally assume that received data are correct, but there can be mistakes in data collection, reporting, copying, and transfer. Undetected and uncorrected, these mistakes can mean permanent error. Sometimes careful examination of attributes reveals obvious inconsistencies or concerns. Note the attribute labels on the illustration: A1 and 13 are inconsistent and could be errors; at least they are worth investigation with the other labels that have X.

◆ *Completeness:* It may be difficult for the user to know if the data are complete, both in area coverage and use of attributes. Some obvious problems may be easily spotted, but others require careful examination of data files and statistical testing—complex and time-consuming tasks for most users. The illustration shows a high level of soil mapping on the lower half, but the top half seems very weak. Either data collection is incomplete or an unusual landscape situation exists. Both are worth investigation.

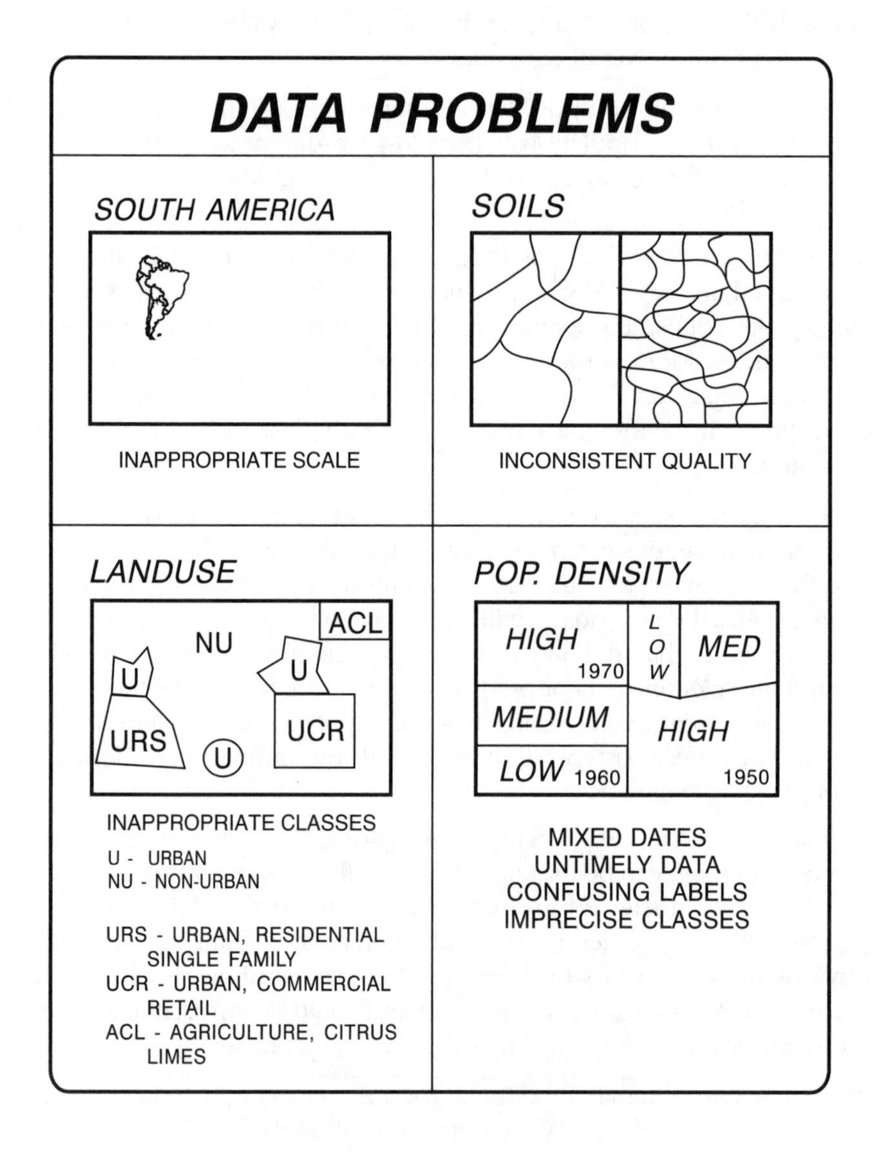

Potential Problems

Similar to the previous discussion, there are other data problems and concerns that must be recognized. Among many are the following (corresponding to the illustration).

- *Inappropriate Scales or Resolutions:* Too much or too little detail may be presented at different scales. If this map (upper left) represents the original data scale, insufficient details exists for each country. If this is a final presentation map, it seems to be inappropriate for use. Although the computer permits zooming to overcome problems of small size, the user must work within the original data and/or final presentation scale formats.

◆ *Inconsistent Quality and Coverage:* Inconsistent levels of detail may raise questions of completeness, as discussed, but there may be valid reasons for differences. The soils data display is from two nations, for which data collection methods differ. Each nation has satisfactory soil detail for itself and the data are accurate, but regional analysis will be uneven and perhaps even misleading because of the differences in data. Special considerations are needed to keep the project believable.

◆ *Inappropriate Classification Scheme:* Choosing the best, or an accurate, classification scheme can be difficult. It is easy to use incorrect or unsuitable classifications, such as too many or too few categories. Illustrated is a poor landuse display due to an inconsistent scheme—only two major classifications and several detailed ones are used. The use of an Urban/Nonurban system could be acceptable, and exclusive use of the more detailed scheme would be preferable, but mixing is not good design. Some very good data sets may be organized and written in a confusing manner, making the data useless.

◆ *Multiple Problems:* The Population Density coverage in the illustration is a good example of a bad coverage. Different dates (a 20-year span) are used—illogical for this theme, as indicated earlier. For a modern study, even the latest date is probably too old (except for a historical study). The date labels are confusing because of their placement. Dates are not known for those polygons not having a date label.

◆ Also, the classes used are rather imprecise. Low, Medium, and High are generally descriptive, but the comparison base is not given (low as compared to what?). A quantitative figure would be better. Clearly, this is a not a good coverage display or map.

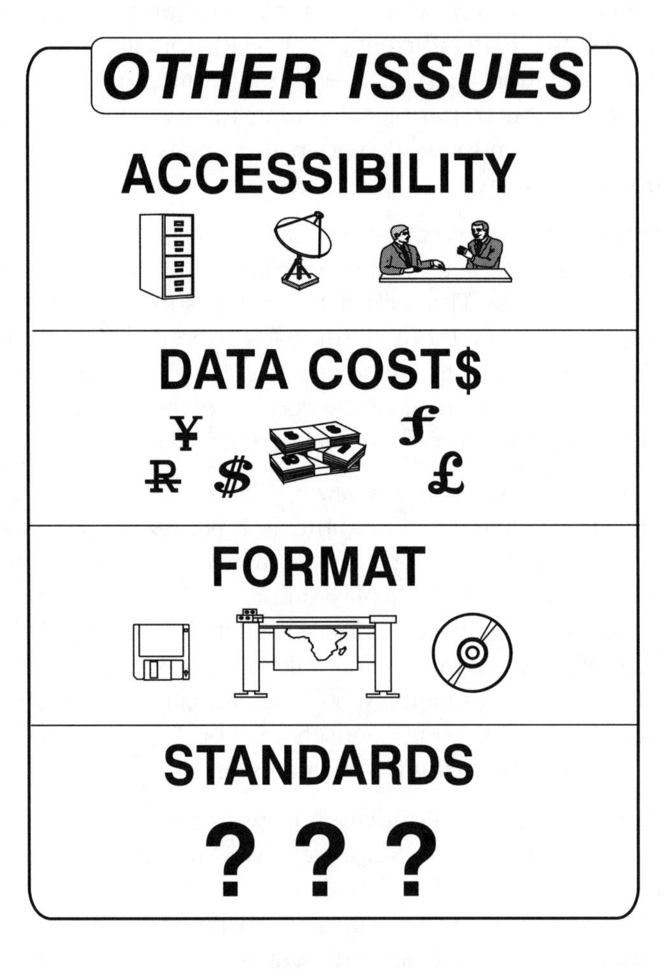

Data Issues

There are many other issues that concern GIS users. The following (corresponding to the illustration) are a few.

◆ *Accessibility:* Getting access to essential data is a major problem for many users, especially for the developing world and regions out of the data mainstream, such as the South Pacific. Users know what is needed, but often it may be out of reach for various reasons, such as ownership, distance, bad format, or cost. Data locked away in an inaccessible filing cabinet is a common problem.

◆ Although access through satellite relay is improving data collection, such expensive high technology represents serious problems for many users. Negotiations and meetings may be necessary to locate and obtain permission for access.

◆ *Data Cost:* There is no guideline on the pricing of data. Some data sets are free, whereas others cost too much for all but the well financed. Satellite data can be expensive by nature, but because of the large imaged area, it can be cost efficient per unit area. However, when the cost is 4,000 U.S. dollars per scene, pricing is beyond most users.

◆ Data is a commodity still seeking an economic "level." The international symbols used in the illustration are appropriate today; many nations are using and selling data.

◆ *Format:* Format can refer to basic media—such as paper, text, digital, and so on—or to the structure of data. There is always a question whether or not a particular set of data will fit into the GIS, and if so, how much work will be needed.

◆ Maps are usually in paper form, but when they are large or in many pieces, they become difficult to handle and use. Maps and other types of data need transfer into GIS, which can create time, effort, and error problems. Even digital formats are different. One computer type or GIS format may be incompatible with another, and data will not transfer without difficulty. Digital tapes come in different sizes and formats. Even transfer over phone lines may not work due to different data structures and transmission incompatibilities.

◆ *Standards:* GIS users would like to have reliable data—data that can be trusted. They need to have consistent formats, interchangeable classifications, dependable accuracy, and compatible media for exchanging data. Users and their data need to talk the same "language"; otherwise, chaos reigns. The development of standards is very difficult because there is little agreement what should be included and what measures should apply. The development of standards is a major current issue.

◆ The illustrated question marks mean that we sometimes don't even know what standards are for some parts of GIS, let alone have agreement of the best way to proceed or format to be used. But progress is being made and the GIS world is becoming more organized.

Glossary

Algorithm

A computer operation used to solve a particular problem or carry out a set of steps.

Application

The practical, real-world use of Geographic Information Systems (GISs) and their data. For example, a forestry service might use vegetation data for ecological mapping.

Artificial intelligence (AI)

Advanced computer programs that simulate human intelligence using logical rules. For example, an AI program could help decide which GIS operation is best for a particular task. (*See also* "Expert systems.")

Aspect

The direction a slope faces. Important in determining sun exposure as it relates to such considerations as soil moisture and effects on vegetation. GIS can generate and use aspect data in ecological and resource evaluation.

Attribute

Data description, characteristic, or quality. Describes or explains the data. For example, "Vegetation = brush, 10 ha, and owned by Mrs. Jones" contains three attributes: identification, size, and ownership. Some GISs store attribute data as category or class numbers, whereas others use descriptive text.

Binary

Having only two possible values or states. For example, 0 = Off, 1 = On; or Yes/No. Computer instructions normally use binary data.

Boolean operation

A multiple-condition query that represents relationships based on criteria specified by "logical operators" such as AND, OR, NOT, and AND/OR. For example, "Show all forests AND/OR brush areas." (*See also* "Relational database.")

Buffer

The zone (corridor, or area) on each side of or around a feature as defined by GIS operations. For example, "A 2-km environmental-protection buffer zone is established around a pollution site." Other names for this, such as *spread,* are used by various software.

CAD

Computer-aided drafting, or computer-aided design; typically used for engineering and design work, and now being adopted for mapping and integration with GIS. (Alternate: CADD, computer-aided drafting and design.)

Cartography

The science and art (both are important) of making maps.

CD-ROM

Compact disk read-only memory. Small compact storage disk read by a laser device. Slightly larger than a 3.5-inch computer disk, it can hold more than 500 times the amount of information and is becoming an increasingly popular medium for data and software.

Cell

See "Grid cell."

Chain

The connection between vertices and nodes. Single part of a line or polygon. A two-part line might consist of two chains. A square has four chains. The word *line* is avoided as a synonym because of the confusion with the same-named GIS feature. Sometimes called an *arc.*

Coordinate System

Location system using an X-Y (horizontal and vertical) grid to permit accurate Earth location of an item. For example, Latitude-Longitude is a world coordinate system (*see also* "UTM"). Might also include a third dimension, usually elevation or depth, termed the Z coordinate.

Coverage

Map data file. The digital form of data for a map. Usually devoted to a single theme or type of data. Also termed *layer.*

CPU

Central processing unit. The "brains" of a computer that computes and directs functions.

Database

Collection of organized data in a manageable structure, usually resembling a table. In GIS, this normally means a computer program that efficiently stores, manages, and retrieves spatial and associated nonspatial data for either quick query (selection) or incorporation into graphics. (*See also* "Relational database" *for a special type.*)

Database management system (DBMS)

A computer system used to store and manage information. In GIS, DBMS software is usually included as part of the system.

Data entry

The process of loading data directing into a database, either manually or automatically. (*See also* "Digitize" *and* "Digitizer.")

Data set

A collection of related data, sometimes referring to a single file but usually to a set of related files. Provides data for the database.

Data structure

The organization and format used to define data so that the software can use it. For example, a raster structure uses a grid format to define GIS data. (*See* "Raster" *and* "Vector" *for examples.*)

Digital Elevation Model (DEM)

Database of gridded digital surface data that can be shown in either 2D (flat view) or 3D (realistic perspective). Digital Terrain Model (DTM) is a special version that uses heights above mean sea level.

Digitize

To convert maps or images into digital format. Normally used in context of employing a digitizer, but can refer to any instrument that transforms analog data to digital format.

Digitizer

An electronically sensitive table that uses a tracing device for copying paper map data into a computer for conversion to digital data. *Manual digitizing:* human tracing of map features with a small hand device, which relays the locations of features (points, lines, polygons) into the GIS. *Automatic digitizing:* television-type scanning of map images (like a photograph converted into digital format) or light (laser) beam tracing of lines and features, which is performed faster and more accurately than through a human process.

Distributed system

Set of computers and devices that may be widely separated by location but linked through a network. For example, a central database located in one city, with employees with computers in other cities linked to that database. The system is "distributed."

Expert system

An advanced computer program that uses rules to simulate a human expert's reasoning to make specific decisions. For example, an expert system for identifying air-photo features might use rules that describe possible choices among them, which eventually lead to a logical answer or solution to a particular focus.

Fractal

A graphics format that uses advanced mathematics to define the shape of a feature or entire image. Technically based on the natural curves of objects. Represents a new way of designating data structures (in addition to vector and raster formats).

Geocoding

Translating or transferring geographic coordinates into X-Y digital format for use in a GIS. Might also include assigning locations to features not based on a world coordinate system, such as street addresses.

Georeferenced

Properly aligned or registered to a fixed coordinate system. A GIS coverage should be fixed in correct cartographic space so that other data can be associated or connected.

GIS

Geographic Information System. A computer-based system that enters, stores, manages, analyzes, and presents spatial (and associated nonspatial) data, combining databases and graphics operations to make a variety of products, from lists to maps.

GPS

Global Positioning System. A system of satellites that transmits signals used by special receivers on the ground for precise determination of location, sometimes within meters. Receivers might be small, hand-held units or large, highly accurate systems. Data might include elevation and speed.

Grid cell

One cell containing a single GIS (or remote sensing image) value or attribute. The cell is part of a large grid that depicts a scene. The grid is a raster coverage. The cell is also termed a *raster* or *pixel* (picture element).

Hard copy

A map or graphical depiction of a coverage on paper. In effect, something that can be held. It is the opposite form of a digital file.

Hologram

A 3D image made by splitting a laser beam. Holograms can store large amounts of data and represent a new frontier in data storage and display.

Icon

A graphic (picture) used to relate or symbolize a computer operation or data set. Typically, pointing and clicking on an icon (using a mouse) starts the process or selects the data. Using icons and the "point and click" method, rather than typing complex commands, is becoming the preferred method for operating a GIS.

Information

Data combined and integrated to indicate something; meaningful data. Note that data by itself is not necessarily information, but merely a fact or collection of facts.

Interactive

Marked by real-time communication between computer and operator. As the operator puts in instructions, the computer reacts immediately, producing results as soon as input is completed.

Label

Data identity stamp or designation; nonfeature items attached to each coverage feature and data structure to provide an identity. For example, the first digitized polygon of a set of polygons might be labeled Poly 1, as assigned by the user.

Land information system (LIS)

Also called a land records system (LRS). A database system containing physical, quantitative, legal, and other descriptions of land, such as elevation, land value, and ownership. Not necessarily a complete GIS.

Landsat

U.S. Earth resources satellite system that produces images of land in multiple electromagnetic-spectrum bands. Its TM (thematic mapper) senses areas as small as 28 meters on a side (784 square meters). The program began in 1972. (*See also* "Spot.")

Layer

One coverage data file, usually devoted to a single theme. For example, "Layer 3 = vegetation."

Line

A one-dimensional feature on a map having length and direction, but normally no width. A GIS line begins and ends with nodes. For example, a railroad in a GIS is depicted as a line, starting at a node and ending at a node.

Map projection

See "Projection."

Map scale

The depiction or calculation of map distance to its corresponding real-world distance. Scale is expressed as a ratio of map units to real-world units. The nature of units does not matter, but metric or English measurements are normally used. For example, 1:50,000 (also 1/50,000, the mathematical ratio) means that 1 centimeter on a map equals 50,000 centimeters of real distance; or, 1 map inch = 50,000 real-world inches. Note that the units must be the same on each side of the equation.

Model

In GIS, this is (1) a representation of reality, (2) an attempt to duplicate or display a real-world process in a simplified manner, or (3) a set of GIS procedures for accomplishing a specific task; also known as "cookbook" instructions. Typically a GIS model consists of a set of spatial and database operations governed by rules. For example, a model of erosion rates for varying slopes under specified precipitation conditions: "For each millimeter of rain per hour, there is an increase of 0.1 cubic meter of erosion per square meter of surface for each additional 1-percent slope."

Mouse

A small, hand-controlled device attached to a computer that allows directing and aiming of a cursor (point of light) on the monitor screen. It is used to control menus, to draw, and to point to features.

Network

1. Per GIS data: A set of interconnected lines that represent paths of movement. For example, a road system. Supports GIS-network spatial operations such as finding the shortest and fastest routes between two locations.

2. Per computer technology: A system linking computers (and other devices, such as printers) by communications lines. The network might be "local" in an office or building (LAN, local-area network) or cover a wider area (WAN, wide-area network). The Internet is a new global network. (*See also* "Distributed system.")

Node

The point at each end of a chain. Nodes are part of the line structure and should not be confused with the GIS feature termed *point.* Nodes are not landscape features and have no dimension. (*See also* "Line" *and* "Vertex.")

Output

Product constructed for display, printing, or reporting. In GIS, this usually refers to monitor displays, maps, and reports.

Overlay

The combining (or superimposing) of two or more GIS coverages to produce new data or a new coverage. For example, an overlay of vegetation, soils, and slopes to make a new erosion potential coverage.

Parcel

A map feature designating land ownership and rights. A parcel map is devoted to legal property depiction and descriptions.

Peripherals

Hardware linked to computers to provide additional assistance. For example, printers and digitizers are standard GIS peripherals.

Pixel

Modified abbreviation of "picture element," the smallest portion of an image to contain a data value. For example, a raster grid cell is a pixel.

Plotter

A graphics drawing device normally used to produce large maps. Typically higher quality than standard printers, a plotter produces continuous lines (as opposed to a printer), and may use colored pens, ink jet spray, or electromagnetic means of printing. A common GIS peripheral.

Point

The most simple GIS spatial feature; a single-location object or event (occurrence), having only one X-Y coordinate location and no length or width (no spatial dimensions). In GIS, a point locates something that is either too small to show realistically or has no real spatial measurement. For example, a house (too small to show in detail in the image area) or an accident site (an event having no measurement).

Polygon

An enclosed area on a map or coverage, having length and width; a feature with three or more sides that completely surrounds an area. For example, representation of a political district or forest site.

Projection

A special shape used to fit a portion of the globe onto a flat view; converting spherical data to 2D presentation. There are many projections, each with specific parameters for particular application. GIS coverages usually need a projection in order to calculate georeferencing accurately.

Query

"Asking" a question (making a request) of a database, typically in the form of a command for specific data.

Raster

A cell that contains a single GIS or remote sensing image value. The cell is part of a larger grid system (rows and columns) constituting an entire coverage or image. A raster system is a data structure that uses the gridded arrangement. (*See also* "Grid cell" *and* "Pixel.")

Relational database

A powerful and flexible type of database that allows multiple linkages of data so that each field can be related to all other fields. Typically, a relational query consists of defining specific conditions from one or more fields to find records meeting all or many of those needed conditions. For example, "Find (1) all properties valued over $50,000 AND (2) that are used for farming AND (3) that cover over 50 ha."

Remote sensing

Gathering data some distance from the target. In GIS, this usually implies the imaging of land (or ocean, atmosphere) from above, such as by aircraft or satellite. It could include sensing from ground level (or a raised platform). The Landsat satellite provides Earth images.

Site suitability analysis

A GIS process typically involving several operations (such as overlays and buffers) to identify the best or worst sites for a particular purpose. For ecological analysis, often termed *site sensitivity analysis.*

Spatial data

Data that occupies cartographic (mappable) space and usually has specific location according to some georeference system (such as Latitude-Longitude). In GIS, may be associated with nonspatial data. For example, the location of a political area is spatial, as is its area; the name and demographics are nonspatial. Also known as geographic data.

Spot

French remote sensing satellite similar to the U.S. Landsat. It can produce stereo images and can gather data at ten-meter resolution.

Thematic map

A map or coverage devoted to a single theme. For example, a geologic map.

Topology

In GIS, a powerful data structure used to maintain spatial relationships between and among features. In effect, each point, line, and polygon "knows" where it is, what is connected to it, and what is around it. Used with vector systems.

UTM (Universal Transverse Mercator)

Commonly used with GIS, a higher-resolution alternative coordinate system to Latitude-Longitude. Based on sixty global zones of 16 degrees longitude width.

Vector

In cartography, a line, having direction and length. In GIS, a chain. Refers also in GIS to a data structure that defines points, lines, and polygons by their true position and dimensions. Each point on a chain (node and vertex) is a coordinate label rather than a graphic item. For example, a square is defined by the coordinates of its four corners rather than by four drawn lines. This allows high accuracy at any scale and is more "maplike" than a raster system.

Vertex

The directional turning point on a chain; a point on a chain given a coordinate label. Distinct from a node (end point). The plural is vertices.

Viewshed

The area visible from a given location, considering the visual obstructions that may interfere, such as mountains. Viewshed coverages are constructed to show paths of visibility.

Workstation

A high-performance computer (and sometimes its connected equipment, such as printers). The term is imprecise and changes often, but typically refers to a system consisting of more than an individual microcomputer (PC) and less than a large, "mainframe" computer. In GIS, typically refers to a central, powerful system holding the primary software, with satellite PCs or terminals performing individual tasks. Sometimes confused with a satellite terminal acting as a "working station."

X-Y coordinates (values)

A coordinate system for precise location of a feature. For example, "Longitude = X and Latitude = Y."

Z value

The third coordinate, after X and Y, which locates a point in space in terms of elevation (height or depth). For example, "The Z value of the mountain is 3,000 meters." In earlier GISs, the Z value also referred to the attribute measure, but that is a declining reference.

Index

A

B

C

D

E

F

G

H

I

L

M

N

O

P

Q

R

S

T

U

V

Z

SunSoft Solaris Series

SunSoft Solaris 2. User's Guide*
Book $29.95 Includes Disk

SunSoft Solaris 2. for Managers and Administrators*
Book $34.95

SunSoft Solaris 2. Quick Reference*
Book $18.95

*Five Steps to SunSoft Solaris 2.**
Book $24.95 Includes Disk

SunSoft Solaris 2. for Windows Users*
Book $24.95

The Hewlett Packard HP-UX Series

HP-UX User's Guide
Book $29.95 Includes Disk

HP-UX Quick Reference
Book $18.95

Five Steps to HP-UX
Book $24.95 Includes Disk

Softdesk

Softdesk Civil 1 Certified Courseware
Book $34.95 Includes CD-ROM

Softdesk Civil 2 Certified Courseware
Book $34.95 Includes CD-ROM

Softdesk Architecture 1 Certified Courseware
Book $34.95 Includes CD-ROM

Softdesk Architecture 2 Certified Courseware
Book $34.95 Includes CD-ROM

INSIDE Softdesk Architecture
Book $49.95 Includes Disk

INSIDE Softdesk Civil
Book $49.95 Includes Disk

Other CAD

Manager's Guide to Computer-Aided Engineering
Book $49.95

Fallingwater in 3D Studio: A Case Study and Tutorial
Book $39.95 Includes Disk

Geographic Information Systems (GIS)

INSIDE ARC/INFO
Book $74.95 Includes CD-ROM

ARC/INFO Quick Reference
Book $24.95

INSIDE ArcView
Book $39.95 Includes CD-ROM

ArcView Developer's Guide
Book $49.95

ArcView/Avenue Programmer's Reference
Book $49.95

101 ArcView/Avenue Scripts: The Disk
Disk $101.00

ArcView Exercise Book
Book $49.95 Includes CD-ROM

INSIDE ArcCAD
Book $39.95 Includes Disk

The GIS Book, 3d ed.
Book $34.95

INSIDE MapInfo Professional
Book $49.95

Raster Imagery in Geographic Information Systems
Book $59.95

GIS: A Visual Approach
Book $39.95

Interleaf Books

INSIDE Interleaf (v. 6)
Book $49.95 Includes Disk

Adventurer's Guide to Interleaf Lisp
Book $49.95 Includes Disk

Interleaf Exercise Book
Book $39.95 Includes Disk

Interleaf Quick Reference (v. 6)
Book $24.95

Interleaf Tips and Tricks
Book $49.95 Includes Disk

OnWord Press Distribution

End Users/User Groups/Corporate Sales

OnWord Press books are available worldwide to end users, user groups, and corporate accounts from your local bookseller or computer/software dealer, or from Softstore/CADNEWS Bookstore: call 1-800-CADNEWS (1-800-223-6397) or 505-474-5120; fax 505-474-5020; write to CADNEWS Bookstore, 2530 Camino Entrada, Santa Fe, NM 87505-4835, or e-mail ORDERS@HMP.COM. CADNEWS Bookstore is a division of SoftStore, Inc., a High Mountain Press Company.

Wholesale, Including Overseas Distribution

High Mountain Press distributes OnWord Press books internationally. For terms call 1-800-4-ONWORD (1-800-466-9673) or 505-474-5130; fax to 505-474-5030; e-mail ORDERS@HMP.COM, or write to High Mountain Press, 2530 Camino Entrada, Santa Fe, NM 87505-4835, USA. Outside North America, call 505-474-5130.

On the Internet: http://www.hmp.com

OnWord Press 2530 Camino Entrada, Santa Fe, NM 87505-4835 USA